MATHEMATICAL METHODS IN ENGINEERING AND PHYSICS

DAVID E. JOHNSON, Ph.D. Applied Mathematics, Auburn University, is Professor of Electrical Engineering at Louisiana State University. He did his undergraduate work at Louisiana Polytechnic Institute, receiving degrees in both mathematics and electrical engineering. Dr. Johnson was also Associate Professor of Mathematics at Louisiana Polytechnic Institute, where he taught for many years. He is a member of Sigma Xi, Tau Beta Pi, and Pi Mu Epsilon.

JOHNNY R. JOHNSON, Ph.D. Applied Mathematics, Auburn University, is Professor of Electrical Engineering at Louisiana State University. He was formerly Associate Professor of Mathematics at Appalachian State Teachers College and Assistant Professor of Mathematics at Louisiana Polytechnic Institute. Dr. Johnson is a member of Sigma Xi, Tau Beta Pi, and Pi Mu Epsilon.

MATHEMATICAL METHODS IN ENGINEERING AND PHYSICS

DAVID E. JOHNSON
LOUISIANA STATE UNIVERSITY

JOHNNY R. JOHNSON
LOUISIANA STATE UNIVERSITY

PRENTICE-HALL, INC., ENGLEWOOD CLIFFS, NJ 07632

Library of Congress Cataloging in Publication Data

JOHNSON, DAVID E.
Mathematical methods in engineering and physics.

Bibliography: p.
Includes index.
1. Engineering mathematics. 2. Mathematical physics. I. Johnson, Johnny Ray. II. Title.
TA330.J59 510'.2453 81–13922
ISBN 0–13–561126–1 AACR2
(previously ISBN 65-12753)

Production supervision by Karen Skrable
Manufacturing buyer: Joyce Levatino
Cover design by Infield/D'Astolfo Associates

Printed in the United States of America

10 9 8 7 6 5 4 3 2 1

ISBN 0-13-561126-1

Prentice-Hall International, Inc., *London*
Prentice-Hall of Australia Pty. Limited, *Sydney*
Prentice-Hall of Canada, Ltd., *Toronto*
Prentice-Hall of India Private Limited, *New Delhi*
Prentice-Hall of Japan, Inc., *Tokyo*
Prentice-Hall of Southeast Asia Pte. Ltd., *Singapore*
Whitehall Books Limited, *Wellington, New Zealand*

Contents

Preface

This textbook is intended for courses in applied mathematics covering special functions and their applications, particularly those concerned with boundary-value problems of mathematical physics and engineering.

In order to make this book as self-contained as possible, we have included a chapter on Fourier series and one on series solution of differential equations. For students already possessing a knowledge of these subjects the chapters may serve as a review or may be omitted in a short course. Topics from advanced calculus are considered in the book as the need for them arises; others are included in the Appendix.

The book is divided into a number of general categories according to the types of functions and their applications. First, orthogonal functions are considered in general, and terms used throughout the remainder of the book are defined. Following this, the more common special functions are discussed. In order to avoid the monotony of a relentless examination of one set of functions after another, we have included in these chapters the applications in which the functions arise. Hermite, Laguerre, Chebyshev, Mathieu, and other special functions are considered next, along with their applications, most of which are boundary-value problems. Integral transforms, especially Laplace and Fourier transforms, are then developed, and applied to boundary-value problems, particularly those not amenable to the separation-of-variables method. The Sturm-Liouville transforms are considered primarily as a method of solving problems, and also to exhibit further properties of the special functions. Finally we consider a general method for developing sets of polynomials orthogonal on any given interval. This last method is based on a general Rodrigues' formula, from which general orthogonality relations and general differential equations are obtained.

Tables of the more important properties of the special functions and short tables of the integral transforms are included in the Appendix for ready reference. References to other works are made throughout the text by abbreviations, a list of which is given in the Bibliography.

Although the topics in this book are normally considered to be on an advanced level, they are presented in a form suitable for students of less extensive backgrounds. Depending on the previous training of the students, the book may be used for a one-semester course at the senior or senior-graduate level. Chapters 14 and 15 on the transform method could be omitted without loss of continuity for a more leisurely paced course in which the primary interest is in the special functions that occur in boundary-value problems.

We wish to acknowledge the many helpful suggestions made by Professor Robert E. Zink of Purdue University, in his review of the manuscript. In particular, we are indebted to him for his proof that the Bessel functions of integral order, J_n and J_{-n}, are linearly dependent.

DAVID E. JOHNSON
JOHNNY R. JOHNSON

Baton Rouge, Louisiana

MATHEMATICAL METHODS
IN
ENGINEERING AND PHYSICS

1

Orthogonal Functions

1.1 INTRODUCTION

In mathematical physics and engineering many phenomena may be described by a function ψ which satisfies an ordinary or partial differential equation, valid throughout a given region, together with prescribed conditions at the boundary of the region. The problem of determining ψ meeting these requirements constitutes a *boundary-value problem*, the solution of which is generally undertaken in two major steps. The first of these consists of obtaining a general solution of the differential equation, which, because it generally contains arbitrary constants or arbitrary functions, may be thought of as a family of solutions. The second step is concerned with selecting the member or members of this family which satisfy the boundary conditions. If the solution is unique, there will be, of course, only one such member.

In this book the boundary-value problems with which we shall be primarily concerned are those whose differential equations are partial differential equations such as *Laplace's* equation,

$$\nabla^2\psi = 0, \tag{1.1.1}$$

the *wave* equation,

$$\nabla^2\psi = \frac{1}{c^2}\frac{\partial^2\psi}{\partial t^2}, \tag{1.1.2}$$

the *diffusion* equation,

$$\nabla^2\psi = \frac{1}{k}\frac{\partial\psi}{\partial t}, \tag{1.1.3}$$

Poisson's equation,

$$\nabla^2\psi = f(x, y, z), \tag{1.1.4}$$

and the *Helmholtz* equation,

$$\nabla^2\psi + K^2\psi = 0, \tag{1.1.5}$$

where in rectangular coordinates the *Laplacian* operator ∇^2 is defined by

$$\nabla^2\psi = \frac{\partial^2\psi}{\partial x^2} + \frac{\partial^2\psi}{\partial y^2} + \frac{\partial^2\psi}{\partial z^2}. \tag{1.1.6}$$

The function ψ will generally be a function of the space coordinates x, y, z and time. These are but a few of the partial differential equations which occur in applications, but they appear in a relatively large number of boundary-value problems. For example, problems in gravitation, electrostatics, magnetostatics, fluid flow, and steady-state heat flow lead to Laplace's equation.

In general, the physical boundary of the region under consideration determines the coordinate system which can best be used. For example if the region were a rectangular solid, it would be desirable to use rectangular coordinates so that each surface could be represented by an equation in which a coordinate is equated to a constant. On the other hand, if the region were a cylinder, the equations of its boundaries would be unnecessarily complicated in rectangular coordinates but would be of the form $r = a$ and $z = b$ in cylindrical coordinates.

Since $\nabla^2\psi$ is involved in the differential equation, it is desirable to obtain its representation in the other coordinate systems of interest. The coordinate systems which we shall consider are orthogonal curvilinear coordinates which are related to the rectangular coordinates by the relations $x = x(q_1, q_2, q_3)$, $y = y(q_1, q_2, q_3)$, $z = z(q_1, q_2, q_3)$. In these coordinate systems it may be shown that

$$\nabla^2\psi = \frac{1}{h_1h_2h_3}\left[\frac{\partial}{\partial q_1}\left(\frac{h_2h_3}{h_1}\frac{\partial\psi}{\partial q_1}\right) + \frac{\partial}{\partial q_2}\left(\frac{h_1h_3}{h_2}\frac{\partial\psi}{\partial q_2}\right) + \frac{\partial}{\partial q_3}\left(\frac{h_1h_2}{h_3}\frac{\partial\psi}{\partial q_3}\right)\right], \tag{1.1.7}$$

where the *scale factors* h_i are given by

$$h_i^2 = \left(\frac{\partial x}{\partial q_i}\right)^2 + \left(\frac{\partial y}{\partial q_i}\right)^2 + \left(\frac{\partial z}{\partial q_i}\right)^2. \tag{1.1.8}$$

(See, for example, [D]*, Chapter 3.)

There are two generally accepted procedures which are used to solve a wide variety of partial differential equations. These procedures are the method of *integral representation*, which embraces such subjects as integral transforms and Green's functions, and the method of *separation of variables*. Although we shall consider the transform method briefly in Chapters 14 and 15, the method of separation of variables will be our main concern.

Separation of variables consists of seeking solutions of the form

$$\psi(q_1, q_2, q_3, t) = F_1(q_1)F_2(q_2)F_3(q_3)T(t) \tag{1.1.9}$$

* References, thus cited in brackets, are listed in the Bibliography.

and splitting the partial differential equation into four (or fewer) ordinary differential equations. The solutions to these equations are the *special functions* that will engage a major part of our interest in the book.

As an example of the separation-of-variables process consider the diffusion equation in one space coordinate,

$$k \frac{\partial^2 \psi}{\partial x^2} = \frac{\partial \psi}{\partial t}.$$

If the solution is of the form $\psi = X(x)T(t)$, we must have

$$kX''T = XT',$$

where the primes indicate differentiation with respect to the arguments. Upon division by XT this equation becomes

$$\frac{kX''(x)}{X(x)} = \frac{T'(t)}{T(t)}.$$

This expression, requiring that a function of x alone be identical with a function of t alone, can be satisfied only if each function is constant. That is, the function on the left cannot vary with t and the function on the right cannot vary with x. Setting the two functions equal to a common constant λ, known as the *separation constant*, we have the desired separated equations,

$$kX'' - \lambda X = 0, \quad T' - \lambda T = 0.$$

A limitation of the method is that the separation is not always possible to effect. For example, separation is possible in Eqs. (1.1.1)–(1.1.5) only for certain coordinate systems. For a thorough discussion of the subject of separability, the interested reader is referred to [MF], Chapter 5 and [MS].

We will postpone further discussion of the separation-of-variables method until we have considered some preliminary definitions pertinent to all the special functions, and have studied in detail the properties of the functions individually.

EXERCISES

1.1.1. Show that for cylindrical coordinates, $x = r \cos \theta$, $y = r \sin \theta$, and $z = z$, the Laplacian is given by

$$\nabla^2 \psi(r, \theta, z) = \frac{\partial^2 \psi}{\partial r^2} + \frac{1}{r}\frac{\partial \psi}{\partial r} + \frac{1}{r^2}\frac{\partial^2 \psi}{\partial \theta^2} + \frac{\partial^2 \psi}{\partial z^2}.$$

1.1.2. Show that for spherical coordinates, $x = r \sin \theta \cos \phi$, $y = r \sin \theta \sin \phi$, and $z = r \cos \theta$, the Laplacian becomes

$$\nabla^2 \psi(r, \theta, \phi) = \frac{\partial^2 \psi}{\partial r^2} + \frac{2}{r}\frac{\partial \psi}{\partial r} + \frac{1}{r^2 \sin^2 \theta}\frac{\partial^2 \psi}{\partial \phi^2} + \frac{1}{r^2}\frac{\partial^2 \psi}{\partial \theta^2} + \frac{\cot \theta}{r^2}\frac{\partial \psi}{\partial \theta}.$$

1.1.3. Show that for elliptical cylinder coordinates, $x = a \cosh \xi \cos \eta$, $y = a \sinh \xi \sin \eta$, and $z = z$, the Laplacian is

$$\nabla^2 \psi(\xi, \eta, z) = \frac{2}{a^2(\cosh 2\xi - \cos 2\eta)}\left[\frac{\partial^2 \psi}{\partial \xi^2} + \frac{\partial^2 \psi}{\partial \eta^2}\right] + \frac{\partial^2 \psi}{\partial z^2}.$$

1.1.4. Show that Laplace's equation, the wave equation, and the diffusion equation are separable in rectangular coordinates.

1.1.5. Work Exercise 1.1.4 in cylindrical coordinates.

1.1.6. Work Exercise 1.1.4 in spherical coordinates.

1.2 PRELIMINARY DEFINITIONS

The *inner product* of two real functions $f(x)$ and $g(x)$ on the closed interval $[a, b]$ with *weight function* $w(x) > 0$, is defined by

$$(f, g) = \int_a^b w(x) f(x) g(x)\, dx, \tag{1.2.1}$$

with the *norm* of $f(x)$ given by

$$\|f\| = \sqrt{(f, f)}. \tag{1.2.2}$$

(We are assuming that the norms of both f and g exist, in which case f and g are said to be *square-integrable.* Under these conditions (1.2.1) will exist.) If $(f, g) = 0$, then $f(x)$ and $g(x)$ are said to be *orthogonal* on $[a, b]$ with weight function $w(x)$. We note from (1.2.1) that for a *non-trivial function* (that is, one whose norm is not zero), the norm is real and positive. Also we note that if the inner product is to exist, it is sufficient, for finite limits, that the functions involved in the integral be *sectionally continuous,* a term which we define as follows:

> **Definition.** A function $f(x)$ is sectionally continuous on a finite interval $a \le x \le b$ if the interval can be divided into a finite number of subintervals on each of which $f(x)$ is continuous and has finite limits as x approaches either end point of the subinterval from the interior.

Unless otherwise stated we shall restrict our attention to non-trivial functions which are sectionally continuous.

A *set* or *sequence* of functions $\{\phi_n(x)\}$, $n = 0, 1, 2, \ldots$, is an *orthogonal sequence* if

$$(\phi_m, \phi_n) = \|\phi_n\|^2 \, \delta_{mn}, \tag{1.2.3}$$

where δ_{mn} is the *Kronecker delta,* defined as 1 if $m = n$ and as 0 if $m \ne n$. In particular, if $\|\phi_n\| = 1$, the set $\{\phi_n(x)\}$ is said to be *orthonormal* on $[a, b]$. An orthonormal sequence can be constructed from any orthogonal sequence

$\{\psi_n(x)\}$ by setting $\phi_n(x) = \psi_n(x)/\|\psi_n\|$, in which case the sequence $\{\psi_n\}$ is said to be *normalized.*

A well-known example of an orthonormal sequence is the sequence $\{\phi_k\}$, where

$$\phi_{2n-1}(x) = \frac{1}{\sqrt{L}} \sin \frac{n\pi x}{L}, \quad \phi_{2n}(x) = \frac{1}{\sqrt{L}} \cos \frac{n\pi x}{L}, \quad n = 1, 2, 3, \ldots, \tag{1.2.4}$$

$$\phi_0(x) = \frac{1}{\sqrt{2L}},$$

with interval $(-L, L)$ and $w(x) = 1$. More generally the interval can be taken as any interval of length $2L$, and may be open or closed.

A more general definition of inner product in which the functions f and g may be complex is the *hermitian* inner product

$$(f, g) = \int_a^b w(x) f(x) \overline{g(x)}\, dx, \tag{1.2.5}$$

where $\overline{g(x)}$ is the complex conjugate of g, and $w(x)$ is real and positive. We note that by this definition the norm of f given by

$$\|f\|^2 = \int_a^b w(x) f(x) \overline{f(x)}\, dx = \int_a^b w(x)\, |f(x)|^2\, dx \tag{1.2.6}$$

is a real number. In (1.2.5) we are considering the integral of a complex function, $f = u + iv$ (u and v real), to be defined as

$$\int_a^b f(x)\, dx = \int_a^b u\, dx + i \int_a^b v\, dx. \tag{1.2.7}$$

EXERCISES

1.2.1. Show that for f and g real

(a) $(f, g) = (g, f)$.

(b) $(cf, g) = (f, cg) = c(f, g)$, where c is a constant.

(c) $(f + g, h) = (f, h) + (g, h)$.

(d) $(f, \sum_{k=1}^{n} g_k) = \sum_{k=1}^{n} (f, g_k)$, $n \geq 1$.

(e) For $f(x)$ a continuous function on (a, b), $\|f\| = 0$ if and only if $f \equiv 0$. Does $\|f\| = 0$ imply $f \equiv 0$?

(f) If $(\phi_n, \phi_m) = \delta_{mn}$, then $\|\phi_n - \phi_m\|^2 = 2(1 - \delta_{mn})$.

1.2.2. Consider the double integral

$$\int_a^b \int_a^b w(x) w(y) [f(x) g(y) - f(y) g(x)]^2\, dx\, dy \geq 0$$

to obtain the *Cauchy-Schwarz inequality*, $(f, g) \leq \|f\| \cdot \|g\|$.

1.2.3. Derive the *Minkowski inequality*, $\|f + g\| \leq \|f\| + \|g\|$. Show that equality holds if $f = g$, and that if $(f, g) = 0$, then $\|f + g\|^2 = \|f\|^2 + \|g\|^2$.

1.2.4. Show that the sequence of functions defined by (1.2.4) is orthonormal over $(\alpha, \alpha + 2L)$, where α is any real constant.

1.2.5. Show that if the sequence $\{\phi_n(x)\}$ is orthogonal then the sequence $\{\psi_n(x)\}$, with $\psi_n(x) = \phi_n(x)/\|\phi_n\|$, is orthonormal.

1.2.6. Show that $\{\sin nx\}$, $n = 1, 2, 3, \ldots$, is an orthogonal sequence over $(0, \pi)$, with $w(x) = 1$, and from this sequence construct an orthonormal sequence.

1.2.7. Show that the functions $L_0(x) = 1$, $L_1(x) = -x + 1$, and $L_2(x) = x^2 - 4x + 2$ are orthogonal over $(0, \infty)$ with $w(x) = e^{-x}$.

1.2.8. Show that if $f(x)$ is sectionally continuous on (a, b) then it has no infinite discontinuities there.

1.2.9. Show that if f and g are sectionally continuous on (a, b) then fg is sectionally continuous on (a, b).

1.2.10. Show that

$$\int_0^a \sin \lambda_n x \sin \lambda_m x \, dx = \frac{\lambda_m \cos a\lambda_m \sin a\lambda_n - \lambda_n \cos a\lambda_n \sin a\lambda_m}{\lambda_n^2 - \lambda_m^2}$$

if $\lambda_n^2 \neq \lambda_m^2$, and

$$\int_0^a \sin^2 \lambda_n x \, dx = \frac{a}{2} - \frac{1}{4\lambda_n} \sin 2\lambda_n a.$$

1.2.11. Show from the results of Exercise 1.2.10 that $\{\sin \lambda_n x\}$, where λ_n, $n = 1, 2, 3, \ldots$, are positive roots of $\tan a\lambda = b\lambda$, is an orthogonal set over $(0, a)$ with $w(x) = 1$. Show also that $\|\sin \lambda_n x\|^2 = \frac{1}{2}(a - b \cos^2 \lambda_n a)$.

1.3 GENERALIZED FOURIER SERIES

Let us suppose that a function $f(x)$ defined on (a, b) is to be represented as a series of orthogonal functions $\{\phi_n(x)\}$, $n = 0, 1, 2, \ldots$, defined by (1.2.3); that is,

$$f(x) = \sum_{n=0}^{\infty} c_n \phi_n(x), \quad a < x < b. \tag{1.3.1}$$

A formal procedure for determining the unknown coefficients c_n consists of multiplying both sides of (1.3.1) by $w(x)\phi_m(x)$ and integrating term by term from a to b, obtaining

$$c_n = \frac{(f, \phi_n)}{\|\phi_n\|^2}. \tag{1.3.2}$$

These are called the *Fourier coefficients*, and (1.3.1) is the *generalized Fourier series* corresponding to $f(x)$ relative to the set $\{\phi_n(x)\}$.

These formal steps in obtaining c_n are justified if the series (1.3.1) converges to $f(x)$ and if after multiplying it through by $\phi_m(x)w(x)$ the resulting series is uniformly convergent [see Appendix A]. It should be noted that if the series (1.3.1) does not converge to $f(x)$, our use of the equality sign is incorrect, and we shall in that case write $f(x) \sim \Sigma\, c_n\phi_n(x)$. Whether it converges to $f(x)$ or not, the series (1.3.1) with coefficients determined by (1.3.2) is defined as the generalized Fourier series corresponding to $f(x)$. As we shall see in the next section, this series is in one sense a better approximation of $f(x)$, regardless of its convergence, than any other linear combination of the $\phi_n(x)$.

1.4 BESSEL'S INEQUALITY

In applications the use, as in (1.3.1), of orthogonal functions to represent an arbitrary function $f(x)$ is of fundamental importance. Accordingly, we now seek to approximate $f(x)$ by a linear combination

$$g_n(x) = \sum_{k=0}^{n} a_k \phi_k(x), \tag{1.4.1}$$

and to show that the *mean-square error*,

$$E_n = \int_a^b w(x)[f(x) - g_n(x)]^2 \, dx, \tag{1.4.2}$$

is a minimum when $a_k = c_k$, given by (1.3.2). If the mean-square error approaches zero as n becomes infinite, we will say that the sequence $\{g_n(x)\}$ *converges in the mean* to $f(x)$.

Substituting (1.4.1) into (1.4.2) and expanding, we have

$$E_n = \|f\|^2 - 2\sum_{k=0}^{n} a_k(f, \phi_k) + \sum_{k=0}^{n} a_k^2 \, \|\phi_k\|^2, \tag{1.4.3}$$

where we have used the orthogonality property (1.2.3). By (1.3.2) this becomes

$$E_n = \|f\|^2 + \sum_{k=0}^{n}(-2a_k c_k + a_k^2) \, \|\phi_k\|^2, \tag{1.4.4}$$

which may be written

$$E_n = \|f\|^2 + \sum_{k=0}^{n}(a_k - c_k)^2 \, \|\phi_k\|^2 - \sum_{k=0}^{n} c_k^2 \, \|\phi_k\|^2.$$

Clearly E_n is a minimum when $a_k = c_k$, and

$$\min E_n = \|f\|^2 - \sum_{k=0}^{n} c_k^2 \, \|\phi_k\|^2. \tag{1.4.5}$$

Since by (1.4.2) E_n is non-negative, (1.4.5) is equivalent to

$$\sum_{k=0}^{n} c_k^2 \, \|\phi_k\|^2 \le \|f\|^2. \tag{1.4.6}$$

This is *Bessel's inequality*.

1.5 PARSEVAL'S EQUATION

If the mean-square error E_n converges to zero as n becomes infinite, then the approximation in the mean to $f(x)$ can be made as close as we wish by taking a sufficient number of terms in (1.3.1). The set $\{\phi_n(x)\}$ is said to be *complete* if for every sectionally continuous function $f(x)$ on (a, b), E_n converges to 0 as $n \to \infty$. If the set $\{\phi_n(x)\}$ is complete, Bessel's inequality becomes *Parseval's equation*:

$$\sum_{k=0}^{\infty} c_k^{\,2} \, \|\phi_k\|^2 = \|f\|^2. \tag{1.5.1}$$

Conversely, if Parseval's equation holds, E_n converges to 0, by (1.4.5) and the set $\{\phi_n(x)\}$ is complete. Hence a necessary and sufficient condition for the convergence in the mean of (1.3.1) is the validity of Parseval's equation.

If the set $\{\phi_n(x)\}$ is orthonormal, Parseval's equation becomes

$$\sum_{k=0}^{\infty} c_k^{\,2} = \|f\|^2. \tag{1.5.2}$$

Hence if $\|f\|^2$ exists, the series of $c_k^{\,2}$ converges and it is necessary that

$$\lim_{k\to\infty} c_k = 0. \tag{1.5.3}$$

This important result may also be obtained, even if Parseval's equation does not hold, by considering Bessel's inequality for the case $\|\phi_k\|^2 = A$, independent of k. In this case we have in (1.4.6) a series of positive terms bounded above, and hence the series must converge.

EXERCISES

1.5.1. Find the expansion (1.3.1) for $f(x) = x$, $0 < x < \pi$, using the set $\{\sin nx\}$ given in Exercise 1.2.6. Ans. $x = 2\sum_{n=1}^{\infty}[(-1)^{n+1} \sin nx]/n$.

1.5.2. Show that for the set in Exercise 1.5.1, Parseval's equation is

$$\sum_{n=1}^{\infty} c_n^{\,2} = \frac{2}{\pi}\int_0^{\pi}[f(x)]^2\,dx,$$

where $c_n = (2/\pi)\int_0^{\pi} f(x) \sin nx\,dx$. If $f'(x)$ is continuous, show directly by integrating by parts that $\lim_{n\to\infty} c_n = 0$.

1.5.3. Given that Parseval's equation holds in Exercise 1.5.1, find the sum of the series $\sum_{n=1}^{\infty} 1/n^2$. Ans. $\pi^2/6$.

1.5.4. An orthonormal set $\{\phi_n(x)\}$ is *closed* on (a, b) if, for any $\psi(x)$ such that $\int_a^b w(x)\psi^2(x)\,dx$ exists and $(\psi, \phi_k) = 0$ for all k, $\psi(x) \equiv 0$ (or a trivial function). Show that if $\{\phi_n(x)\}$ is complete, then it is closed.

1.5.5. Show that the set $\{\sqrt{2/\pi}\cos nx\}$, $n = 1, 2, 3, \ldots$, is orthonormal on $(0, \pi)$ with $w(x) = 1$, but it is not closed. Is $\{\sqrt{2/\pi}\sin nx\}$ closed for these conditions?

1.6 AN ORTHOGONALIZATION PROCESS

A set of functions $\{\phi_n(x)\}$, $n = 1, 2, \ldots, m$, is said to be *linearly independent* on (a, b) if the equation

$$\sum_{n=1}^{m} c_n\phi_n(x) \equiv 0, \quad a < x < b, \tag{1.6.1}$$

is satisfied if and only if $c_n = 0$, for $n = 1, 2, 3, \ldots, m$. A set which is not linearly independent is defined to be *linearly dependent*, and in this case at least one coefficient, say c_k, in (1.6.1) is not zero. We may then divide through by c_k and solve for $\phi_k(x)$ in terms of the other members of the set. Hence for a linearly dependent set, one member can always be expressed as a linear combination of the others. As an example, the set $\{\sin^2 x, \cos^2 x, \cos 2x\}$ is not linearly independent because

$$c_1 \sin^2 x + c_2 \cos^2 x + c_3 \cos 2x \equiv 0$$

is satisfied for $c_1 = 1$, $c_2 = -1$, $c_3 = 1$.

A method for checking for the linear independence of the set $\{\phi_n(x)\}$, if the ϕ_n's are differentiable $m - 1$ times, is as follows. Differentiate Eq. (1.6.1) $m - 1$ times, obtaining a system of m homogeneous equations in the m unknown c_n's:

$$\begin{aligned}
&c_1\phi_1 + c_2\phi_2 + \cdots + c_m\phi_m \equiv 0,\\
&c_1\phi'_1 + c_2\phi'_2 + \cdots + c_m\phi'_m \equiv 0,\\
&\cdots\cdots\cdots\cdots\cdots\cdots\cdots\cdots\cdots\cdots,\\
&c_1\phi_1^{(m-1)} + c_2\phi_2^{(m-1)} + \cdots + c_m\phi_m^{(m-1)} \equiv 0.
\end{aligned}$$

In order that there exist a non-trivial solution (at least one $c_n \neq 0$), the determinant of the system,

$$W(x) = \begin{vmatrix} \phi_1 & \phi_2 & \cdots & \phi_m \\ \phi'_1 & \phi'_2 & \cdots & \phi'_m \\ \cdots & \cdots & \cdots & \cdots \\ \phi_1^{(m-1)} & \phi_2^{(m-1)} & \cdots & \phi_m^{(m-1)} \end{vmatrix}, \tag{1.6.2}$$

must vanish. If $W(x) \not\equiv 0$, the only solution of the system is the trivial one and the set $\{\phi_n(x)\}$ is linearly independent. The determinant in (1.6.2) is called the *Wronskian* of the functions $\{\phi_n\}$.

If the set under consideration is an infinite set, it is linearly independent if every finite subset is linearly independent. Then $W(x) \not\equiv 0$ in (1.6.2) for any $m = 1, 2, 3, \ldots$.

If the set $\{\phi_n(x)\}$ is an orthogonal set, it is linearly independent. This can be shown by taking a termwise inner product in (1.6.1) of $\phi_n(x)$ and

$\phi_k(x)$, resulting in

$$\sum_{n=1}^{m} c_n(\phi_k, \phi_n) \equiv 0, \quad k = 1, 2, \ldots, m;\ m = 1, 2, 3, \ldots,$$

or

$$\sum_{n=1}^{m} c_n \, \|\phi_k\|^2 \, \delta_{kn} = c_k \, \|\phi_k\|^2 = 0.$$

Hence $c_k = 0$, $k = 1, 2, 3, \ldots, m$. (We have assumed that the m members of the finite subset have been renumbered, if necessary, $\phi_1, \phi_2, \ldots, \phi_m$.)

Conversely, any set of linearly independent functions can be transformed into an orthogonal set, as we shall now show. The method we give is known as the *Gram-Schmidt process*.

Consider the set of linearly independent functions $\{f_n(x)\}$, $n = 0, 1, 2, \ldots$. We shall replace this set by an orthogonal set $\{\phi_n(x)\}$, each member of which is a linear combination of the f_n. We begin by setting

$$\phi_0(x) = f_0(x).$$

Next we write

$$\phi_1(x) = f_1(x) - a_{01}\phi_0(x),$$

and determine a_{01} by the condition $(\phi_0, \phi_1) = 0$. The next function is determined by the conditions

$$\phi_2(x) = f_2(x) - a_{02}\phi_0(x) - a_{12}\phi_1(x),$$

$$(\phi_m, \phi_2) = 0, \text{ for } m = 0, 1.$$

Proceeding in this manner, we write in general

$$\phi_0(x) = f_0(x),$$

$$\phi_n(x) = f_n(x) - \sum_{k=0}^{n-1} a_{kn}\phi_k(x), \quad n = 1, 2, 3, \ldots, \tag{1.6.3}$$

where the a_{kn} are determined by the n equations

$$(\phi_m, \phi_n) = 0, \quad m = 0, 1, 2, \ldots, n - 1,$$

which may be written, using (1.6.3), as

$$(\phi_m, \phi_n) = (\phi_m, f_n) - \sum_{k=0}^{n-1} a_{kn}(\phi_m, \phi_k)$$

$$= (\phi_m, f_n) - a_{mn} \, \|\phi_m\|^2 = 0.$$

Therefore

$$a_{kn} = \frac{(\phi_k, f_n)}{\|\phi_k\|^2}, \quad k = 0, 1, 2, \ldots, n - 1;\ n = 1, 2, 3, \ldots. \tag{1.6.4}$$

The linear independence of the $f_n(x)$ ensures the non-vanishing of the denominators. That is, $\phi_n(x)$ is not identically zero, for then (1.6.3) would imply that $f_n(x)$ is a linear combination of $\phi_0, \phi_1, \ldots, \phi_{n-1}$, and consequently is a linear combination of $f_0, f_1, \ldots, f_{n-1}$. This contradicts the linear independence of the set $\{f_n(x)\}$.

Transforming from a non-orthogonal to an orthogonal set of functions is of considerable importance in approximating a function $f(x)$ by a linear combination of linearly independent functions. While it may be possible to make the approximation by means of a set of non-orthogonal functions, in general the constants are not as readily obtainable as those given in (1.3.2).

1.7 AN EXAMPLE

As an illustration of the orthogonalization process we consider the set of linearly independent functions $\{x^n\}$, $n = 0, 1, 2, \ldots,$ as our set $\{f_n(x)\}$. A variety of sets of orthogonal functions can be obtained by considering various weight functions and intervals. In this example we take $w(x) = 1$ and the interval as $(-1, 1)$.

According to (1.6.3) and (1.6.4), $\phi_0(x) = 1$, and

$$\begin{aligned}\phi_1(x) &= x - a_{01}\\ &= x - \frac{(1, x)}{\|1\|^2} = x.\end{aligned}$$

Similarly,

$$\begin{aligned}\phi_2(x) &= x^2 - \tfrac{1}{3},\\ \phi_3(x) &= x^3 - \tfrac{3}{5}x,\\ \phi_4(x) &= x^4 - \tfrac{6}{7}x^2 + \tfrac{3}{35},\\ &\ldots\end{aligned}$$

A well-known orthogonal set, the *Legendre polynomials* $\{P_n(x)\}$, can be obtained from this set by writing

$$P_n(x) = b_n\phi_n(x), \quad n = 0, 1, 2, \ldots,$$

and determining b_n so that $P_n(1) = 1$. The resulting set is

$$\begin{aligned}P_0(x) &= 1,\\ P_1(x) &= x,\\ P_2(x) &= \tfrac{1}{2}(3x^2 - 1),\\ P_3(x) &= \tfrac{1}{2}(5x^3 - 3x),\\ P_4(x) &= \tfrac{1}{8}(35x^4 - 30x^2 + 3),\\ &\ldots\end{aligned} \tag{1.7.1}$$

EXERCISES

1.7.1. Show that the Wronskian for the set $\{\sin^2 x, \cos^2 x, \cos 2x\}$ is identically zero. Does this imply that the set is linearly dependent? See Exercises 1.7.7 and 1.7.10.

1.7.2. Show that the set $\{x^k\}$, $k = 0, 1, 2, \ldots, n$, is linearly independent.

1.7.3. Use the set $\{x^n\}$, $n = 0, 1, 2, \ldots$, $w(x) = 1$, and the interval $(0, 1)$ to obtain an orthogonal set $\{\phi_n(x)\}$.

1.7.4. Using $\{x^n\}$, $n = 0, 1, 2, \ldots$, $w(x) = e^{-x^2}$, on $(-\infty, \infty)$, obtain an orthogonal set $\{\phi_n(x)\}$. Show that $2^n\phi_n(x) = H_n(x)$, where

$$\begin{aligned}
H_0(x) &= 1, \\
H_1(x) &= 2x, \\
H_2(x) &= 4x^2 - 2, \\
H_3(x) &= 8x^3 - 12x, \\
H_4(x) &= 16x^4 - 48x^2 + 12, \\
H_5(x) &= 32x^5 - 160x^3 + 120x, \\
&\cdots
\end{aligned}$$

These are *Hermite polynomials* (see Chapter 9).

1.7.5. Obtain the *Laguerre polynomials*, $\{L_n(x)\}$, by using $\{x^n\}$, $n = 0, 1, 2, \ldots$, $w(x) = e^{-x}$, on $(0, \infty)$, taking $L_n(x) = [(-1)^n/n!]\phi_n(x)$. As we shall see in Chapter 10, these are given by

$$L_n(x) = \sum_{k=0}^{n} \frac{(-1)^k n! x^k}{(k!)^2(n-k)!}.$$

1.7.6. Obtain an orthogonal set based on the set $\{e^{-nx}\}$, $n = 0, 1, 2, \ldots$, for $w(x) = e^{-x}$, on $(0, \infty)$. Note that $(e^{-ax}, e^{-bx}) = 1/(a + b + 1)$.

Ans. $\phi_0(x) = 1$, $\phi_1(x) = e^{-x} - \frac{1}{2}$, $\phi_2(x) = e^{-2x} - e^{-x} + \frac{1}{6}$,
$\phi_3(x) = e^{-3x} - \frac{3}{2}e^{-2x} + \frac{3}{5}e^{-x} - \frac{1}{20}, \ldots$

1.7.7. The vanishing of the Wronskian is not sufficient to ensure the linear dependence of the set of functions. To show this, prove that the set $\{x^3, |x|^3\}$ in any interval including the origin is linearly independent, but $W(x) \equiv 0$.

1.7.8. Extend the result of Exercise 1.7.7 to the set $\{x^n, |x|^n\}$, where n is an odd positive integer greater than 1. What if $n = 1$?

1.7.9. Show that the set $\{e^{m_k x}\}$, $k = 1, 2, \ldots, n$, with $m_i \neq m_j$, is linearly independent over any finite interval. *Hint*: The *Vandermonde determinant* V has the value

$$V = \begin{vmatrix} 1 & 1 & 1 & \cdots & 1 \\ x_1 & x_2 & x_3 & \cdots & x_n \\ x_1^2 & x_2^2 & x_3^2 & \cdots & x_n^2 \\ \cdots & \cdots & \cdots & \cdots & \cdots \\ x_1^{n-1} & x_2^{n-1} & x_3^{n-1} & \cdots & x_n^{n-1} \end{vmatrix} = \prod_{i>j}(x_i - x_j).$$

1.7.10. Show that if the Wronskian of two functions vanishes identically in $[a, b]$ and if the functions do not vanish simultaneously in $[a, b]$, then they are linearly dependent in $[a, b]$.

1.8 GENERATING FUNCTIONS AND RECURRENCE RELATIONS

A function $g(x, t)$ whose power series in t is given by

$$g(x, t) = \sum_n \phi_n(x)t^n = G(\{\phi_n\}) \tag{1.8.1}$$

is a *generating function* of the set $\{\phi_n(x)\}$. The summation is taken over all values of n appearing in the set. The notation $G(\{\phi_n\})$ is used to indicate that the set of functions is being generated rather than a single member. For convenience we will henceforth use $G(\phi_n)$.

If n is non-negative, $\phi_n(x)$ may be obtained by differentiating (1.8.1) n times with respect to t, and equating t to zero. That is,

$$\phi_n(x) = \frac{1}{n!}\frac{\partial^n}{\partial t^n} g(x, t)\bigg]_{t=0}. \tag{1.8.2}$$

This result is, of course, not surprising since (1.8.1) is the Maclaurin series of $g(x, t)$ in powers of t.

As an example, the generating function of the set $\{x^n\}$ considered in Sec. 1.7 is

$$\frac{1}{1 - xt} = \sum_{n=0}^{\infty} x^n t^n, \quad |xt| < 1. \tag{1.8.3}$$

As another example, the generating function for the set $\{\cos n\theta\}$ may be obtained from (1.8.3) by letting $x = e^{i\theta}$ and equating real parts:

$$\frac{1 - t\cos\theta}{1 - 2t\cos\theta + t^2} = \sum_{n=0}^{\infty} (\cos n\theta)t^n. \tag{1.8.4}$$

Recurrence relations for the set $\{\phi_n(x)\}$ are equations relating two or more members of the set or their derivatives. For example, the set $\{\cos n\theta\}$ in (1.8.4) has as a recurrence relation the trigonometric identity,

$$\cos (n + 1)\theta + \cos (n - 1)\theta = 2 \cos \theta \cos n\theta. \tag{1.8.5}$$

As another example, the *Fibonacci numbers*, $\{1, 1, 2, 3, 5, 8, \ldots\}$, a set of constant functions, satisfy the relations,

$$\phi_{n+1} = \phi_n + \phi_{n-1}; \quad \phi_0 = 1; \quad \phi_n = 0, \quad n < 0. \tag{1.8.6}$$

The recurrence relations are thus special types of difference equations, and usually connect consecutive members of the set. If no derivatives are present, the recurrence relation is called *pure*.

Frequently it is possible to determine a generating function from a given recurrence relation. If the recurrence relation has constant coefficients, the problem can be very simple. To illustrate a formal procedure that can be used, we consider (1.8.6) in the form

$$\phi_{n+2} = \phi_{n+1} + \phi_n, \quad n = 0, 1, 2, \ldots .$$

Multiplying by t^n and summing, we obtain

$$G(\phi_{n+2}) = G(\phi_{n+1}) + G(\phi_n), \tag{1.8.7}$$

where, for example,

$$G(\phi_{n+2}) = \sum_{n=0}^{\infty} \phi_{n+2} t^n.$$

We proceed to express $G(\phi_{n+2})$ and $G(\phi_{n+1})$ in terms of $G(\phi_n)$ as follows:

$$\begin{aligned} G(\phi_{n+2}) &= t^{-2} \sum_{n=2}^{\infty} \phi_n t^n \\ &= t^{-2}\left[\sum_{n=0}^{\infty} \phi_n t^n - \phi_0 - \phi_1 t\right] \\ &= t^{-2} G(\phi_n) - t^{-2} - t^{-1}. \end{aligned} \tag{1.8.8}$$

Similarly,

$$G(\phi_{n+1}) = t^{-1}[G(\phi_n) - \phi_0] = t^{-1}G(\phi_n) - t^{-1}. \tag{1.8.9}$$

Substituting (1.8.8) and (1.8.9) into (1.8.7) and solving for $G(\phi_n)$ yields the generating function for the Fibonacci numbers,

$$G(\phi_n) = (1 - t - t^2)^{-1} = \sum_{n=0}^{\infty} \phi_n t^n. \tag{1.8.10}$$

In the example above the coefficients in the difference equation were independent of n. We shall now consider the problem of determining the generating function when the recurrence relation has coefficients which are functions of n. For this purpose we need to express $G(n\phi_n)$ in terms of $G(\phi_n)$. Recalling that

$$G(\phi_n) = \sum_{n=0}^{\infty} \phi_n t^n,$$

we see that the term $n\phi_n$ can be obtained by differentiating with respect to t. Assuming that termwise differentiation is valid, we have

$$\frac{d}{dt} G(\phi_n) = \frac{1}{t} \sum_{n=0}^{\infty} n\phi_n t^n = \frac{1}{t} G(n\phi_n),$$

from which we obtain the relation

$$G(n\phi_n) = t \frac{d}{dt} G(\phi_n). \tag{1.8.11}$$

We leave as an exercise for the reader to show that

$$G[(n+1)\phi_{n+1}] = \frac{1}{t} G(n\phi_n) = \frac{d}{dt} G(\phi_n), \tag{1.8.12}$$

and

$$G[(n-1)\phi_{n-1}] = tG(n\phi_n) = t^2 \frac{d}{dt} G(\phi_n). \tag{1.8.13}$$

We have assumed that $\phi_n = 0$ if $n < 0$, and that the operations on the series are valid.

As an example let us consider the recurrence relation,

$$(n+1)L_{n+1} + (x-1-2n)L_n + nL_{n-1} = 0, \quad n \geq 0. \tag{1.8.14}$$

Multiplying by t^n and summing over all non-negative integers, we have

$$G[(n+1)L_{n+1}] + G[(x-1-2n)L_n] + G[nL_{n-1}] = 0. \tag{1.8.15}$$

By using the derived expressions above, this equation is transformed into the first-order differential equation,

$$(1-t)^2 \frac{dy}{dt} + (x-1+t)y = 0, \tag{1.8.16}$$

where $y = G(L_n)$. Its solution is

$$G(L_n) = C(1-t)^{-1} \exp\left(-\frac{x}{1-t}\right), \tag{1.8.17}$$

where C is an arbitrary function of x. Suppose that $L_0(x) = 1$. This condition requires that $C = e^x$ and, consequently,

$$G(L_n) = (1-t)^{-1} \exp\left(-\frac{xt}{1-t}\right) = \sum_{n=0}^{\infty} L_n(x)t^n. \tag{1.8.18}$$

This is the generating function for the simple Laguerre polynomials given in Exercise 1.7.5.

Conversely, recurrence relations may be obtained from a generating function. For example, writing (1.8.4) as

$$1 - t\cos\theta = (1 - 2t\cos\theta + t^2) \sum_{n=0}^{\infty} (\cos n\theta)t^n$$

and equating coefficients of t^n, we obtain the recurrence relation,

$$\cos n\theta - 2\cos\theta\cos(n-1)\theta + \cos(n-2)\theta = 0, \quad n > 1.$$

If n is replaced by $n+1$, this relation becomes the addition formula (1.8.5).

Interesting recurrence relations may also be obtained by differentiating the generating function and comparing coefficients of t^n. For example, if we differentiate (1.8.18) with respect to x, we obtain

$$-\frac{t}{(1-t)^2}\exp\left(-\frac{xt}{1-t}\right) = \sum_{n=0}^{\infty} L'_n(x)t^n. \tag{1.8.19}$$

This can be written as

$$-\frac{t}{1-t}\sum_{n=0}^{\infty} L_n(x)t^n = \sum_{n=0}^{\infty} L'_n(x)t^n.$$

Multiplying by $1 - t$ and comparing coefficients, we have

$$L'_n(x) - L'_{n-1}(x) = -L_{n-1}(x). \tag{1.8.20}$$

Finally, in connection with recurrence relations, we state the following theorem:

Theorem 1.8.1. Any orthogonal set $\{\phi_n\}$, $n = 0, 1, 2, \ldots$, where ϕ_n is a polynomial of degree n, satisfies a recurrence relation of the type

$$\phi_{n+1} + (a_n x + b_n)\phi_n + c_n\phi_{n-1} = 0, \quad n \geq 1, \tag{1.8.21}$$

where a_n, b_n, c_n are constants, provided there is one and only one polynomial of each degree present in the set.

To establish the theorem, we first take the inner product of both sides of (1.8.21) with ϕ_{n+1}, obtaining

$$\|\phi_{n+1}\|^2 + a_n(x\phi_n, \phi_{n+1}) = 0. \tag{1.8.22}$$

Since $x\phi_n$ is a polynomial of degree $n + 1$, we know from Sec. 1.6 that it may be written as the linear combination

$$x\phi_n = \sum_{k=0}^{n+1} A_k\phi_k,$$

where $A_{n+1} \neq 0$. (If $A_{n+1} = 0$, $x\phi_n$ is not of degree $n + 1$.) Thus $(x\phi_n, \phi_{n+1}) = A_{n+1}\|\phi_{n+1}\|^2 \neq 0$, and hence we may solve for a_n in (1.8.22). Taking the inner products of (1.8.21) with ϕ_n and with ϕ_{n-1}, we obtain

$$a_n(x\phi_n, \phi_n) + b_n\|\phi_n\|^2 = 0,$$
$$a_n(x\phi_{n-1}, \phi_n) + c_n\|\phi_{n-1}\|^2 = 0,$$

from which we may determine b_n and c_n. It may be verified that these indicated values of a_n, b_n, c_n satisfy (1.8.21).

EXERCISES

1.8.1. Derive Eq. (1.8.4).

1.8.2. Obtain the generating function for the set $\{\sin n\theta\}$, $n = 1, 2, \ldots$.

Ans. $g(x, t) = t\sin\theta/(1 - 2t\cos\theta + t^2)$.

1.8.3. Using (1.8.2) show that for $\phi_n(x) = e^{nx}$,

$$\frac{1}{1 - te^x} = \sum_{n=0}^{\infty} \phi_n(x)t^n.$$

1.8.4. Find the first few Legendre polynomials from the relation

$$(1 - 2xt + t^2)^{-1/2} = \sum_{n=0}^{\infty} P_n(x)t^n.$$

1.8.5. Given

$$g(x, t) = \exp(2xt - t^2) = \sum_{n=0}^{\infty} \frac{H_n(x)}{n!} t^n.$$

Obtain the first few Hermite polynomials by expanding $g(x, t)$ in its Maclaurin series.

1.8.6. Obtain the first few Fibonacci numbers from (1.8.10) by (a) the binomial expansion, and (b) Eq. (1.8.2). In (b), note that if $Q = (1 - t - t^2)^{-1}$, $Q' = (2t + 1)Q^2$, etc., and $Q(0) = 1$.

1.8.7. Obtain a general expression for the Fibonacci numbers by expanding the generating function (1.8.10).

Ans.

$$\phi_n = \sum_{k=0}^{[n/2]} \binom{n-k}{k}, \text{ where } [n/2] = \text{largest integer} \le \frac{n}{2}, \text{ and } \binom{n}{r} = \frac{n!}{r!(n-r)!}.$$

1.8.8. Prove that the Fibonacci numbers are given by

$$\phi_n = \frac{1}{\sqrt{5}}\left[\left(\frac{1 + \sqrt{5}}{2}\right)^{n+1} - \left(\frac{1 - \sqrt{5}}{2}\right)^{n+1}\right].$$

Use partial fractions to write the generating function as a sum of two fractions.

1.8.9. Beginning with ϕ_n in Exercise 1.8.8, obtain the generating function $G(\phi_n)$ by multiplying by t^n and summing over n. Note that

$$(1 - xt)^{-1} = \sum_{n=0}^{\infty} x^n t^n.$$

1.8.10. For the set $\{\phi_n(x)\} = \{e^{-nx}\}$ obtain a recurrence relation that holds for any three consecutive functions.

1.8.11. Use the method of Exercise 1.8.9 to obtain a generating function for $\{e^{-nx}\}$.

1.8.12. Verify (1.8.12) and (1.8.13).

1.9 STURM-LIOUVILLE SYSTEMS

A general second-order linear equation

$$P(x)y'' + Q(x)y' + R(x)y = 0, \tag{1.9.1}$$

may be put in *self-adjoint form*,

$$\frac{d}{dx}[r(x)y'] + g(x)y = 0, \tag{1.9.2}$$

by multiplying through by the expression,

$$u(x) = \frac{1}{P(x)} \exp\left[\int \frac{Q(x)}{P(x)}\,dx\right] = \exp\left[\int \left(\frac{Q - P'}{P}\right)dx\right], \quad P(x) \neq 0. \quad (1.9.3)$$

In (1.9.2) we have

$$r(x) = \exp\left[\int \frac{Q(x)}{P(x)}\,dx\right]; \quad g(x) = \frac{R(x)}{P(x)} \exp\left[\int \frac{Q(x)}{P(x)}\,dx\right]. \quad (1.9.4)$$

The boundary-value problem of finding solutions valid for $a \leq x \leq b$ of the special case of (1.9.2),

$$\frac{d}{dx}[r(x)y'] + [p(x) + \lambda w(x)]y = 0, \quad (1.9.5)$$

subject to the *homogeneous boundary conditions*,

$$\begin{aligned} Ay'(a) + By(a) &= 0, \\ Cy'(b) + Dy(b) &= 0, \end{aligned} \quad (1.9.6)$$

is known as the *Sturm-Liouville problem*, and Eq. (1.9.5) is known as the Sturm-Liouville equation. The constants A, B, C, D are real and not all zero, the functions, $r(x)$, $r'(x)$, $p(x)$, and $w(x)$ are real and continuous, with $r(x) > 0$, $w(x) > 0$, and λ is a parameter. We are seeking solutions y which are continuous and have continuous first derivatives on (a, b).

The system (1.9.5) and (1.9.6) obviously has the trivial solution $y = 0$ for any value of λ, but in general, non-trivial solutions will depend on λ. The notation we will use is

$$y_n = y(x, \lambda_n), \quad (1.9.7)$$

where λ_n, $n = 0, 1, 2, \ldots$, are values of λ known as *eigenvalues*. The corresponding y_n are *eigenfunctions* of the system.

We may show that the set of eigenfunctions $\{y_n\}$ is an orthogonal set over (a, b) by considering the system for two distinct eigenvalues,

$$\begin{aligned} [r(x)y'_n]' + (p + \lambda_n w)y_n &= 0, \\ [r(x)y'_m]' + (p + \lambda_m w)y_m &= 0. \end{aligned}$$

Multiplying the first equation by y_m and the second by y_n, subtracting, and rearranging, we have

$$\frac{d}{dx}\{r(x)[y'_n y_m - y'_m y_n]\} = (\lambda_m - \lambda_n)w(x)y_n y_m.$$

Integrating both members from a to b we have, noting that $w(x)$ is the weight function defined in Sec. 1.2,

$$\begin{aligned} (\lambda_m - \lambda_n)\int_a^b w(x)y_n y_m\,dx &= (\lambda_m - \lambda_n)(y_n, y_m) \\ &= [r(x)\{y'_n y_m - y'_m y_n\}]_a^b. \end{aligned} \quad (1.9.8)$$

Equations (1.9.6) must hold for both eigenfunctions y_n and y_m, yielding two pairs of homogeneous equations for A and B and for C and D. If these are to have non-zero solutions we must have the determinants of the coefficients vanish. That is,

$$\begin{vmatrix} y'_n(a) & y_n(a) \\ y'_m(a) & y_m(a) \end{vmatrix} = \begin{vmatrix} y'_n(b) & y_n(b) \\ y'_m(b) & y_m(b) \end{vmatrix} = 0. \tag{1.9.9}$$

This makes the right member of (1.9.8) zero and, since we have assumed that $\lambda_m \neq \lambda_n$ for $m \neq n$, we have the orthogonality of y_n and y_m established:

$$(y_n, y_m) = \int_a^b w(x) y_n y_m \, dx = 0, \quad n \neq m. \tag{1.9.10}$$

If $r(a)$ or $r(b)$ or both vanish, we do not need both conditions in (1.9.9) in order for $\{y_n\}$ to be orthogonal. For example, the eigenfunctions for

$$\frac{d}{dx}[(1 - x^2)y'_n] + \lambda_n y_n = 0 \tag{1.9.11}$$

are orthogonal on $(-1, 1)$ independently of (1.9.9) because $r(x)$ vanishes at both end points.

The special case that will be of the most interest to us is

$$r(a) = 0, \quad Cy'(b) + Dy(b) = 0, \tag{1.9.12}$$

in which case (1.9.10) still holds.

We now state without proof (see [C-2], pp. 267–279) an important property of Sturm-Liouville systems, in the form of a theorem.

Theorem 1.9.1. If r, r', p, w, and $(rw)''$ are real and continuous, $w(x) > 0$, and $r(x) > 0$ on $a \leq x \leq b$, and A, B, C, D are real constants independent of λ, then the system (1.9.5) and (1.9.6) has an infinite set of eigenvalues λ_n corresponding to a set of real eigenfunctions $\{\phi_n(x)\}$, orthogonal on (a, b) with weight function $w(x)$, where $\phi_n(x)$ and $\phi'_n(x)$ are continuous. The series $\Sigma_n C_n \phi_n(x) = f(x)$, with C_n the Fourier coefficients, converges at every point where $f(x)$ is continuous on (a, b) if f and f' are sectionally continuous functions.

EXERCISES

1.9.1. Show that the functions $\{\sin(n\pi x/L)\}$ are the eigenfunctions of the Sturm-Liouville system,

$$y'' + \lambda y = 0, \quad y(0) = y(L) = 0,$$

and find the eigenvalues.

1.9.2. Find the eigenvalues and functions of the following systems: $y'' + \lambda y = 0$ and

(a) $y(0) = y'(L) = 0$.
(b) $y'(0) = y'(L) = 0$.
(c) $y(0) = 0$, $y(1) + y'(1) = 0$.

1.9.3. Show that $y_n(x) = (c_n/\sqrt{x}) \sin[(n\pi/2) \ln x]$, $n = 1, 2, \ldots$, for the system,

$$(x^2y')' + \lambda y = 0, \quad y(1) = y(e^2) = 0.$$

This differential equation is of the form $\Sigma_{k=0}^{n} a_k x^k y^{(k)}(x) = 0$, and is known by the names, Euler's equation, Cauchy's equation, the homogeneous equation, and the equidimensional equation. It can be solved by assuming a solution of the type $y = x^p$.

1.9.4. Put *Chebyshev's equation*, $(1 - x^2)y'' - xy' + n^2y = 0$, into self-adjoint form.

1.9.5. Put *Laguerre's equation*, $xy'' + (1 - x)y' + ay = 0$ (a constant), into self-adjoint form.

1.9.6. Obtain from (1.9.1) the differential equation that the function $u(x)$ in (1.9.3) must satisfy, and from this equation show that (1.9.3) is true.

1.9.7. Show that $\bar{y}$, the complex conjugate of y, is an eigenfunction corresponding to $\bar{\lambda}$ in (1.9.5) and (1.9.6). Using this fact, show that the eigenvalues are real.

1.9.8. Show that if conditions (1.9.6) are replaced by the *periodic conditions*, $r(a) = r(b)$, $y(a) = y(b)$, $y'(a) = y'(b)$, then (1.9.10) still holds.

1.10 ORTHOGONAL FUNCTIONS OF TWO VARIABLES

In this section we shall extend some of the definitions and results obtained in Secs. 1.2–1.5 to the case of functions $f(x, y)$ defined over a region S in the x-y plane. The class of functions we shall consider are those which are *continuous by subregions*. That is, we may subdivide the region into polygons within each of which the function $f(x, y)$ is continuous except possibly for finite jumps on the boundaries.

As before, a set $\{\phi_{nm}(x, y)\}$, $n, m = 0, 1, 2, \ldots$, is defined to be orthogonal over the region if the inner product of any two different members is zero. For functions of two variables this assumes the form

$$(\phi_{nm}, \phi_{rs}) = \iint_S w(x, y)\phi_{nm}\phi_{rs}\, dx\, dy = \|\phi_{nm}\|^2\, \delta_{nm}^{rs}, \qquad (1.10.1)$$

where the norm is defined by

$$\|\phi_{nm}\| = \sqrt{(\phi_{nm}, \phi_{nm})}, \qquad (1.10.2)$$

and $\delta_{nm}^{rs} = 1$ if $r = n$ and $s = m$, and is zero otherwise. If $\|\phi_{nm}\| = 1$, $n, m = 0, 1, 2, \ldots$, the set $\{\phi_{nm}\}$ is orthonormal over S.

The generalized double Fourier series corresponding to a function $f(x, y)$ is given by

$$f(x, y) \sim \sum_{n} \sum_{m} C_{nm} \phi_{nm}(x, y), \tag{1.10.3}$$

where by a formal manipulation corresponding to the derivation of Eq. (1.3.2), we obtain the Fourier coefficients

$$C_{nm} = \frac{(f, \phi_{nm})}{\|\phi_{nm}\|^2}. \tag{1.10.4}$$

This procedure is valid if the series (1.10.3) converges to $f(x, y)$ and if after multiplying through by $w\phi_{rs}$ the resulting series converges uniformly [see Appendix A].

An example of an orthonormal set is $\{\phi_{nm}(x, y)\}$, where

$$\phi_{nm}(x, y) = \frac{2 \sin (n\pi x/a) \sin (m\pi y/b)}{\sqrt{ab}}, \tag{1.10.5}$$

where $w(x, y) = 1$ and the region S is the rectangle $0 < x < a$, $0 < y < b$.

If a function $f(x, y)$ is to be expanded in a series of type (1.10.3) where the orthonormal set is defined by (1.10.5), the reader may verify that the coefficients are given by

$$C_{nm} = \frac{4}{ab} \int_0^b \int_0^a f(x, y) \sin \frac{n\pi x}{a} \sin \frac{m\pi y}{b} \, dx \, dy. \tag{1.10.6}$$

The extension to functions of three or more variables of the ideas of orthogonal sets, generalized Fourier series, and coefficients is analogous to the foregoing procedure and will not be given here. Special cases will be considered as they arise later in the book.

EXERCISES

1.10.1. We shall see in Chapter 4 that for the set $\{P_n(x)\}$ of Legendre polynomials obtained in (1.7.1) we have $\int_{-1}^{1} P_n(x)P_k(x) \, dx = [2/(2n + 1)]\delta_{nk}$. Using this fact, show that the set $\{P_n(x) \cos my\}$, $n, m = 0, 1, 2, \ldots$, is orthogonal over the region $-1 < x < 1$, $0 < y < \pi$, with $w(x, y) = 1$. Calculate the norm and obtain from this set an orthonormal set $\{\phi_{nm}\}$.

1.10.2. If $f(x) = \sum_{n=0}^{\infty} \sum_{m=0}^{\infty} C_{nm}\phi_{nm}(x, y)$, with ϕ_{nm} given in Exercise 1.10.1, find

(a) C_{00}, C_{01}, C_{10} for $f(x) = e^x$.
(b) All C_{nm} for $f(x) = 1$.

1.10.3. Show that the mean-square error defined as

$$E_N = \iint_S w(x, y)\left[f(x, y) - \sum_{n=0}^{N}\sum_{m=0}^{N} C_{nm}\phi_{nm}\right]^2 dx\,dy$$

is a minimum when the C_{nm} are given by (1.10.4).

1.10.4. Show that Bessel's inequality and Parseval's equation for the two-variable case are, respectively,

$$\sum_{n=0}^{N}\sum_{m=0}^{N} C_{nm}^2 \,\|\phi_{nm}\|^2 \leq \|f(x, y)\|^2,$$

and

$$\sum_{n=0}^{\infty}\sum_{m=0}^{\infty} C_{nm}^2 \,\|\phi_{nm}\|^2 = \|f\|^2.$$

1.10.5. Show that for the set $\{\sin(n\pi x/a)\sin(m\pi y/b)\}$, $n, m = 1, 2, 3, \ldots$, for $0 \leq x \leq a$, $0 \leq y \leq b$, and $w(x, y) = 1$, Bessel's inequality takes the form

$$\sum_{n=1}^{N}\sum_{m=1}^{N} C_{nm}^2 \leq \frac{4}{ab}\,\|f(x, y)\|^2.$$

1.11 RODRIGUES' FORMULA

Thus far we have seen that we may obtain explicit expressions for the members of a set of special functions $\{\phi_n\}$ in various ways. For example, in (1.8.2) $\phi_n(x)$ was given explicitly by means of its generating function, in (1.6.3) the orthogonalization process was employed, and in Exercise 1.7.5 the set $L_n(x)$ was given in the form of a summation. In this section we give another explicit form, applicable when $\phi_n(x)$ is a polynomial, known as the *Rodrigues' formula*. Actually the formula discovered by Rodrigues was restricted to the case of the Legendre polynomials considered in (1.7.1). This formula, published by Rodrigues in 1816 in his "Mémoire sur l'attraction des sphéroides," is given for $D = d/dx$, by

$$P_n(x) = \frac{1}{2^n n!} D^n(x^2 - 1)^n, \tag{1.11.1}$$

which we consider in detail in Chapter 4. Similar formulas exist for the other polynomial sets and have become known as Rodrigues' formulas. In this section we derive a more general expression of type (1.11.1) which is applicable to all the polynomial sets that we consider.

We begin by considering the second-order differential equation

$$f(x)y'' + g(x)y' + ky = 0, \tag{1.11.2}$$

where $f(x)$ is a polynomial of degree 2 at most, $g(x)$ is a polynomial of degree 1 at most, and k is a constant, which we restrict later. We recall from Sec. 1.9 that (1.11.2) may be put in self-adjoint form by multiplying it through by the integrating factor

$$\mu = \frac{1}{f} \exp\left[\int \frac{g\,dx}{f}\right], \quad f \neq 0. \tag{1.11.3}$$

Now let us consider the function

$$w = [f(x)]^n \mu = f^{n-1} \exp\left[\int \frac{g\,dx}{f}\right]. \tag{1.11.4}$$

The differential equation for w may be obtained by one differentiation, and is found to be

$$fw' = [g + (n-1)f']w. \tag{1.11.5}$$

Using *Leibnitz's rule* for differentiating a product,

$$D^n uv = \sum_{k=0}^{n} \binom{n}{k} D^k u D^{n-k} v \tag{1.11.6}$$

(see Exercise 1.11.1), where $\binom{n}{k}$ is the binomial coefficient $n!/k!(n-k)!$, we may write

$$D^{n+1}(fw') = \sum_{k=0}^{n+1} \binom{n+1}{k} D^k f D^{n+1-k} w'.$$

Since f is a second-degree polynomial, we have only three non-zero terms,

$$D^{n+1}(fw') = fD^{n+2}w + (n+1)f'D^{n+1}w + \frac{n(n+1)}{2} f'' D^n w.$$

Similarly, we may apply Leibnitz's rule to obtain

$$D^{n+1}\{[g + (n-1)f']w\}$$
$$= [g + (n-1)f']D^{n+1}w + (n+1)[g' + (n-1)f'']D^n w.$$

Equating these last two expressions, which is equivalent to differentiating (1.11.5) $n+1$ times, we have the differential equation for $D^n w$,

$$f(D^n w)'' + (2f' - g)(D^n w)' + \frac{n+1}{2}[(2-n)f'' - 2g'](D^n w) = 0.$$

If we now make the substitution

$$D^n w = \mu y = \frac{1}{f} y \exp\left[\int \frac{g\,dx}{f}\right] \tag{1.11.7}$$

in the last equation, we have, after some simplification,

$$f(x)y'' + g(x)y' + \left[\frac{n}{2}(1-n)f'' - ng'\right]y = 0. \tag{1.11.8}$$

If we restrict k to be the constant in the brackets, this equation is identical with (1.11.2).

Collecting our results from (1.11.7) and (1.11.4), we have, for $\phi_n(x)$ a polynomial solution of (1.11.8),

$$\phi_n(x) = \frac{C_n}{\mu} D^n[f^n \mu], \tag{1.11.9}$$

or, substituting for μ, we have

$$\phi_n(x) = C_n f \exp\left(-\int \frac{g\,dx}{f}\right) D^n\left[f^{n-1} \exp\left(\int \frac{g\,dx}{f}\right)\right]. \tag{1.11.10}$$

The constant C_n is an arbitrary multiplier, making ϕ_n a more general solution of (1.11.8).

To see that $\phi_n(x)$ is a polynomial, we note that applying the operator D to the bracket expression in (1.11.10) results in

$$D\left[f^{n-1} \exp\left(\int \frac{g\,dx}{f}\right)\right] = [f^{n-2}g + (n-1)f^{n-2}f'] \exp\left(\int \frac{g\,dx}{f}\right),$$

and hence reduces the degree of the polynomial coefficient of $\exp[\int (g/f)dx]$ by 1 from $2n - 2$ to $2n - 3$. Applying the operator n times, we have a polynomial coefficient of degree $n - 2$, which when multiplied by f in (1.11.10) is of degree n. The exponentials cancel, leaving $\phi_n(x)$ a polynomial of degree n. The coefficient C_n may be determined for the special cases by comparing (1.11.10) with other known expressions for $\phi_n(x)$.

EXERCISES

1.11.1. Establish (1.11.6) by induction on n.

1.11.2. Supply the details in deriving (1.11.8).

1.11.3. Determine a Rodrigues' formula for the case $f(x) = 1 - x^2$ and $g(x) = -2x$. Obtain the differential equation for $\phi_n(x)$ for this case. Determine C_n by requiring that the highest-degree coefficient be $(2n)!/2^n(n!)^2$. The answer should be (1.11.1).

1.11.4. Determine a Rodrigues' formula and the differential equation for $\phi_n(x)$ if

(a) $f(x) = 1$, $g(x) = -2x$.
(b) $f(x) = x$, $g(x) = 1 - x$.
(c) $f(x) = 1 - x^2$, $g(x) = -x$.
(d) $f(x) = 1 - x^2$, $g(x) = -3x$.
(e) $f(x) = x(1 - x)$, $g(x) = \gamma - (\alpha + \beta + 1)x$, where $-\alpha = n = 0, 1, 2, \ldots$.
(f) $f(x) = x^2$, $g(x) = 2x + 2$.

Ans. (f) $\phi_n = C_n e^{2/x} D^n(x^{2n} e^{-2/x})$, $x^2 y'' + (2x + 2)y' = n(n + 1)y$.

1.11.5. We have shown that $\phi(x)$ in (1.11.10) is a polynomial of degree n for the case $f(x) = ax^2 + bx + c$, $g(x) = dx + e$, with $a, d \neq 0$. Show that this result is still true for the other possibilities, f and g of degrees less than or equal to 2 and 1 respectively, except for the special case f and g constants.

2

Fourier Series

2.1 FOURIER TRIGONOMETRIC SERIES

A special case of (1.3.1) is the series,

$$f(x) \sim \frac{a_0}{2} + \sum_{n=1}^{\infty}\left(a_n \cos\frac{n\pi x}{L} + b_n \sin\frac{n\pi x}{L}\right), \quad -L < x < L. \tag{2.1.1}$$

To show this, we consider the orthonormal set $\{\phi_k(x)\}$ given in (1.2.4),

$$\phi_0(x) = \frac{1}{\sqrt{2L}},$$

$$\phi_{2n-1}(x) = \frac{1}{\sqrt{L}}\sin\frac{n\pi x}{L}, \quad \phi_{2n}(x) = \frac{1}{\sqrt{L}}\cos\frac{n\pi x}{L}, \quad n = 1, 2, 3, \ldots,$$

and let

$$\begin{aligned} c_0 &= \sqrt{\frac{L}{2}}\, a_0, \\ c_{2k-1} &= \sqrt{L}\, b_k, \quad c_{2k} = \sqrt{L}\, a_k, \quad k = 1, 2, 3, \ldots. \end{aligned} \tag{2.1.2}$$

Then (2.1.1) reduces to

$$f(x) = \sum_{n=0}^{\infty} c_n \phi_n(x).$$

It follows from (1.3.2) that $c_n = (f, \phi_n)$, from which we obtain

$$\begin{aligned} a_n &= \frac{1}{L}\int_{-L}^{L} f(x) \cos\frac{n\pi x}{L}\, dx, \\ b_n &= \frac{1}{L}\int_{-L}^{L} f(x) \sin\frac{n\pi x}{L}\, dx, \quad n = 0, 1, 2, \ldots. \end{aligned} \tag{2.1.3}$$

The series defined by (2.1.1) and (2.1.3) is called the *Fourier trigonometric series* corresponding to $f(x)$. If the series converges to $f(x)$ on $(-L, L)$, it will still not represent $f(x)$ outside the interval unless $f(x)$ is periodic of period $2L$; that is, unless

$$f(x + 2L) = f(x). \tag{2.1.4}$$

For non-periodic $f(x)$, series (2.1.1) corresponds to the *periodic extension* of the portion of $f(x)$ on $(-L, L)$, whose graph will consist of the graph of $f(x)$ on $(-L, L)$ repeated periodically every $2L$ radians.

Since all functions are not expandable in a valid Fourier trigonometric series, we now state some sufficient conditions that the series (2.1.1) converge. These are known as the *Dirichlet conditions* and may be summarized as follows.

> If $f(x)$ is a periodic function of period $2L$ such that, for $-L \le x \le L$, $f(x)$ is bounded and has a finite number of maxima and minima and a finite number of discontinuities, then the Fourier series (2.1.1) converges to
>
> $$S_f(x) = \frac{f(x + 0) + f(x - 0)}{2}, \tag{2.1.5}$$
>
> where
>
> $$f(x \pm 0) = \lim_{h \to 0} f(x \pm h).$$

The Dirichlet conditions are general enough for the applications we shall consider, and they are usually easy to apply. Another set of conditions, which we show in Appendix B to be sufficient, is the following:

> If $f(x)$ is sectionally continuous on $(-L, L)$ and periodic of period $2L$, then the Fourier series corresponding to $f(x)$ converges to $S_f(x)$ given in (2.1.5) at every point on $(-L, L)$ at which $f(x)$ has a right-hand and a left-hand derivative.

Right- and left-hand derivatives of $f(x)$ at a point x are defined, respectively, as the limits, as $h \to 0$ through positive values, of

$$\frac{f(x + h) - f(x + 0)}{h} \quad \text{and} \quad \frac{f(x - 0) - f(x - h)}{h}.$$

For the special case $L = \pi$, the Fourier series assumes the simpler form,

$$f(x) = \frac{a_0}{2} + \sum_{n=1}^{\infty} (a_n \cos nx + b_n \sin nx), \tag{2.1.6}$$

where

$$\begin{aligned} a_n &= \frac{1}{\pi} \int_{-\pi}^{\pi} f(x) \cos nx \, dx, \\ b_n &= \frac{1}{\pi} \int_{-\pi}^{\pi} f(x) \sin nx \, dx. \end{aligned} \tag{2.1.7}$$

The use of the term $a_0/2$ rather than a_0 enables us to write one equation for all the a_n, for $n = 0, 1, 2, \ldots$.

2.2 AN EXAMPLE

To illustrate the method of the Fourier series expansion, we consider the function defined by

$$\begin{aligned} f(x) &= 1, \qquad 0 < x \leq \pi, \\ &= -1, \qquad -\pi < x \leq 0. \end{aligned}$$

We note that this function satisfies the Dirichlet conditions and hence it has a Fourier expansion. By Eq. (2.1.7) we have

$$a_n = -\frac{1}{\pi}\int_{-\pi}^{0} \cos nx\, dx + \frac{1}{\pi}\int_{0}^{\pi} \cos nx\, dx = 0,$$

$$\begin{aligned} b_n &= -\frac{1}{\pi}\int_{-\pi}^{0} \sin nx\, dx + \frac{1}{\pi}\int_{0}^{\pi} \sin nx\, dx \\ &= \frac{2}{n\pi}[1 - \cos n\pi] = \frac{2}{n\pi}[1 - (-1)^n]. \end{aligned}$$

Thus all the coefficients with even subscripts are zero. The series is therefore

$$f(x) = \frac{4}{\pi}\sum_{n\,\text{odd}} \frac{1}{n}\sin nx,$$

which may also be written in the form,

$$f(x) = \frac{4}{\pi}\sum_{n=1}^{\infty} \frac{\sin (2n-1)x}{2n-1}, \tag{2.2.1}$$

where we have replaced the summation index n by $2n - 1$.

EXERCISES

2.2.1. If $f(x)$ is periodic of period T, show that

(a) $f'(x)$ (if it exists) is periodic of period T.

(b) $f(x)$ is also periodic of period nT, where n is an integer.

(c) If in addition $g(x)$ is periodic of period T', then $f(x)g(x)$ is periodic of period P where P is the least common multiple of T and T'.

2.2.2. Verify Eqs. (2.1.3).

2.2.3. Find the Fourier series for

$$\begin{aligned} f(x) &= 0, \qquad -\pi < x < 0 \\ &= \pi, \qquad 0 < x < \pi. \end{aligned}$$

2.2.4. Find the Fourier series for

$$\begin{aligned} f(x) &= -1, \qquad -1 < x < 0 \\ &= x, \qquad 0 < x < 1, \end{aligned}$$

and use (2.1.5) to show that

$$\sum_{n=1}^{\infty} \frac{1}{(2n-1)^2} = \frac{\pi^2}{8}.$$

2.2.5. Show that the Fourier series for $f(x) = x$, $-\pi < x < \pi$, is

$$x = 2 \sum_{n=1}^{\infty} (-1)^{n+1} \frac{\sin nx}{n}.$$

(Compare with Exercise 1.5.1.) Plot the sum of the first four terms for $0 < x < \pi$, and compare with the graph of $f(x) = x$.

2.2.6. Show that Bessel's inequality for the set $\{\phi_n(x)\}$, $(-L, L)$, defined in (1.2.4), with c_n given by (2.1.2), is

$$\frac{a_0^2}{2} + \sum_{n=1}^{m} (a_n^2 + b_n^2) \leq \frac{1}{L} \int_{-L}^{L} [f(x)]^2 \, dx, \quad \text{for any } m = 1, 2, 3, \ldots,$$

and hence, if the integral exists, $a_n, b_n \to 0$ as $n \to \infty$. Show that Parseval's equality for this case with $L = \pi$ is

$$\frac{a_0^2}{2} + \sum_{n=1}^{\infty} (a_n^2 + b_n^2) = \frac{1}{\pi} \int_{-\pi}^{\pi} [f(x)]^2 \, dx.$$

2.2.7. Find the Fourier series for

$$\begin{aligned} f(x) &= x + 1, \quad && 0 < x < \pi \\ &= x - 1, && -\pi < x < 0. \end{aligned}$$

Ans. $f(x) = \dfrac{2}{\pi} \displaystyle\sum_{n=1}^{\infty} \left[\frac{1 - (-1)^n(1 + \pi)}{n}\right] \sin nx.$

2.2.8. Find the Fourier series for

$$\begin{aligned} f(x) &= 1, \quad && 0 < x < \pi \\ &= 2, && \pi < x < 2\pi. \end{aligned}$$

Ans. $f(x) = \dfrac{3}{2} - \dfrac{2}{\pi} \displaystyle\sum_{n=1}^{\infty} \frac{\sin (2n-1)x}{2n-1}.$

2.2.9. Find the Fourier series for

$$f(x) = e^x, \quad -\pi < x < \pi.$$

Ans. $f(x) = \dfrac{2 \sinh \pi}{\pi} \left\{\dfrac{1}{2} + \displaystyle\sum_{n=1}^{\infty} \frac{(-1)^n}{n^2 + 1} (\cos nx - n \sin nx)\right\}.$

2.2.10. Show, by integration of (2.1.3) by parts, that if $f(x)$ is continuous and $f'(x)$ sectionally continuous on $(-L, L)$, then

$$a_n = -\frac{1}{n\pi} \int_{-L}^{L} f'(x) \sin \frac{n\pi x}{L} \, dx, \quad n = 1, 2, 3, \ldots,$$

and hence $|a_n| < M/n$, M a positive constant. Thus $\lim_{n\to\infty} a_n = 0$. Show similarly that $\lim_{n\to\infty} b_n = 0$.

2.2.11. Show that if f and f' are continuous and f'' sectionally continuous on $(-L, L)$, then $|a_n| < 2ML^2/n^2\pi^2$ and $|b_n| < 2ML^2/n^2\pi^2$, where M is an upper bound of $|f''|$ on $(-L, L)$.

2.2.12. Show that the Fourier series corresponding to the function $f(x)$ in Exercise 2.2.11 converges uniformly. Use the Weierstrass M-test [see Appendix A].

2.2.13. Show from Exercises 1.9.1 and 1.9.2(b) with the aid of Theorem 1.9.1 that, for f and f' sectionally continuous, (2.1.1) converges to $f(x)$ at points of continuity on $-L \leq x \leq L$.

2.3 EVEN AND ODD FUNCTIONS

A function $f(x)$, defined for all x, is said to be *even* if $f(-x) = f(x)$, and to be *odd* if $f(-x) = -f(x)$, for all x. For example, the functions x^2 and $\cos nx$ are even and the functions x and $\sin nx$ are odd. Graphs of even and odd functions are shown in Fig. 2–1. We note that the even function displayed

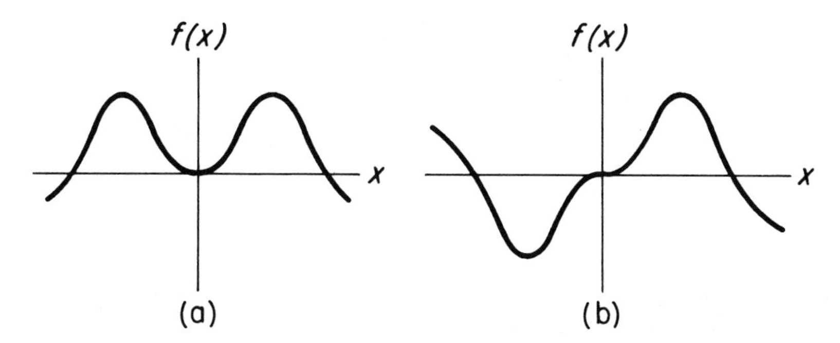

Fig. 2–1. (a) $f(x)$ Even; (b) $f(x)$ Odd

in Fig. 2–1(a) is symmetric about the vertical axis and the odd function in Fig. 2–1(b) is symmetric about the origin.

By consideration of the algebraic area under the curves in Fig. 2–1, it is clear that

$$\begin{aligned} &\int_{-a}^{a} f(x)\,dx = 2\int_0^a f(x)\,dx, \quad f(x) \text{ even}, \\ &\int_{-a}^{a} f(x)\,dx = 0, \quad f(x) \text{ odd}. \end{aligned} \tag{2.3.1}$$

An analytical proof of (2.3.1) will be left to the exercises (see Exercise 2.4.1).

Products of two even functions, or of two odd functions, are even, and products of an even function and an odd function are odd. For example, consider the function $F(x) = f(x)g(x)$, where $f(x)$ and $g(x)$ are both odd.

Then

$$F(-x) = f(-x)g(-x) = [-f(x)][-g(x)] = F(x).$$

This property together with (2.3.1) may be used to advantage in obtaining the Fourier series of functions which are even or odd, as we shall see in the next section.

The function $f(x) = x^2 + x$ is obviously neither even nor odd but is a sum of two functions, one of which is even and the other odd. It is not as obvious that the function $f(x) = xe^x$ can be written as the sum of an even and an odd function, as can be seen from

$$xe^x = \frac{xe^x - xe^{-x}}{2} + \frac{xe^x + xe^{-x}}{2}.$$

The first term on the right is even, and the second is odd. In general we may write

$$f(x) = f_e(x) + f_o(x), \tag{2.3.2}$$

where $f_e(x)$ is an even function and $f_o(x)$ is an odd function. To see this, we begin by replacing x by $-x$ in (2.3.2), which gives

$$f(-x) = f_e(x) - f_o(x).$$

Solving this equation with (2.3.2) gives the even and odd components,

$$f_e(x) = \frac{f(x) + f(-x)}{2}, \quad f_o(x) = \frac{f(x) - f(-x)}{2}. \tag{2.3.3}$$

2.4 EXPANSIONS OF EVEN AND ODD FUNCTIONS; HALF-RANGE SERIES

When the function to be expanded is either odd or even, the properties developed in Sec. 2.3 enable us to simplify the calculation of the Fourier coefficients. If $f(x)$ is even, the coefficients are

$$a_n = \frac{2}{L}\int_0^L f(x)\cos\frac{n\pi x}{L}\,dx, \quad b_n = 0, \tag{2.4.1}$$

and the resulting series is the *Fourier cosine series*,

$$f(x) = \frac{a_0}{2} + \sum_{n=1}^{\infty} a_n \cos\frac{n\pi x}{L}. \tag{2.4.2}$$

Similarly, if $f(x)$ is odd, we have

$$a_n = 0, \quad b_n = \frac{2}{L}\int_0^L f(x)\sin\frac{n\pi x}{L}\,dx, \tag{2.4.3}$$

and the *Fourier sine series*,

$$f(x) = \sum_{n=1}^{\infty} b_n \sin \frac{n\pi x}{L}. \tag{2.4.4}$$

We may note, as an example, that the function considered in Sec. 2.2 is an odd function and that its Fourier coefficients could have been obtained more readily from (2.4.3).

In many applied problems the function $f(x)$ may be defined only in the *half-range* $(0, L)$, and it could be expanded in a Fourier series as a periodic function of period L. However, in many cases it is more convenient to extend the definition to the interval $(-L, 0)$ so that the Fourier series represents a periodic function of period $2L$. Such a series is called a *half-range* series. If the extension of the interval is done in such a way as to make $f(x)$ either odd or even, the half-range series will take the form of either the sine or cosine series and will in either case converge to the same function $f(x)$ on $(0, L)$ if the extended function has a valid Fourier series. Sometimes, as will be seen in a later chapter, the nature of the problem determines the type of extension. If we have a choice of extensions, the rapidity of convergence of the two series may determine which is preferable.

An example of a half-range series involving both sine and cosine terms arises when on $(-L, 0)$ the function satisfies the condition, $f(x \pm L) = -f(x)$. For this case, as the reader may verify (see Exercise 2.4.10), the expansion for $f(x)$ contains only *odd harmonics*; that is,

$$a_n = b_n = 0, \text{ for } n \text{ even}. \tag{2.4.5}$$

We may also show that for n odd,

$$\begin{aligned} a_n &= \frac{2}{L} \int_0^L f(x) \cos \frac{n\pi x}{L}\, dx, \\ b_n &= \frac{2}{L} \int_0^L f(x) \sin \frac{n\pi x}{L}\, dx. \end{aligned} \tag{2.4.6}$$

As an example of the various ways to represent a given function defined on $(0, L)$, we take the function

$$f(x) = \pi - x, \quad 0 < x < \pi. \tag{2.4.7}$$

If we are only interested in a series which converges to $f(x)$ on $(0, \pi)$, we may do this in a variety of ways. We show seven of these ways in Fig. 2–2. The series for the functions in Figs. 2–2(f) and 2–2(g) are, of course, the half-range series of the function in Fig. 2–2(d).

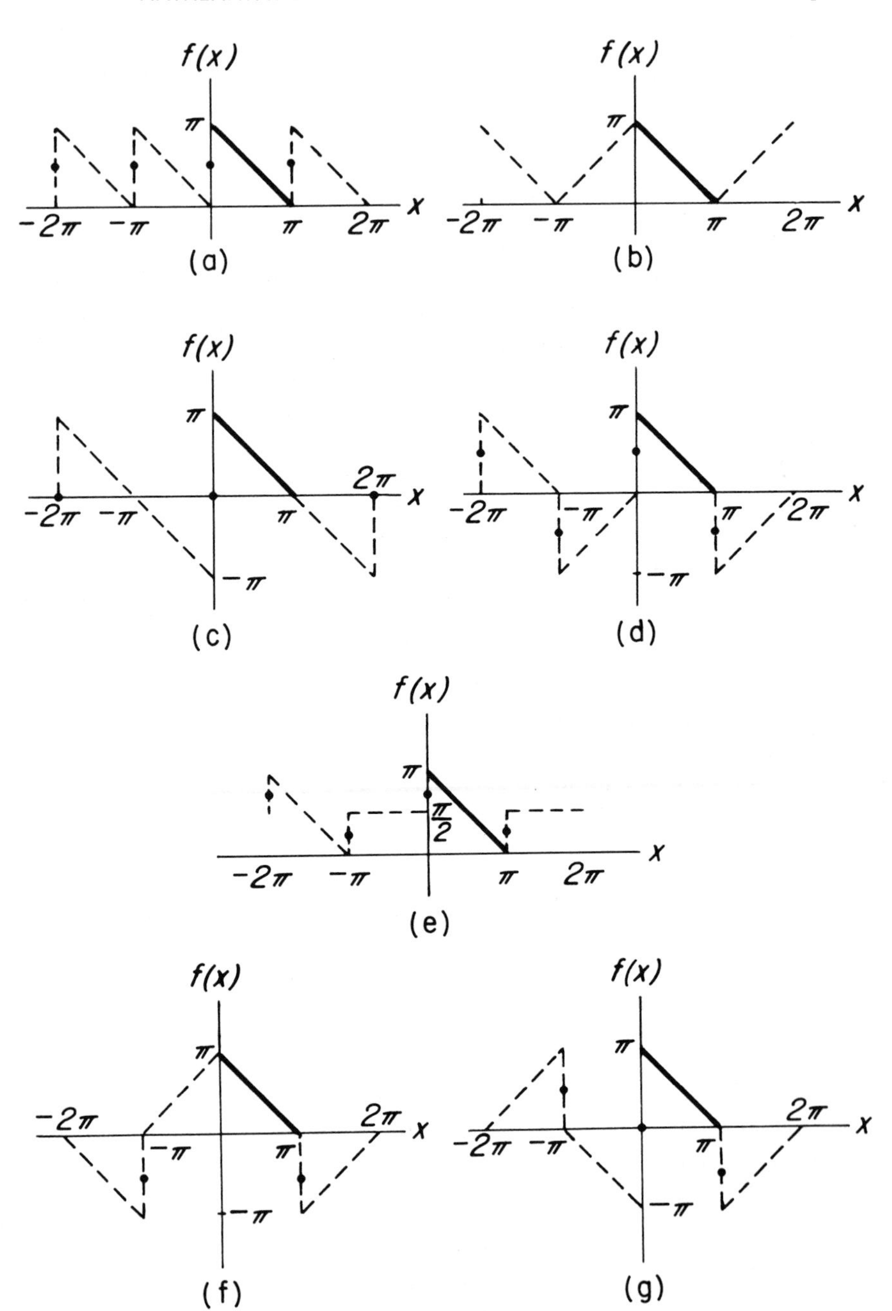

Fig. 2–2. (a) Sines and Cosines, Period π; (b) Half-Range Cosine, Period 2π; (c) Half-Range Sine, Period 2π; (d) Odd Harmonics, Period 2π; (e) Sines and Cosines, Period 2π; (f) Half-Range Cosine, Period 4π; (g) Half-Range Sine, Period 4π

EXERCISES

2.4.1. Prove analytically that if $f(x)$ is integrable on $(-a, a)$, then

$$\int_{-a}^{a} f(x)\,dx = 0, \quad f(x) \text{ odd,}$$
$$= 2\int_{0}^{a} f(x)\,dx, \quad f(x) \text{ even.}$$

2.4.2. Find (a) the sine series, (b) the cosine series, of $f(x) = \sin x$, on $0 < x < \pi$.

Ans. (a) $\sin x$; (b) $2/\pi - (4/\pi)\sum_{n=1}^{\infty} (\cos 2nx)/(4n^2 - 1)$.

2.4.3. Find the Fourier series of $f(x) = x^2$, $-1 < x < 1$.

Ans. $\frac{1}{3} + (4/\pi^2)\sum_{n=1}^{\infty} [(-1)^n \cos n\pi x]/n^2$.

2.4.4. Find the sine series of $f(x) = \frac{1}{8}\pi x(\pi - x)$, $0 < x < \pi$.

Ans. $\sum_{n=1}^{\infty} [\sin(2n-1)x]/(2n-1)^3$.

2.4.5. Expand $f(x) = x$ on $(0, \pi)$ in a cosine and a sine half-range series, and compare rapidity of convergence. Compare this result with that of Exercise 2.2.5.

2.4.6. Expand in a Fourier series

$$f(x) = 0, \quad -\pi < x < 0,$$
$$= x, \quad 0 < x < \pi.$$

Compare values on $(0, \pi)$ with similar points of the two series in Exercise 2.4.5.

2.4.7. Show that Eqs. (2.1.3) may be written

$$a_n = \frac{2}{L}\int_0^L f_e(x) \cos\frac{n\pi x}{L}\,dx,$$

$$b_n = \frac{2}{L}\int_0^L f_o(x) \sin\frac{n\pi x}{L}\,dx,$$

where $f(x) = f_e(x) + f_o(x)$ as defined by (2.3.2).

2.4.8. By the method of Exercise 2.4.7, expand

$$f(x) = 0, \quad -\pi < x < 0,$$
$$= 1, \quad 0 < x < \frac{\pi}{2},$$
$$= 2, \quad \frac{\pi}{2} < x < \pi.$$

2.4.9. Find the sine series on $(0, \pi)$ for

(a) $f(x) = x(\pi - x)$.
(b) $f(x) = x(x^2 - \pi^2)$.
(c) $f(x) = 2x - 1$.
(d) $f(x) = x, \quad 0 < x < \dfrac{\pi}{2},$
$\quad = \pi - x, \quad \dfrac{\pi}{2} < x < \pi.$
(e) $f(x) = x^2$.

2.4.10. Derive Eqs. (2.4.5) and (2.4.6).

2.4.11. Show that if $f(x \pm L) = f(x)$ in (2.1.1), we obtain only even harmonics.

2.4.12. Find the Fourier series for the functions shown in (d), (f), and (g) of Fig. 2–2. Compare with Exercises 2.4.5 and 2.4.6.

2.4.13. Find C_n and θ_n so that Eq. (2.1.1) may be written

$$f(x) = \sum_{n=0}^{\infty} C_n \cos\left(\frac{n\pi x}{L} + \theta_n\right).$$

Ans. $C_0 = a_0/2$; $C_n = \sqrt{a_n^2 + b_n^2}$, $n = 1, 2, \ldots$; $\theta_n = \arctan(-b_n/a_n)$.

2.4.14. Express the trigonometric functions in terms of complex exponentials in (2.1.1) and define $A_n = \frac{1}{2}(a_n - ib_n)$, $A_{-n} = \frac{1}{2}(a_n + ib_n)$, $A_0 = \frac{1}{2}a_0$, to obtain the *complex Fourier series*

$$f(x) = \sum_{n=-\infty}^{\infty} A_n e^{in\pi x/L}, \quad \text{with } A_n = \frac{1}{2L}\int_{-L}^{L} f(x)e^{-in\pi x/L}\,dx.$$

2.4.15. Show that if, in general, we define (see Sec. 1.2)

$$(\phi_n, \phi_m) = \int_a^b w(x)\phi_n(x)\overline{\phi_m(x)}\,dx,$$

where $\overline{\phi_m(x)}$ is the complex conjugate of $\phi_m(x)$, then the set $\{e^{in\pi x/L}\}$ is orthogonal over $(-L, L)$ with $w(x) = 1$. Use this to obtain the A_n given in Exercise 2.4.14.

2.4.16. Put the Fourier series obtained in Exercise 2.2.4 in the form given in Exercise 2.4.13.

2.4.17. Find the complex Fourier series for $f(x) = e^{-x}$ on $(-1, 1)$. Convert this to the real form (2.1.1).

2.4.18. Show that Bessel's inequality for the set $\{\phi_n(x)\}$, $(-L, L)$, defined in (1.2.4) with c_n given by (2.1.2), takes the respective forms for the sine and cosine series,

$$\sum_{n=1}^{m} b_n^2 \le \frac{2}{L}\int_0^L [f(x)]^2\,dx,$$

$$\frac{1}{2}a_0^2 + \sum_{n=1}^{m} a_n^2 \le \frac{2}{L}\int_0^L [f(x)]^2\,dx, \quad \text{for any } m = 1, 2, 3, \ldots.$$

Hence if the integral exists, $a_n, b_n \to 0$ as $n \to \infty$.

2.5 OPERATIONS WITH FOURIER SERIES

In this section we wish to consider the conditions under which the Fourier series for $f(x)$ given by (2.1.1) may be differentiated or integrated term by term, resulting in a Fourier series for the derivative or integral of $f(x)$ respectively. That these operations are not always valid may be seen by considering the example of Sec. 2.2 in which we had for

$$f(x) = 1, \qquad 0 < x \le \pi,$$
$$= -1, \quad -\pi < x \le 0,$$

the series

$$f(x) = \frac{4}{\pi} \sum_{n=1}^{\infty} \frac{\sin (2n - 1)x}{2n - 1}, \qquad -\pi < x < \pi.$$

Differentiating term by term, we have the series

$$\frac{4}{\pi} \sum_{n=1}^{\infty} \cos (2n - 1)x,$$

which is not a Fourier series because $c_n = 4/\pi$ and $\lim_{n\to\infty} c_n \ne 0$ [see Eq. (1.5.3)]. The function $f(x)$ that we have considered is not a continuous function on $(-\pi, \pi)$, nor does it have a derivative at every point. This is the difficulty, as we shall see.

We now state some sufficient conditions for differentiability of a Fourier series, which are as follows:

Theorem 2.5.1. If $f(x)$ is continuous on $-L \le x \le L$, $f(L) = f(-L)$, periodic of period $2L$, and $f'(x)$ satisfies the Dirichlet conditions, then the Fourier series (2.1.1) for $f(x)$ may be differentiated term by term to give the Fourier series

$$f'(x) = \sum_{n=1}^{\infty} \frac{n\pi}{L}\left(-a_n \sin \frac{n\pi x}{L} + b_n \cos \frac{n\pi x}{L}\right), \tag{2.5.1}$$

which converges to $\frac{1}{2}[f'(x + 0) + f'(x - 0)]$.

To establish (2.5.1) we note that if $f'(x)$ satisfies the Dirichlet conditions, then it has a Fourier series given by

$$f'(x) = \frac{\alpha_0}{2} + \sum_{n=1}^{\infty} \left[\alpha_n \cos \frac{n\pi x}{L} + \beta_n \sin \frac{n\pi x}{L}\right], \tag{2.5.2}$$

where

$$\alpha_n = \frac{1}{L}\int_{-L}^{L} f'(x) \cos \frac{n\pi x}{L}\, dx,$$
$$\beta_n = \frac{1}{L}\int_{-L}^{L} f'(x) \sin \frac{n\pi x}{L}\, dx. \tag{2.5.3}$$

If $f(x)$ is continuous on $(-L, L)$, we may integrate the expressions (2.5.3) by parts, and by substituting for a_n and b_n from (2.1.3) obtain

$$\alpha_0 = \frac{1}{L}[f(L) - f(-L)], \quad \alpha_n = \frac{(-1)^n}{L}[f(L) - f(-L)] + \frac{n\pi}{L} b_n,$$
$$\beta_n = \frac{n\pi}{L} a_n. \tag{2.5.4}$$

If $f(L) = f(-L)$ in (2.5.4), then (2.5.2) becomes (2.5.1), and the theorem is proved.

Since termwise integration of a Fourier series will improve its convergence because of the added factor n in its denominators, the conditions for integrating term by term are weaker than those in Theorem 2.5.1. The integrated series will converge regardless of the convergence of the original series, though it will not be a Fourier series if $a_0 \neq 0$, because of the term $\frac{1}{2}a_0 x$ resulting from the integration. The sufficient conditions for integrating term by term are as follows:

Theorem 2.5.2. If $f(x)$ is sectionally continuous on $-L \leq x \leq L$ with a Fourier series

$$f(x) \sim \sum_{n=1}^{\infty} \left(a_n \cos \frac{n\pi x}{L} + b_n \sin \frac{n\pi x}{L} \right),$$

and

$$F(x) = \int_{-L}^{x} f(t)\, dt,$$

then it is true that

$$F(x) = \sum_{n=1}^{\infty} \int_{-L}^{x} \left(a_n \cos \frac{n\pi t}{L} + b_n \sin \frac{n\pi t}{L} \right) dt, \tag{2.5.5}$$

or

$$F(x) = \frac{-1}{2L} \int_{-L}^{L} x f(x)\, dx + \sum_{n=1}^{\infty} \frac{L}{n\pi} \left(a_n \sin \frac{n\pi x}{L} - b_n \cos \frac{n\pi x}{L} \right) \tag{2.5.6}$$

is the Fourier series for $F(x)$ and converges to $F(x)$.

To show that (2.5.6) holds, we note that if $f(x)$ is sectionally continuous on $(-L, L)$ then $F(x)$ is continuous, and for $F(x + 2L) = F(x)$ we must have $F(L) = F(-L)$. From the definition of $F(x)$ in Theorem 2.5.2, we see that $F(-L) = 0$ and hence $F(L) = 0$. Let the Fourier series for $F(x)$ be written

$$F(x) = \frac{A_0}{2} + \sum_{n=1}^{\infty} \left(A_n \cos \frac{n\pi x}{L} + B_n \sin \frac{n\pi x}{L} \right), \tag{2.5.7}$$

with

$$\begin{aligned} A_n &= \frac{1}{L} \int_{-L}^{L} F(x) \cos \frac{n\pi x}{L}\, dx, \\ B_n &= \frac{1}{L} \int_{-L}^{L} F(x) \sin \frac{n\pi x}{L}\, dx. \end{aligned} \tag{2.5.8}$$

Integrating by parts in (2.5.8), noting that $F'(x) = f(x)$, and again substituting from (2.1.3), we have

$$A_0 = -\frac{1}{L} \int_{-L}^{L} x f(x)\, dx,$$

$$A_n = \frac{-Lb_n}{n}, \quad B_n = \frac{La_n}{n},$$

which with (2.5.7) establishes (2.5.6).

Since it is true that

$$\int_a^b f(x)\,dx = \int_{-L}^b f(x)\,dx - \int_{-L}^a f(x)\,dx,$$

we may use (2.5.5) to write

$$\int_a^b f(t)\,dt = \sum_{n=1}^{\infty} \int_a^b \left(a_n \cos \frac{n\pi t}{L} + b_n \sin \frac{n\pi t}{L}\right) dt. \tag{2.5.9}$$

If $a_0 \neq 0$, we may still use Theorem 2.5.2 and (2.5.9) if we replace $f(x)$ by $f(x) - a_0/2$.

If we carry out the operations indicated in (2.5.5), we have

$$\int_{-L}^{x} f(t)\,dt = \sum_{n=1}^{\infty} \frac{L}{n\pi}\left(a_n \sin \frac{n\pi x}{L} - b_n \cos \frac{n\pi x}{L}\right) + \sum_{n=1}^{\infty} \frac{Lb_n(-1)^n}{n\pi}.$$

Comparing this with (2.5.6) we may equate the two values of $A_0/2$ to obtain

$$\frac{1}{L}\int_{-L}^{L} xf(x)\,dx = \sum_{n=1}^{\infty} \frac{2Lb_n(-1)^{n+1}}{n\pi}. \tag{2.5.10}$$

If $L = \pi$, this becomes

$$\frac{1}{\pi}\int_{-\pi}^{\pi} xf(x)\,dx = \sum_{n=1}^{\infty} \frac{2b_n(-1)^{n+1}}{n}. \tag{2.5.11}$$

EXERCISES

2.5.1. Show that if the series

$$S = \frac{a_0}{2} + \sum_{n=1}^{\infty}\left[a_n \cos \frac{n\pi x}{L} + b_n \sin \frac{n\pi x}{L}\right]$$

converges uniformly to $f(x)$ on $-L \leq x \leq L$, then the two series obtained by term-by-term multiplication of S by $\cos(m\pi x/L)$ and by $\sin(m\pi x/L)$ are uniformly convergent. From this obtain (2.1.3).

2.5.2. Show in two ways that $\Sigma_{n=1}^{\infty} \sin nx$, $0 \leq x \leq \pi$, is not a Fourier series.

2.5.3. Show that $\Sigma_{n=1}^{\infty} (\sin nx)/[\log(1 + n)]$ is not a Fourier series. This is a classic example of a convergent series of sine terms which is not a Fourier series.

2.5.4. Use the functions defined in Exercise 2.4.9(a) and (b) with Eq. (2.5.11) to obtain

$$\sum_{n=1}^{\infty} \frac{1}{n^4} = \frac{\pi^4}{90}; \quad \sum_{n=1}^{\infty} \frac{1}{(2n-1)^4} = \frac{\pi^4}{96}.$$

2.5.5. Use in (2.5.11) the function defined in Exercise 2.2.9 to obtain the sum

$$\sum_{n=1}^{\infty} \frac{1}{n^2+1} = \frac{\pi \coth \pi - 1}{2}.$$

2.5.6. Show that the result in Exercise 2.5.5 can be obtained directly from the series in Exercise 2.2.9 when $x = \pi$.

2.5.7. Show that (2.5.4) is not valid if there exists a number x_0 such that $-L < x_0 < L$ and $f(x_0 + 0) - f(x_0 - 0) = k, k \neq 0$. Obtain correct expressions corresponding to (2.5.4) for this case.

2.6 PARSEVAL'S THEOREM

The result obtained in the preceding section and given in (2.5.11) could have been obtained from the sine series for x given in Exercise 2.2.5, by multiplying by $f(x)$ and integrating term by term. In general, if the Fourier coefficients exist for two functions $f_1(x)$ and $f_2(x)$ and are denoted by a_{n1}, b_{n1} and a_{n2}, b_{n2} respectively, we may represent $f_1(x)$ by its Fourier series and multiply through by $f_2(x)/L$ to obtain

$$\frac{1}{L} f_1(x) f_2(x) \sim \frac{a_{01}}{2L} f_2(x) + \sum_{n=1}^{\infty} \left(a_{n1} \frac{f_2(x)}{L} \cos \frac{n\pi x}{L} + b_{n1} \frac{f_2(x)}{L} \sin \frac{n\pi x}{L} \right).$$

Integrating term by term from $-L$ to L, and recognizing the Fourier coefficients for $f_2(x)$, we have

$$\frac{1}{L} \int_{-L}^{L} f_1(x) f_2(x)\, dx = \frac{a_{01} a_{02}}{2} + \sum_{n=1}^{\infty} (a_{n1} a_{n2} + b_{n1} b_{n2}), \tag{2.6.1}$$

where

$$a_{ni} = \frac{1}{L} \int_{-L}^{L} f_i(x) \cos \frac{n\pi x}{L}\, dx, \quad b_{ni} = \frac{1}{L} \int_{-L}^{L} f_i(x) \sin \frac{n\pi x}{L}\, dx, \quad i = 1, 2.$$

We note that for $f_1(x) = f_2(x)$ and $L = \pi$, this reduces to Parseval's equality given in Exercise 2.2.6. Equation (2.6.1) is known as *Parseval's theorem* and can be established rigorously if $f_1(x)$ and $f_2(x)$ are bounded on $(-L, L)$ and $\int_{-L}^{L} f_i(x)\, dx$ exists for $i = 1, 2$. (See [WW], page 182.)

Equation (2.6.1) assumes simpler forms when we restrict $f_1(x)$ and $f_2(x)$ to both even or both odd functions. If both functions are even (that is, they are represented by their cosine series), we have

$$\begin{aligned} \frac{2}{L} \int_0^L f_1(x) f_2(x)\, dx &= \frac{a_{01} a_{02}}{2} + \sum_{n=1}^{\infty} a_{n1} a_{n2}, \\ a_{ni} &= \frac{2}{L} \int_0^L f_i(x) \cos \frac{n\pi x}{L}\, dx, \end{aligned} \tag{2.6.2}$$

and if both functions are odd,

$$\begin{aligned} \frac{2}{L} \int_0^L f_1(x) f_2(x)\, dx &= \sum_{n=1}^{\infty} b_{n1} b_{n2}, \\ b_{ni} &= \frac{2}{L} \int_0^L f_i(x) \sin \frac{n\pi x}{L}\, dx. \end{aligned} \tag{2.6.3}$$

These results may be used to sum a wide variety of infinite series. For example, let $f_1(x) = f_2(x) = 1$ on $(0, \pi)$, the sine series of which was obtained in Sec. 2.2 as

$$1 = \frac{4}{\pi} \sum_{n=1}^{\infty} \frac{\sin (2n-1)x}{2n-1}, \quad 0 < x < \pi.$$

For this case, (2.6.3) yields the result previously found in Exercise 2.2.4,

$$\sum_{n=1}^{\infty} \frac{1}{(2n-1)^2} = \frac{\pi^2}{8}. \tag{2.6.4}$$

For a more complicated example, consider $f_1(x) = f_2(x) = e^x$ on $(0, \pi)$, for which we have

$$e^x = \sum_{n=1}^{\infty} \frac{2n}{\pi(n^2+1)} [1 - (-1)^n e^{\pi}] \sin nx, \quad 0 < x < \pi. \tag{2.6.5}$$

Again using (2.6.3), we have

$$\sum_{n=1}^{\infty} \frac{n^2}{(n^2+1)^2} [1 - (-1)^n e^{\pi}]^2 = \frac{\pi}{4} (e^{2\pi} - 1). \tag{2.6.6}$$

EXERCISES

In Exercises 2.6.1 through 2.6.4, using Eq. (2.6.1), find $\Sigma_{n=1}^{\infty} A_n$ for $f_1(x), f_2(x)$ defined as indicated.

2.6.1. Equation (2.2.1) and Exercise 2.4.9(d); $A_n = (-1)^{n+1}/(2n-1)^3$.

2.6.2. Exercise 2.4.9(a) and (e); $A_n = 1/(2n-1)^6$. Also show that this result may be obtained by using Exercise 2.4.9(a) in Eq. (2.5.11).

2.6.3. Exercise 2.4.9(b); with $f_1(x) = f_2(x)$; $A_n = 1/n^6$. Ans. $\pi^6/945$.

2.6.4. Exercises 2.2.8 and 2.2.9; $A_n = 1/[(2n-1)^2 + 1]$.

2.6.5. Obtain the result given in (2.6.4) by using $f_1(x) = f_2(x)$ as the function in Exercise 2.2.8.

2.6.6. For an orthonormal set $\{\phi_n(x)\}$, using (1.3.1) and (1.3.2) obtain the most general form of Parseval's theorem:

$$(f_1, f_2) = \sum_{n=1}^{\infty} c_{n1} c_{n2}; \quad c_{ni} = (f_i, \phi_n).$$

2.7 FOURIER INTEGRAL

If the function $f(x)$ considered in Sec. 2.1 is non-periodic and defined on an infinite interval, we can no longer represent it by a Fourier series of the type (2.1.1). However, by considering (2.1.1) as $L \to \infty$, we can obtain an analogous representation which will be valid if $f(x)$ satisfies the Dirichlet conditions and if, in addition,

$$\int_{-\infty}^{\infty} |f(x)|\, dx < M, \text{ a constant.} \tag{2.7.1}$$

Assuming that $f(x)$ satisfies these conditions, and writing (2.1.1) as

$$f(x) = \frac{1}{2L}\int_{-L}^{L} f(\xi)\, d\xi + \sum_{n=1}^{\infty} \frac{1}{L}\int_{-L}^{L} f(\xi)\left[\cos\frac{n\pi\xi}{L}\cos\frac{n\pi x}{L} + \sin\frac{n\pi\xi}{L}\sin\frac{n\pi x}{L}\right] d\xi,$$

we note that for large L, we have the approximation,

$$f(x) \doteq \sum_{n=1}^{\infty} \frac{1}{L}\int_{-L}^{L} f(\xi)\cos\frac{n\pi}{L}(\xi - x)\, d\xi, \tag{2.7.2}$$

since the constant term tends to zero.

If we make the substitution $L = \pi/\Delta\nu$, we may write (2.7.2) as

$$f(x) \doteq \frac{1}{\pi}\sum_{n=1}^{\infty} F(n\,\Delta\nu, x)\,\Delta\nu, \tag{2.7.3}$$

where

$$F(\nu, x) = \int_{-L}^{L} f(\xi)\cos[\nu(\xi - x)]\, d\xi. \tag{2.7.4}$$

Equation (2.7.3) and the definition of the definite integral suggest that as $\Delta\nu \to 0$ $(L \to \infty)$,

$$f(x) \to \frac{1}{\pi}\int_0^{\infty} F(\nu, x)\, d\nu,$$

or

$$f(x) = \frac{1}{\pi}\int_0^{\infty}\int_{-\infty}^{\infty} f(\xi)\cos[\nu(\xi - x)]\, d\xi\, d\nu. \tag{2.7.5}$$

Equation (2.7.5) is known as the *Fourier integral* and can be established rigorously under the conditions we have assumed on $f(x)$. At points of discontinuity the integral converges to $\frac{1}{2}[f(x+0) + f(x-0)]$. For a proof based on the conditions cited in Sec. 2.1 see [C-1], pages 114–117.

An alternative form of (2.7.5) is

$$f(x) = \int_0^{\infty} [A_\nu \cos\nu x + B_\nu \sin\nu x]\, d\nu, \tag{2.7.6}$$

where

$$A_\nu = \frac{1}{\pi}\int_{-\infty}^{\infty} f(\xi)\cos\nu\xi\, d\xi,$$
$$B_\nu = \frac{1}{\pi}\int_{-\infty}^{\infty} f(\xi)\sin\nu\xi\, d\xi, \tag{2.7.7}$$

which is analogous to Eqs. (2.1.6) and (2.1.7).

EXERCISES

2.7.1. Show that (2.7.6) and (2.7.7) become, for $f(x)$ even,

$$f(x) = \int_0^\infty A_\nu \cos \nu x \, d\nu,$$

$$A_\nu = \frac{2}{\pi} \int_0^\infty f(\xi) \cos \nu\xi \, d\xi,$$

$$B_\nu = 0,$$

and for $f(x)$ odd,

$$f(x) = \int_0^\infty B_\nu \sin \nu x \, d\nu,$$

$$A_\nu = 0,$$

$$B_\nu = \frac{2}{\pi} \int_0^\infty f(\xi) \sin \nu\xi \, d\xi.$$

These are the cosine and sine integrals respectively.

2.7.2. Represent by a sine integral the function

$$f(x) = 1, \quad 0 < x < 1$$
$$= 0, \quad x > 1$$
$$f(1) = \tfrac{1}{2}.$$

2.7.3. Show that for $f(x) = e^{-x}$, $x > 0$, the sine and cosine integrals are, respectively,

$$e^{-x} = \frac{2}{\pi} \int_0^\infty \frac{\nu}{\nu^2 + 1} \sin \nu x \, d\nu \quad \text{and} \quad e^{-x} = \frac{2}{\pi} \int_0^\infty \frac{1}{\nu^2 + 1} \cos \nu x \, d\nu.$$

2.7.4. Show by a process similar to that used in deriving (2.7.5) that the complex Fourier series given in Exercise 2.4.14 can be used to obtain the *complex Fourier integral*,

$$f(x) = \frac{1}{2\pi} \int_{-\infty}^\infty \int_{-\infty}^\infty f(\xi) \, e^{i\nu(x-\xi)} \, d\xi \, d\nu.$$

3

Series Solution of Differential Equations

3.1 PRELIMINARY EXAMPLE

As we have seen in Chapter 1, the method of separation of variables leads to a set of ordinary differential equations whose solutions are the special functions that are of interest to us. As we shall see, all of these equations involving the space coordinates take the general form,

$$y'' + P(x)y' + Q(x)y = 0, \tag{3.1.1}$$

a second-order linear differential equation which, because of its importance to us and because we wish to make the book as self-contained as possible, will be considered in some detail in this chapter.

In general, a solution of this equation in closed form cannot be found, but under certain conditions on the coefficients P and Q, solutions in the form of infinite series may be obtained for certain values of x. The method we will consider is known as the *method of Frobenius*. The reader who is already familiar with it may omit this chapter without loss of continuity.

As an example illustrating the method of Frobenius, let us solve the equation

$$y'' + xy' + 2y = 0$$

by assuming a solution of the type

$$y = \sum_{n=0}^{\infty} a_n x^n, \tag{3.1.2}$$

and determining the coefficients a_n by making (3.1.2) satisfy the differential equation identically. Substituting the assumed series into the equation, we have

$$\sum_{n=0}^{\infty} n(n-1)a_n x^{n-2} + \sum_{n=0}^{\infty} na_n x^n + 2\sum_{n=0}^{\infty} a_n x^n \equiv 0.$$

Shifting the index in the first summation and combining terms yields

$$\sum_{n=0}^{\infty} [(n + 1)a_{n+2} + a_n](n + 2)x^n = 0.$$

Since this is to be an identity, the coefficient of each power of x must vanish. This requirement leads to the *recurrence relation*,

$$a_{n+2} = \frac{-a_n}{n + 1}, \quad n = 0, 1, 2, \ldots. \tag{3.1.3}$$

Since this relation places no restriction on a_0 and a_1, their values are arbitrary. The recurrence relation yields, for a few values of n,

$$a_2 = -\frac{a_0}{1},$$

$$a_3 = -\frac{a_1}{2},$$

$$a_4 = -\frac{a_2}{3} = \frac{a_0}{1 \cdot 3},$$

$$a_5 = -\frac{a_3}{4} = \frac{a_1}{2 \cdot 4}.$$

In general

$$a_{2n} = \frac{(-1)^n a_0}{1 \cdot 3 \cdot 5 \cdots (2n - 1)} = \frac{(-2)^n n!\, a_0}{(2n)!}, \quad n = 0, 1, 2, \ldots,$$

and

$$a_{2n+1} = \frac{(-1)^n a_1}{2 \cdot 4 \cdot 6 \cdots (2n)} = \frac{(-1)^n a_1}{2^n n!}, \quad n = 0, 1, 2, \ldots.$$

Substituting these values into (3.1.2) we have the solution

$$y = a_0 y_1 + a_1 y_2, \tag{3.1.4}$$

where

$$y_1 = \sum_{n=0}^{\infty} \frac{(-2)^n n!\, x^{2n}}{(2n)!}$$

and

$$y_2 = \sum_{n=0}^{\infty} \frac{(-1)^n x^{2n+1}}{2^n n!}.$$

Since y_1 and y_2 are obviously linearly independent, (3.1.4) represents the general solution, provided the steps we have taken to obtain it are justifiable. These steps, termwise differentiation and rearrangement of terms, are permissible in the case of a power series for all values of x within the interval of convergence.

The ratio test [see Appendix A] yields, using (3.1.3),

$$\left|\frac{a_{n+2}x^{n+2}}{a_n x^n}\right| \sim {}_0{}^2.$$

Since $x^2 < 1$ if $|x| < 1$, it follows that (3.1.4) is valid for all values of x satisfying $|x| < 1$.

If a solution valid for $|x| > 1$ is desired, we may assume a solution of descending powers,

$$y = \sum_{n=0}^{\infty} a_n x^{m-n},$$

where m is a constant which must be determined. An example of this type will be given in Exercise 4.1.2.

EXERCISES

Use the method of Frobenius to solve the differential equations in Exercises 3.1.1 through 3.1.5.

3.1.1. $y'' + y = 0$. Show that the answer can be put in the form, $y = c_1 \cos x + c_2 \sin x$.

3.1.2. $y'' + 4xy' + 4y = 0$.

Ans. $y = c_1 y_1 + c_2 y_2$, where
$y_1 = \Sigma_{n=0}^{\infty} (-1)^n 2^n x^{2n}/n!$, $y_2 = \Sigma_{n=0}^{\infty} (-1)^n 2^{3n} n!\, x^{2n+1}/(2n+1)!$.

3.1.3. $(x^2 + 1)y'' + 6xy' + 6y = 0$.

Ans. $y = c_1 y_1 + c_2 y_2$, where
$y_1 = \Sigma_{n=0}^{\infty} (-1)^n (2n+1) x^{2n}$, $y_2 = \Sigma_{n=0}^{\infty} (2n+2) x^{2n+1}$.

3.1.4. $x^2 y'' - 4xy' + 4y = 0$. Ans. $y = c_1 x + c_2 x^4$.

3.1.5. $y'' - 2xy' + 2\alpha y = 0$. (This equation is known as *Hermite's equation*, and the polynomial solutions are the Hermite polynomials given in Exercise 1.7.4.)

Ans. $y = c_1 y_1 + c_2 y_2$, where

$$y_1 = 1 - \frac{2\alpha x^2}{2!} + \frac{2^2\alpha(\alpha - 2)x^4}{4!} - \frac{2^3\alpha(\alpha - 2)(\alpha - 4)x^6}{6!} + \cdots,$$

$$y_2 = x - \frac{2(\alpha - 1)x^3}{3!} + \frac{2^2(\alpha - 1)(\alpha - 3)x^5}{5!} - \cdots.$$

3.1.6. Note in Exercise 3.1.5 that if α is a positive integer, one of the two solutions becomes a polynomial. Obtain from y the set of Hermite polynomials given in Exercise 1.7.4 by proper choice of α, c_1, and c_2.

3.2 GENERALIZED SERIES SOLUTION

Many second-order equations do not lend themselves to the method of Sec. 3.1. For example, substitution of (3.1.2) into the equation

$$4x^2y'' + 4xy' + (4x^2 - 1)y = 0 \tag{3.2.1}$$

leads to the recurrence relations

$$(4n^2 - 1)a_n = 0, \quad n = 0, 1,$$

$$a_n = -\frac{4a_{n-2}}{4n^2 - 1}, \quad n = 2, 3, 4, \ldots,$$

and hence to

$$a_n = 0, \quad \text{for } n = 0, 1, 2, \ldots.$$

This method therefore fails to yield a non-trivial solution. However, (3.2.1) may be solved readily by letting $u = \sqrt{x}\, y$, obtaining

$$u'' + u = 0,$$

and finally

$$y = c_1 \frac{\cos x}{\sqrt{x}} + c_2 \frac{\sin x}{\sqrt{x}}. \tag{3.2.2}$$

Obviously (3.1.2) will always fail to yield solutions involving fractional powers of x as is required in (3.2.2). Consequently we are led to consider the more general series

$$y = \sum_{n=0}^{\infty} a_n x^{n+k}, \tag{3.2.3}$$

where the determination of k gives more latitude in the type of solutions available.

We will now show that the method of (3.2.3) will yield a solution to Eq. (3.2.1). After substituting for y and simplifying, we have

$$\sum_{n=0}^{\infty} a_n[4(n + k)^2 - 1]x^{n+k} + 4\sum_{n=0}^{\infty} a_n x^{n+k+2} = 0.$$

Equating to zero the coefficients of x^k and x^{k+1}, which occur only in the first summation, we have respectively

$$(4k^2 - 1)a_0 = 0 \tag{3.2.4}$$

and

$$[4(k + 1)^2 - 1]a_1 = 0. \tag{3.2.5}$$

Equation (3.2.4) is called the *indicial equation* and in general determines k if a_0 is taken as arbitrary. When the roots (sometimes called the *exponents*) of the indicial equation do not differ by an integer, two independent solutions are obtained by taking each root separately. Frequently, when the roots differ by an integer, this method fails to yield two independent solutions. We defer the discussion of this case until later.

Returning to our example we see that the value $k = -\frac{1}{2}$ makes both a_0 and a_1 arbitrary and will enable us to get the two independent solutions at once. Using this value of k and equating the remaining coefficients to zero, we obtain

$$a_{n+2} = -\frac{a_n}{(n+1)(n+2)}.$$

This relation gives us all the coefficients in terms of a_0 and a_1, and upon substituting the values of the coefficients into (3.2.3) we have

$$y = x^{-1/2}\left[a_0 \sum_{n=0}^{\infty} \frac{(-1)^n x^{2n}}{(2n)!} + a_1 \sum_{n=0}^{\infty} \frac{(-1)^n x^{2n+1}}{(2n+1)!}\right],$$

which reduces to the result given in (3.2.2).

3.3 EXISTENCE OF THE SERIES SOLUTION

Although the series in (3.2.3) is more general than the one in (3.1.2), it still fails in many cases to produce a solution. For example, as the reader may verify, the equation

$$x^4 y'' + y = 0$$

has no solution of the type (3.2.3).

Quite evidently it is desirable to know beforehand whether an equation of the type (3.1.1) has a solution of the form (3.1.2) or (3.2.3). Before presenting sufficient conditions for the existence of the solution, we require some preliminary definitions.

Definitions. A real number x_0 is said to be an *ordinary point* of the differential equation

$$y'' + P(x)y' + Q(x)y = 0$$

if and only if $P(x)$ and $Q(x)$ possess Taylor's series expansions about $x = x_0$. If x_0 is not an ordinary point, then it is a *singular point*; that is, $P(x)$ or $Q(x)$ or both fail to possess a Taylor's series expansion about $x = x_0$. A singular point x_0 is a *regular singular point* if $(x - x_0)P(x)$ and $(x - x_0)^2Q(x)$ both have Taylor's series expansions about $x = x_0$. A singular point x_0 which is not a regular singular point is an *irregular singular point*.

We now state, without proof, the following existence theorem due to *Fuchs* (for proof, see [WW], page 197).

Theorem 3.3.1. If x_0 is an ordinary point of (3.1.1), then a convergent series solution exists of the form

$$y = \sum_{n=0}^{\infty} a_n(x - x_0)^n.$$

If x_0 is a regular singular point, then a convergent series solution exists of the form

$$y = \sum_{n=0}^{\infty} a_n(x - x_0)^{n+k}. \tag{3.3.1}$$

It will be noted that Eqs. (3.1.2) and (3.2.3) represent the case $x_0 = 0$. For the example considered in Sec. 3.1,

$$y'' + xy' + 2y = 0,$$

$x = 0$ is an ordinary point, since $P(x) = x$ and $Q(x) = 2$. For Eq. (3.2.1), $x = 0$ is a regular singular point, since $xP(x) = 1$ and $x^2Q(x) = x^2 - \frac{1}{4}$. The example considered in this section,

$$x^4y'' + y = 0,$$

has an irregular singular point $x = 0$ since $xP(x) = 0$, but $x^2Q(x) = 1/x^2$.

Even though the conditions of Fuchs' theorem are satisfied, it may not be possible to obtain two independent solutions. If the roots of the indicial equation differ by an integer, this process may yield only one solution. In this event there are certain techniques that may be applied to the given series solution to obtain a second solution. For our purpose it is convenient to use a method which we describe in the next section.

To see why the method may fail to yield both solutions, we obtain the indicial equation in the general case. Assuming that the conditions of Fuchs' theorem are satisfied, we may write

$$\begin{aligned}(x - x_0)P(x) &= \sum_{m=0}^{\infty} p_m(x - x_0)^m,\\ (x - x_0)^2Q(x) &= \sum_{m=0}^{\infty} q_m(x - x_0)^m.\end{aligned} \tag{3.3.2}$$

Multiplying (3.1.1) through by $(x - x_0)^2$ and making the substitutions indicated in (3.3.1) and (3.3.2), we obtain

$$\begin{aligned}&\sum_{n=0}^{\infty} a_n(n + k)(n + k - 1)(x - x_0)^n\\ &\qquad + \sum_{m=0}^{\infty} p_m(x - x_0)^m \sum_{n=0}^{\infty} a_n(n + k)(x - x_0)^n\\ &\qquad\qquad + \sum_{m=0}^{\infty} q_m(x - x_0)^m \sum_{n=0}^{\infty} a_n(x - x_0)^n \equiv 0.\end{aligned}$$

Using the *Cauchy product*

$$\sum_{n=0}^{\infty} a_n t^n \sum_{n=0}^{\infty} b_n t^n = \sum_{n=0}^{\infty} \sum_{k=0}^{n} a_{n-k} b_k t^n, \tag{3.3.3}$$

we may finally write

$$\sum_{n=0}^{\infty} \left[a_n(n + k)(n + k - 1) + \sum_{m=0}^{n} p_{n-m} a_m (m + k) + \sum_{m=0}^{n} q_{n-m} a_m \right] (x - x_0)^n \equiv 0 \tag{3.3.4}$$

Equating to zero the coefficient of $(x - x_0)^0$ and dividing out $a_0 \neq 0$, we have the indicial equation,

$$F(k) = k^2 + (p_0 - 1)k + q_0 = 0. \tag{3.3.5}$$

In general, equating to zero for $n = 1, 2, 3, \ldots$, the coefficient of $(x - x_0)^n$, which is the expression in brackets in (3.3.4), we have the recurrence relation. Using (3.3.4) and (3.3.5) we may express this relation as

$$a_n F(k + n) = -\sum_{m=0}^{n-1} p_{n-m} a_m (m + k) - \sum_{m=0}^{n-1} q_{n-m} a_m. \tag{3.3.6}$$

Thus we see that if the two roots of (3.3.5) differ by an integer, then for some value of n the quantity $F(k + n)$ will be zero, and (3.3.6) cannot be used to solve for that particular a_n. The method for this case therefore breaks down in general, although there are many exceptions. For example, Eq. (3.2.1), which we solved in a previous section, is an exception.

3.4 A SECOND SOLUTION

We will now consider a method for finding the second solution when the roots of the indicial equation differ by an integer and the method of Frobenius yields only one solution, $y = y_1$, to the equation

$$y'' + P(x)y' + Q(x)y = 0.$$

We shall seek a second solution $y = y_2$ in the form

$$y_2 = v(x)y_1, \tag{3.4.1}$$

where the function $v(x)$ is to be determined so that y_2 satisfies identically the given equation.

Substituting for y the expression (3.4.1) and using the fact that y_1 is a solution, the differential equation becomes

$$v''y_1 + 2v'y'_1 + Pv'y_1 = 0.$$

This may be written

$$\frac{v''}{v'} = -P - \frac{2y'_1}{y_1},$$

which when integrated becomes

$$v' = \frac{\exp\left[-\int P(x)\,dx\right]}{y_1^{\,2}}.$$

Since we are interested only in a particular solution, we omit constants of integration.

Completing the solution for $v(x)$ and substituting in (3.4.1), we have

$$y_2 = y_1 \int \frac{\exp\left[-\int P(x)\,dx\right] dx}{y_1^{\,2}}, \quad y_1 \neq 0, \tag{3.4.2}$$

which is a solution of the given equation.

It remains to show that y_1 and y_2 are linearly independent. We recall from Sec. 1.6 that this is the case if the Wronskian of y_1 and y_2,

$$W(x) = \begin{vmatrix} y_1 & y_2 \\ y'_1 & y'_2 \end{vmatrix}, \tag{3.4.3}$$

is not identically zero. Substituting for y_2 and y'_2 from (3.4.2), we have

$$W(x) = \exp\left[-\int P(x)\,dx\right], \tag{3.4.4}$$

and hence y_1 and y_2 are linearly independent.

EXERCISES

Use the method of Frobenius to solve the equations of Exercises 3.4.1 through 3.4.6.

3.4.1. $xy'' + y' - 4y = 0$ (about $x = 0$).

3.4.2. $(x^3 - x)y'' + (4x^2 - 2)y' + 2xy = 0$ (about $x = 0$).

Ans. $y = c_1x^{-1} + c_2 \sum_{n=0}^{\infty} x^{2n}/(2n+1)$.

3.4.3. $(x^2 - 1)y'' + (2x - 1)y' - y = 0$ (about $x = 1$).

3.4.4. The equation of Exercise 3.4.1 about $x = 1$.

3.4.5. $xy'' + (1 - x)y' - y = 0$ (about $x = 0$).

Ans. $y = e^x\{c_1 + c_2[\log x + \sum_{n=1}^{\infty} (-1)^n x^n/n(n!)]\}$.

3.4.6. $y'' + e^x y = 0$ (about $x = 0$). Find the first five terms of the general solution.

3.4.7. Given that $y_1 = x^{-1}$ is a solution for Exercise 3.4.2, find another solution by the method of Sec. 3.4.

3.4.8. Show that p_0 and q_0 in (3.3.5) are given by

$$p_0 = \lim_{x \to x_0} (x - x_0)P(x),$$
$$q_0 = \lim_{x \to x_0} (x - x_0)^2 Q(x).$$

3.4.9. Obtain the indicial equation and the recurrence relation for Eq. (3.2.1) using (3.3.5) and (3.3.6) directly.

3.4.10. Verify Eq. (3.4.4).

3.4.11. Show directly from (3.1.1) that if $P(x)$ and $Q(x)$ are continuous over $a \leq x \leq b$, then the Wronskian of any two solutions is given by $W(x) = W_0 \exp\left[-\int_{x_0}^{x} P\, dx\right]$, where x_0 is any point in $[a, b]$ and W_0 is constant. Hence the Wronskian of any two solutions has the same sign or is identically zero throughout $[a, b]$.

4

Legendre Functions

4.1 LEGENDRE POLYNOMIALS

The differential equation

$$(1 - x^2)y'' - 2xy' + n(n + 1)y = 0, \tag{4.1.1}$$

where n is a constant, occurs frequently in engineering problems, and is known as *Legendre's equation*. In this section we shall apply the method of Frobenius to obtain solutions about the point $x = 0$.

Since $x = 0$ is an ordinary point, we may assume a solution of the form

$$y = \sum_{k=0}^{\infty} a_k x^k,$$

which leads to the recurrence relation

$$a_{k+2} = -\frac{(n - k)(n + k + 1)a_k}{(k + 1)(k + 2)}, \tag{4.1.2}$$

and to the two independent solutions,

$$y_1 = c_1\left[1 + \sum_{k=1}^{\infty} \frac{(-1)^k n(n - 2)\cdots(n - 2k + 2) \times (n + 1)(n + 3)\cdots(n + 2k - 1)x^{2k}}{(2k)!}\right], \tag{4.1.3}$$

and

$$y_2 = c_2 x\left[1 + \sum_{k=1}^{\infty} \frac{(-1)^k (n - 1)(n - 3)\cdots(n - 2k + 1) \times (n + 2)(n + 4)\cdots(n + 2k)x^{2k}}{(2k + 1)!}\right]. \tag{4.1.4}$$

We note by applying the ratio test, making use of (4.1.2), that these solutions are valid for $|x| < 1$. Also, if n is an integer, positive, zero, or negative, one or the other of the two solutions (4.1.3) and (4.1.4) will be a polynomial. Thus, in such a case, we have a polynomial solution and an infinite series solution. As we shall see later, the case n a non-negative integer gives rise to a solution which is of the most interest to us; so we will consider it in some detail.

For the set of eigenvalues

$$\lambda = n(n+1), \quad n = 0, 1, 2, \ldots, \tag{4.1.5}$$

it is easier to obtain the polynomial solution for both cases, n even and n odd, by returning to the recurrence relation (4.1.2). If we solve for a_k and replace k by $k - 2$, the result is the recurrence relation

$$a_{k-2} = -\frac{k(k-1)a_k}{(n-k+2)(n+k-1)}, \quad k = 2, 3, 4, \ldots, n. \tag{4.1.6}$$

From (4.1.3) if n is even, or (4.1.4) if n is odd, we see that $a_k = 0$ for $k > n$.

Using (4.1.6) (which is equivalent to choosing a_n as arbitrary), we will obtain $a_{n-2}, a_{n-4}, \ldots$. The coefficients $a_{n-1}, a_{n-3}, \ldots$ will not appear in the polynomial solution since, by (4.1.3) if n is even, we will have only even powers of x, and by (4.1.4) if n is odd, we will have only odd powers. Hence our polynomial solution will be either an even polynomial containing a_0, $a_2, \ldots, a_n$, or an odd polynomial containing $a_1, a_3, \ldots, a_n$. In either case (4.1.6) yields

$$a_{n-2} = -\frac{n(n-1)a_n}{2(2n-1)},$$

$$a_{n-4} = -\frac{(n-2)(n-3)a_{n-2}}{4(2n-3)} = \frac{n(n-1)(n-2)(n-3)a_n}{2^2(2!)(2n-1)(2n-3)},$$

and in general

$$a_{n-2k} = \frac{(-1)^k n(n-1)(n-2)\cdots(n-2k+1)a_n}{2^k k!\,(2n-1)(2n-3)\cdots(2n-2k+1)}, \quad k = 1, 2, \ldots, \left[\frac{n}{2}\right]. \tag{4.1.7}$$

We may obtain a particular solution by assigning a value to a_n. The value which gives the standard polynomial solution is

$$a_n = \frac{(2n)!}{2^n(n!)^2},$$

which simplifies (4.1.7) to the more compact form

$$a_{n-2k} = \frac{(-1)^k(2n-2k)!}{2^n k!\,(n-2k)!\,(n-k)!}, \quad k = 1, 2, \ldots, \left[\frac{n}{2}\right]. \tag{4.1.8}$$

We may now write the polynomial solution, for which the standard notation is

$$P_n(x) = \sum_{k=0}^{[n/2]} \frac{(-1)^k(2n-2k)!\, x^{n-2k}}{2^n k!\,(n-2k)!\,(n-k)!}. \tag{4.1.9}$$

This solution is known as the *Legendre polynomial* of degree n. The first five have been tabulated previously in (1.7.1).

The other solution of Legendre's equation for n a non-negative integer may be found as the series solution (4.1.3) if n is odd or (4.1.4) if n is even. The notation usually assigned to it, for a particular value of c_1 or c_2, is $y = Q_n(x)$, so that the general solution is

$$y = AP_n(x) + BQ_n(x), \quad n = 0, 1, 2, \ldots. \tag{4.1.10}$$

The function $Q_n(x)$ is known as Legendre's function of the *second kind.*

EXERCISES

4.1.1. Write out the first six Legendre polynomials from (4.1.9).

4.1.2. To obtain a solution of Legendre's equation valid for $|x| > 1$, assume that $y = \sum_{k=0}^{\infty} a_k x^{m-k}$. Show that, except for the cases $n = 0, -1$, if a_0 is arbitrary then $a_1 = 0$ and $m = n$ or $m = -n - 1$. Obtain the two solutions

$$y_3 = c_3 x^n\left[1 + \sum_{k=1}^{\infty} \frac{(-1)^k n(n-1)\cdots(n-2k+1)x^{-2k}}{2^k k!\,(2n-1)(2n-3)\cdots(2n-2k+1)}\right]$$

and

$$y_4 = c_4 x^{-n-1}\left[1 + \sum_{k=1}^{\infty} \frac{(n+1)(n+2)\cdots(n+2k)x^{-2k}}{2^k k!\,(2n+3)(2n+5)\cdots(2n+2k+1)}\right].$$

4.1.3. Note in Exercise 4.1.2 that if n is a non-negative integer, y_3 is the polynomial solution and y_4 is the series solution. Let $c_4 = 2^n(n!)^2/(2n+1)!$ and obtain the particular solution

$$y_4 = \sum_{k=0}^{\infty} \frac{2^n(n+k)!\,(n+2k)!}{k!\,(2n+2k+1)!} x^{-n-2k-1}, \quad |x| > 1.$$

4.1.4. Solve Legendre's equation for the cases $n = 0$ and $n = -1$.

4.1.5. Show that for $x = \cos\theta$, $0 \le \theta \le 2\pi$, Legendre's equation becomes

$$\frac{d^2y}{d\theta^2} + \cot\theta\,\frac{dy}{d\theta} + n(n+1)y = 0,$$

or

$$\frac{1}{\sin\theta}\left[\frac{d}{d\theta}\left(\sin\theta\,\frac{dy}{d\theta}\right)\right] + n(n+1)y = 0.$$

4.1.6. Let $x = \cosh\phi$, and obtain Legendre's equation in ϕ.

4.2 LEGENDRE FUNCTIONS OF THE SECOND KIND

As we have seen in the previous section, the functions $y = Q_n(x)$ are infinite series solutions of Legendre's equation for n a non-negative integer. For the case $|x| < 1$ these are given by (4.1.3) or (4.1.4). The standard forms are, for n even,

$$Q_n = \frac{(-1)^{n/2}2^n[(n/2)!]^2}{n!} \times \left[x + \sum_{k=1}^{\infty} \frac{(-1)^k(n-1)(n-3)\cdots(n-2k+1) \times (n+2)(n+4)\cdots(n+2k)x^{2k+1}}{(2k+1)!}\right] \tag{4.2.1}$$

and for n odd,

$$Q_n = \frac{(-1)^{(n+1)/2}2^{n-1}\{[(n-1)/2]!\}^2}{n!} \times \left[1 + \sum_{k=1}^{\infty} \frac{(-1)^k n(n-2)\cdots(n-2k+2) \times (n+1)(n+3)\cdots(n+2k-1)x^{2k}}{(2k)!}\right] \tag{4.2.2}$$

For $|x| > 1$, by Exercise 4.1.3 we have, for n even or odd,

$$Q_n = \sum_{k=0}^{\infty} \frac{2^n(n+k)!\,(n+2k)!}{k!\,(2n+2k+1)!} x^{-n-2k-1}. \tag{4.2.3}$$

We shall now show that when n is a non-negative integer, $P_n(x)$ is the only solution of Legendre's equation which is finite at $x = \pm 1$. Let $F(x)$ be any other solution such that $F(x)$ and $P_n(x)$ form a linearly independent set. Then the Wronskian of $P_n(x)$ and $F(x)$ is given by [see (3.4.4)]

$$W[P_n, F] = c_1 \exp\left[-\int \frac{-2x\,dx}{1-x^2}\right],$$

where $c_1 \neq 0$. Dividing by $P_n^2(x)$ and integrating, we have

$$F(x) = c_1 P_n(x) \int \frac{dx}{P_n^2(x)(1-x^2)} + c_2 P_n(x). \tag{4.2.4}$$

[Compare with (3.4.2).] When the integrand is expanded by partial fractions, there are terms

$$\frac{1}{2P_n^2(1)}\frac{1}{1-x} \quad \text{and} \quad \frac{1}{2P_n^2(-1)}\frac{1}{1+x}$$

which lead to logarithmic terms with arguments $(1 \pm x)$. Therefore $F(x)$ becomes infinite at $x = \pm 1$.

Thus we see that in any problem in which Legendre's equation arises, a condition that the solution be finite at $x = \pm 1$ requires that $B = 0$ in Eq. (4.1.10).

In fact, the only solutions of the equation

$$(1 - x^2)y'' - 2xy' + \lambda_n y = 0 \tag{4.2.5}$$

which are continuous with continuous first derivatives on $-1 \leq x \leq 1$ are $\{P_n(x)\}$, $n = 0, 1, 2, \ldots,$ occurring when $\lambda_n = n(n + 1)$. This is true because (4.2.5) is a Sturm-Liouville equation with $r(x) = 1 - x^2$ (see Sec. 1.9), and for the case $r(a) = r(b) = 0$, which we have, the set of eigenfunctions is orthogonal; that is, $\int_{-1}^{1} y_n y_m \, dx = 0$. As we shall see in Secs. 4.5 and 4.10, the set $\{P_n(x)\}$ is also orthogonal, and for a sectionally continuous function $f(x)$ the series $\sum_{n=0}^{\infty} c_n P_n(x)$ converges to $f(x)$, where c_n are the Fourier coefficients. Hence the set $\{P_n(x)\}$ is *complete in the sense of pointwise convergence* on $(-1, 1)$ (see Sec. 1.5), and by Exercise 1.5.4 the set is also closed. That is, if $Y \neq P_n(x)$ is an eigenfunction corresponding to $\lambda_Y \neq n(n + 1)$ such that $(Y, P_n) = 0$ for all n, then $Y \equiv 0$. Thus the only continuous eigenfunction with continuous first derivative other than $P_n(x)$ is $Y \equiv 0$.

4.3 GENERATING FUNCTION FOR $P_n(x)$

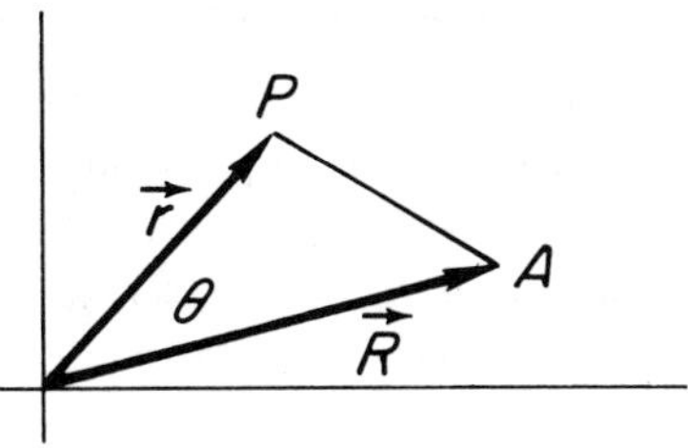

Fig. 4–1

The Legendre polynomials arose from the representation of the quantity $(r^2 + R^2 - 2rR \cos \theta)^{-1/2}$ as a power series in $r/R < 1$ where r and R are as shown in Fig. 4–1. If there is a unit mass at A, then the gravitational potential at point P is $|\vec{r} - \vec{R}|^{-1}$. The quantity $|\vec{r} - \vec{R}|^{-1}$, using the law of cosines, is $(1/R)[1 - (2r/R) \cos \theta + (r/R)^2]^{-1/2}$. We shall now show that this expression, neglecting $1/R$, is the generating function of $P_n(x)$ if we let $r/R = t$ and $\cos \theta = x$; that is,

$$g(x, t) = (1 - 2xt + t^2)^{-1/2} = \sum_{n=0}^{\infty} P_n(x) t^n. \tag{4.3.1}$$

Expanding the left-hand side by the binomial expansion we have

$$g(x, t) = \sum_{n=0}^{\infty} \frac{(\frac{1}{2})_n (2xt - t^2)^n}{n!},$$

where the symbol $(\alpha)_n$ is defined by

$$(\alpha)_n = \alpha(\alpha + 1) \cdots (\alpha + n - 1) = \prod_{k=0}^{n-1} (\alpha + k),$$
$$(\alpha)_0 = 1. \tag{4.3.2}$$

Thus we have

$$g(x, t) = \sum_{n=0}^{\infty} \frac{(\frac{1}{2})_n}{n!} \sum_{k=0}^{n} \frac{n!\,(2x)^{n-k} t^{n-k} (-t^2)^k}{k!\,(n-k)!}, \tag{4.3.3}$$

which can be written as (see Exercise 4.3.6)

$$g(x, t) = (1 - 2xt + t^2)^{-1/2} = \sum_{n=0}^{\infty} \left[\sum_{k=0}^{[n/2]} \frac{(-1)^k (2n-2k)!\, x^{n-2k}}{2^n k!\,(n-2k)!\,(n-k)!} \right] t^n. \tag{4.3.4}$$

The coefficient of t^n is $P_n(x)$ as given by (4.1.9).

EXERCISES

4.3.1. Verify that $P_0(x) = 1$, $P_1(x) = x$, $P_2(x) = \frac{1}{2}(3x^2 - 1)$, $P_3(x) = \frac{1}{2}(5x^3 - 3x)$.

4.3.2. Verify that

$$P_0(\cos\theta) = 1, \quad P_1(\cos\theta) = \cos\theta, \quad P_2(\cos\theta) = \tfrac{1}{4}(3\cos 2\theta + 1),$$
$$P_3(\cos\theta) = \tfrac{1}{8}(5\cos 3\theta + 3\cos\theta).$$

4.3.3. Find the generating function for $P_n(\cos\theta)$.

4.3.4. Obtain the expression for $P_n(\cos\theta)$.

4.3.5. Work Exercises 4.3.2, 4.3.3, and 4.3.4 using $x = \cosh\phi$.

4.3.6. Write the summation in (4.3.3) as $\sum_{n=0}^{\infty} \sum_{k=0}^{n} c(k, n) t^{n+k}$, and make the change of indices $k = j$, $n = m - j$. Noting that $0 \le k \le n$, show that

$$\sum_{n=0}^{\infty} \sum_{k=0}^{n} c(k, n) t^{n+k} = \sum_{m=0}^{\infty} \sum_{j=0}^{[m/2]} c(j, m-j) t^m.$$

(See [R], pages 56–58.)

4.3.7. Note from (4.3.1) that $P_n(x) = (1/n!)(\partial^n/\partial t^n) g(x, 0)$ and that $g(x, 0) = 1$. Use these results with the derivatives expressed in terms of g [for example, $\partial g/\partial t = (x - t)g^3$] to obtain the first five Legendre polynomials.

4.3.8. Suppose a function $\phi_n(x)$ is given by

$$\phi_n(x) = \frac{1}{\pi} \int_0^{\pi} (x + \sqrt{x^2 - 1}\cos t)^n \, dt.$$

Obtain the generating function of $\phi_n(x)$ of the form $\sum_{n=0}^{\infty} \phi_n(x) s^n$ by using the method outlined in Sec. 1.8, and show that $\phi_n(x) = P_n(x)$. (This result is due to Laplace.)

4.3.9. In the known integral,

$$\int_0^{\pi} \frac{d\theta}{a + b\cos\theta} = \frac{\pi}{\sqrt{a^2 - b^2}},$$

let $a = 1 - tx$, $b = -t\sqrt{x^2 - 1}$, and obtain the result in Exercise 4.3.8. Consider $|t(x + \sqrt{x^2 - 1}\cos\theta)| < 1$.

4.4 RODRIGUES' FORMULA

In this section we shall show that $P_n(x)$ is a constant multiple of $D^n v$ where $v = (x^2 - 1)^n$. We first obtain

$$\frac{dv}{dx} = 2nx(x^2 - 1)^{n-1},$$

and multiplying by $(x^2 - 1)$, we have

$$(x^2 - 1)\frac{dv}{dx} - 2nxv = 0.$$

We proceed to differentiate this equation $n + 1$ times using *Leibnitz's rule* for differentiating a product, given in Sec. 1.11 as

$$D^n(uv) = \sum_{k=0}^{n} \binom{n}{k} D^k u D^{n-k} v, \tag{4.4.1}$$

where $\binom{n}{k}$ is the binomial coefficient. This yields

$$(1 - x^2)\frac{d^2}{dx^2}[D^n v] - 2x\frac{d}{dx}[D^n v] + n(n+1)D^n v = 0, \tag{4.4.2}$$

which is Legendre's equation of order n. Since $D^n v$ is a polynomial of degree n, it can differ from $P_n(x)$ only by a constant factor. Hence

$$P_n(x) = AD^n(x^2 - 1)^n.$$

The constant A can be determined by comparing the coefficients of x^n:

$$\frac{(2n)!}{2^n(n!)^2} = A[2n(2n-1)\cdots(n+1)] = A\frac{(2n)!}{n!}$$

Therefore $A = 1/2^n n!$, and we have the *Rodrigues' formula*,

$$P_n(x) = \frac{1}{2^n n!} D^n(x^2 - 1)^n. \tag{4.4.3}$$

We can determine $P_n(\pm 1)$ by using the Rodrigues' formula and Leibnitz's rule for the case

$$D^n(x^2 - 1)^n = \sum_{k=0}^{n} \binom{n}{k} D^k(x+1)^n D^{n-k}(x-1)^n.$$

The only term which does not contain $x - 1$ occurs when $k = 0$. Hence

$$P_n(1) = \frac{1}{2^n n!} 2^n n! = 1. \tag{4.4.4}$$

In a similar manner we have

$$P_n(-1) = (-1)^n. \tag{4.4.5}$$

4.5 ORTHOGONALITY OF THE $P_n(x)$

We shall first show that any two Legendre polynomials of different order are orthogonal over the interval $(-1, 1)$ with weight function $w(x) = 1$. This can be done by using the generating function, as follows. Multiplying the two generating functions,

$$g(x, t) = \sum_{n=0}^{\infty} P_n(x)t^n,$$

$$g(x, s) = \sum_{m=0}^{\infty} P_m(x)s^m,$$

we have

$$g(x, t)g(x, s) = \sum_{n=0}^{\infty} \sum_{m=0}^{\infty} P_n(x)P_m(x)t^n s^m,$$

and integrating both sides with respect to x results in

$$\int_{-1}^{1} g(x, t)g(x, s)\, dx = \sum_{n=0}^{\infty} \sum_{m=0}^{\infty} t^n s^m \int_{-1}^{1} P_n(x)P_m(x)\, dx.$$

Noting that the left member is given by

$$\int_{-1}^{1} [(1 - 2xt + t^2)(1 - 2xs + s^2)]^{-1/2}\, dx = \frac{1}{\sqrt{st}} \log \frac{1 + \sqrt{st}}{1 - \sqrt{st}},$$

we obtain

$$\sum_{n=0}^{\infty} \sum_{m=0}^{\infty} t^n s^m \int_{-1}^{1} P_n(x)P_m(x)\, dx = \frac{1}{\sqrt{st}} \log \frac{1 + \sqrt{st}}{1 - \sqrt{st}}$$
$$= 2\sum_{k=0}^{\infty} \frac{t^k s^k}{2k + 1}.$$

Equating coefficients of $s^m t^n$, we have

$$\int_{-1}^{1} P_n(x)P_m(x)\, dx = \frac{2}{2n + 1}\, \delta_{mn}, \tag{4.5.1}$$

which is the desired orthogonality relation. In the notation of Chapter 1, (4.5.1) can be written

$$(P_n(x), P_m(x)) = \frac{2}{2n + 1}\, \delta_{mn},$$
$$\|P_n(x)\| = \sqrt{\frac{2}{2n + 1}}. \tag{4.5.2}$$

The foregoing is a formal derivation, and its validity depends upon the convergence properties of the series. An alternative method of obtaining (4.5.1) is considered in Exercise 4.5.3.

EXERCISES

4.5.1. Prove (4.4.5).

4.5.2. Obtain $P_n(\pm 1) = (\pm 1)^n$ by using the generating function.

4.5.3. Show by the method of Sec. 1.9 that $(P_n, P_m) = 0$, $n \neq m$; n, m non-negative integers. Use the Rodrigues' formula to obtain $\|P_n\|$.

4.5.4. Show that $P_{2n}(0) = (-1)^n(2n)!/2^{2n}(n!)^2$.

4.5.5. Show that (a) $(P_{2n}, P_{2m}) = [1/(4n+1)]\delta_{mn}$, and (b) $(P_{2n+1}, P_{2m+1}) = [1/(4n+3)]\delta_{mn}$, where the interval is $(0, 1)$, $w(x) = 1$, and $n, m = 0, 1, 2, \ldots$.

4.5.6. It is known (see [1], page 786) that the series (4.3.1) converges for $|x| \leq 1$, $|t| < 1$, and (see [C-1], page 211) that $n^{2k}\,|P_n{}^{(2k)}(x)| \leq 1$, $k = 0, 1, 2, \ldots$, $|x| \leq 1$. Show that (4.3.1) can be differentiated k times termwise with respect to both x and t.

4.6 RECURRENCE RELATIONS FOR $P_n(x)$

As we have seen in Chapter 1, we may obtain relationships between various members of an orthogonal set of functions by considering explicit expressions for the functions themselves or by considering the generating function of the set. In this section we will use the latter method to obtain some recurrence relations for the Legendre polynomials.

The generating function,

$$g(x, t) = (1 - 2xt + t^2)^{-1/2} = \sum_{n=0}^{\infty} P_n(x)t^n,$$

may be differentiated with respect to t (see Exercise 4.5.6) to yield

$$\frac{\partial g}{\partial t} = (1 - 2xt + t^2)^{-3/2}(x - t) = \sum_{n=0}^{\infty} nP_n(x)t^{n-1}, \tag{4.6.1}$$

which may be written

$$(x - t)\sum_{n=0}^{\infty} P_n(x)t^n = (1 - 2xt + t^2)\sum_{n=0}^{\infty} nP_n(x)t^{n-1}.$$

Equating coefficients of t^n, we have the pure recurrence relation

$$(n + 1)P_{n+1}(x) - (2n + 1)xP_n(x) + nP_{n-1}(x) = 0, \quad n > 0. \tag{4.6.2}$$

Next, differentiating the generating function with respect to x, we have

$$\frac{\partial g}{\partial x} = t(1 - 2xt + t^2)^{-3/2} = \sum_{n=0}^{\infty} P'_n(x)t^n,$$

and from this result and (4.6.1) we see that

$$(1 - 2xt + t^2)^{-3/2} = (x - t)^{-1}\frac{\partial g}{\partial t} = t^{-1}\frac{\partial g}{\partial x}.$$

Substituting for g in the last two members we have

$$(x - t)\sum_{n=0}^{\infty} P'_n(x)t^n = t\sum_{n=0}^{\infty} nP_n(x)t^{n-1}, \tag{4.6.3}$$

so that when we equate coefficients of t^n, we obtain

$$xP'_n(x) - P'_{n-1}(x) - nP_n(x) = 0, \quad n > 0. \tag{4.6.4}$$

Equations (4.6.2) and (4.6.4) are the basic recurrence relations for $P_n(x)$. Other relations may be obtained from these two as will be seen in the exercises.

EXERCISES

4.6.1. Starting with

$$(1 - 2xt + t^2)^{-1/2} = \sum_{n=0}^{\infty} A_n(x)t^n,$$

show that $A_n(x)$ satisfies the Legendre differential equation. Since $A_n(x)$ is a polynomial, show that it is $P_n(x)$.

4.6.2. From (4.6.2) and (4.6.4), obtain

$$P'_{n+1}(x) - xP'_n(x) = (n + 1)P_n(x).$$

4.6.3. Derive the relations

(a) $P'_{n+1}(x) - (2n + 1)P_n(x) - P'_{n-1}(x) = 0.$

(b) $(x^2 - 1)P'_n(x) - nxP_n(x) + nP_{n-1}(x) = 0.$

4.6.4. Evaluate $\int_{-1}^{1} P_n(x)\,dx$.

4.6.5. Evaluate $\int_{-1}^{1} (1 - x^2)P'_n(x)\,dx$.

4.6.6. Evaluate $\int_{-1}^{1}(1 - x^2)P'_n(x)P_n(x)\,dx$.

4.6.7. Show that any polynomial $R_m(x)$ of degree m can be expressed as a linear combination of Legendre polynomials of degrees m and less. Generalize this to any set of orthogonal polynomials $\{\phi_n(x)\}$, where $\phi_n(x)$ is non-zero and of degree n, $n = 0, 1, 2, \ldots$.

4.6.8. Show that for $R_m(x)$ in Exercise 4.6.7,

$$\int_{-1}^{1} P_n(x)R_m(x)\,dx = 0, \quad m < n.$$

Generalize to the set $\{\phi_n(x)\}$ in Exercise 4.6.7.

4.6.9. Show that all the zeros of $P_n(x)$ lie in the interval $-1 < x < 1$ and are distinct. [*Hint:* Assume $a_1, \ldots, a_m$ are the zeros of $P_n(x)$ in $-1 < x < 1$, $m < n$. Take $R_m(x) = \Pi_{k=1}^{m}(x - a_k)$ and consider the integral $\int_{-1}^{1} P_n(x)R_m(x)\,dx$.]

4.6.10. Using (4.6.3), express $P'_n(x)$ as a finite sum of Legendre polynomials.

4.6.11. Using (4.6.2) and the procedure in Sec. 1.8, determine the generating function for $P_n(x)$.

4.6.12. In seeking a generating function

$$g(x, t) = \sum_{n=0}^{\infty} \frac{P_n(x)}{n!} t^n,$$

set $u_n(x) = P_n(x)/n!$ and obtain the difference equation

$$(n + 1)^2 u_{n+1} - (2n + 1)xu_n + u_{n-1} = 0.$$

Show that $g(x, t) = e^{xt} J_0(t\sqrt{1 - x^2})$, where $J_0(\alpha x)$ is a solution of $xy'' + y' + \alpha^2 xy = 0$. The function $J_0(x)$ is a Bessel function discussed in Chapter 6.

4.6.13. Obtain the differential equation for $g(x, t)$ if

$$g(x, t) = \sum_{n=0}^{\infty} \frac{P_n(x)t^n}{(n!)^2} .$$

4.6.14. Use Theorem 1.8.1 to obtain a three-term recurrence relation for the set $\{2^n(n!)^2 P_n(x)/(2n)!\}$.

4.6.15. Obtain an integration formula for

$$\int_x^1 P_n(\xi)\, d\xi.$$

4.7 SERIES EXPANSIONS INVOLVING $P_n(x)$

The generalized Fourier series in Legendre polynomials has the form (see Sec. 1.3)

$$f(x) = \sum_{k=0}^{\infty} c_k P_k(x), \quad -1 < x < 1 \tag{4.7.1}$$

where

$$c_k = \frac{(f, P_k)}{\|P_k\|^2}, \quad w(x) = 1, \quad (a, b) = (-1, 1).$$

The square of the norm, $\|P_k\|^2$, is given by (4.5.2) as

$$\|P_k\|^2 = \frac{2}{2k + 1},$$

and therefore the coefficients are

$$c_k = \frac{2k + 1}{2} (f, P_k). \tag{4.7.2}$$

The series converges to $f(x)$ at every point on $(-1, 1)$ where $f(x)$ is continuous if $f(x)$ and $f'(x)$ are sectionally continuous (see Theorem 1.9.1).

As an example, if $f(x) = x^2$, then

$$c_0 = \tfrac{1}{2}\int_{-1}^{1} x^2 \cdot 1\,dx = \tfrac{1}{3},$$

$$c_1 = \tfrac{3}{2}\int_{-1}^{1} x^2 \cdot x\,dx = 0,$$

$$c_2 = \tfrac{5}{2}\int_{-1}^{1} x^2(\tfrac{3}{2}x^2 - \tfrac{1}{2})\,dx = \tfrac{2}{3},$$

and

$$x^2 = \tfrac{2}{3}P_2(x) + \tfrac{1}{3}P_0(x).$$

The Gram-Schmidt orthogonalization process indicates that for any non-negative integer n, x^n can be expressed as a linear combination of Legendre polynomials:

$$x^n = \sum_{k=0}^{n} b_k P_k(x). \tag{4.7.3}$$

The proof is by mathematical induction, as follows: Since $P_0(x) = 1$, the relation is satisfied for $n = 0$. Assuming the relation holds for n and using $x^{n+1} = x^n \cdot x$, we have

$$x^{n+1} = \sum_{k=0}^{n} b_k x P_k(x).$$

Replacing $xP_k(x)$ by its value from the recurrence relation (4.6.2),

$$xP_k(x) = \frac{1}{2k+1}\,[(k+1)P_{k+1}(x) + kP_{k-1}(x)], \quad (P_{-1}(x) = 0), \tag{4.7.4}$$

we obtain

$$\begin{aligned} x^{n+1} &= \sum_{k=0}^{n} \frac{b_k(k+1)}{2k+1} P_{k+1}(x) + \sum_{k=0}^{n} \frac{kb_k}{2k+1} P_{k-1}(x). \\ &= \sum_{k=1}^{n+1} \frac{kb_{k-1}}{2k-1} P_k(x) + \sum_{k=0}^{n-1} \frac{(k+1)b_{k+1}}{2k+3} P_k(x) \\ &= \sum_{k=0}^{n+1} a_k P_k(x), \end{aligned}$$

where

$$a_0 = \frac{b_1}{3},$$

$$a_k = \frac{kb_{k-1}}{2k-1} + \frac{(k+1)b_{k+1}}{2k+3}, \quad k = 1, 2, \ldots, n-1,$$

$$a_k = \frac{kb_{k-1}}{2k-1}, \quad k = n, n+1.$$

Therefore the relation (4.7.3) is valid for all non-negative integers n.

The powers x^n can be computed by repeatedly using (4.7.4). For example,

$$\begin{aligned} x^3 = xx^2 &= \tfrac{2}{3}xP_2(x) + \tfrac{1}{3}xP_0(x) \\ &= \tfrac{2}{3}\{\tfrac{1}{5}[3P_3(x) + 2P_1(x)]\} + \tfrac{1}{3}P_1(x) \\ &= \tfrac{2}{5}P_3(x) + \tfrac{3}{5}P_1(x). \end{aligned}$$

This method is impractical, of course, if n is large. However, a general formula developed by a different method (see [R], page 181) gives, for $n = 0, 1, 2, \ldots,$

$$x^n = \frac{n!}{2^n} \sum_{k=0}^{[n/2]} \frac{(2n - 4k + 1)P_{n-2k}(x)}{k!\,(\tfrac{3}{2})_{n-k}}. \tag{4.7.5}$$

(The notation $(\tfrac{3}{2})_{n-k}$ is defined in (4.3.2).)

As another example, consider $f(x) = e^{-x}$. The coefficients are given by

$$c_k = \frac{2k+1}{2} \int_{-1}^{1} e^{-x}P_k(x)\,dx.$$

We can evaluate the indefinite integral by using the "exponential shift" property of the antiderivative operator $1/D$ which can be found in most elementary differential equations books:

$$\begin{aligned} \frac{1}{D} e^{-x}P_k(x) &= e^{-x} \frac{1}{D-1} P_k(x) \\ &= -e^{-x} \sum_{n=0}^{k} D^n P_k(x). \end{aligned} \tag{4.7.6}$$

Using Rodrigues' formula and then Leibnitz's rule, we have

$$\begin{aligned} \frac{1}{D} e^{-x}P_k(x) &= -\frac{e^{-x}}{2^k k!} \sum_{n=0}^{k} D^{n+k}(x^2-1)^k \\ &= -\frac{e^{-x}}{2^k k!} \sum_{n=0}^{k} \sum_{m=0}^{n+k} \binom{n+k}{m} D^m(x-1)^k D^{n+k-m}(x+1)^k. \end{aligned}$$

When $x = 1$, the only non-zero term on the right occurs for $m = k$:

$$\frac{1}{D} e^{-x}P_k(x)\bigg|_{x=1} = -\frac{e^{-1}}{2^k k!} \sum_{n=0}^{k} \binom{n+k}{k} k!\,(k-n+1)_n 2^{k-n}.$$

Similarly for $x = -1$, the only non-zero term occurs for $m = n$:

$$\frac{1}{D} e^{-x}P_k(x)\bigg|_{x=-1} = -\frac{e}{2^k k!} \sum_{n=0}^{k} \binom{n+k}{k} k!\,(k-n+1)_n(-2)^{k-n}.$$

Combining these two results, we find in general that

$$c_k = \frac{2k+1}{2} \sum_{n=0}^{k} \binom{n+k}{k} \frac{(k-n+1)_n}{2^n} [(-1)^{k-n}e - e^{-1}].$$

Using this result for $k = 0, 1, 2$, the expansion of e^{-x} can be written

$$e^{-x} = -\tfrac{1}{2}(e^{-1} - e)P_0(x) - 3e^{-1}P_1(x) - \tfrac{5}{2}[7e^{-1} - e]P_2(x) + \cdots.$$

EXERCISES

4.7.1. Obtain the generalized Fourier series expansion of a function $f(\theta)$ in $P_k(\cos\theta)$.

4.7.2. Obtain the first three non-zero terms in the expansion of $f(\theta) = \sin\theta$.

4.7.3. Express x^4 as a linear combination of Legendre polynomials by two methods other than (4.7.5).

4.7.4. Show that if $f(x)$, $-1 < x < 1$, is even, then $c_{2k+1} = 0$ for $k = 0, 1, 2, \ldots$, and if $f(x)$ is odd, $c_{2k} = 0$ for $k = 0, 1, 2, \ldots$.

4.7.5. Obtain the expansion of

$$f(x) = \begin{cases} 0, & -1 < x < 0 \\ 1, & 0 \le x < 1 \end{cases}$$

in Legendre polynomials.

4.7.6. Expand $(d/dx)P_n(x)$ in a series of Legendre polynomials.

$$\text{Ans.}\quad P'_n(x) = (2n-1)P_{n-1}(x) + (2n-5)P_{n-3}(x)$$
$$+ \cdots \begin{cases} + 3P_1(x), & n \text{ even} \\ + P_0(x), & n \text{ odd} \end{cases} \quad n \ge 1.$$

4.7.7. Determine c_n where (a) $f(x) = \sum_{n=0}^{\infty} c_n P_{2n}(x)$, (b) $f(x) = \sum_{n=0}^{\infty} c_n P_{2n+1}$, for $0 \le x \le 1$. (See Exercise 4.5.5.)

4.8 ASSOCIATED LEGENDRE FUNCTIONS

Suppose we differentiate Legendre's differential equation m times by Leibnitz's rule. The terms will be given by

$$D^m[(1-x^2)v''] = (1-x^2)v^{(m+2)} + m(-2x)v^{(m+1)} + \frac{m(m-1)}{2}(-2)v^{(m)},$$
$$D^m(-2xv') = -2xv^{(m+1)} + m(-2)v^{(m)},$$
$$D^m[n(n+1)v] = n(n+1)v^{(m)},$$

which, when combined, results in the differential equation

$$(1-x^2)D^2v^{(m)} - 2x(m+1)Dv^{(m)} + [n(n+1) - m(m+1)]v^{(m)} = 0. \tag{4.8.1}$$

If we eliminate $v^{(m)}$ by the substitution $y = (1-x^2)^r v^{(m)}$, we have

$$(1-x^2)^2y'' + 2x(2r-m-1)(1-x^2)y'$$
$$+ \{4rx^2[(r+1)-(m+1)] + (1-x^2)[2r + n(n+1) - m(m+1)]\}y = 0$$

Finally, letting $2r = m$, we have *Legendre's associated equation*,

$$(1-x^2)y'' - 2xy' + \left[n(n+1) - \frac{m^2}{1-x^2}\right]y = 0. \tag{4.8.2}$$

Since $v = AP_n(x) + BQ_n(x)$, we must have

$$y = A(1 - x^2)^{m/2}P_n^{(m)}(x) + B(1 - x^2)^{m/2}Q_n^{(m)}(x),$$

the standard notation for which is

$$y = AP_n{}^m(x) + BQ_n{}^m(x). \tag{4.8.3}$$

The first solution,

$$P_n{}^m(x) = (1 - x^2)^{m/2}P_n^{(m)}(x), \quad m,n = 0, 1, 2, \ldots, \tag{4.8.4}$$

is defined as the *associated Legendre function* of the *first kind*. As a consequence of the definition we see that $P_n{}^m(x) = 0$ for $m > n$ and $P_n{}^m(\pm 1) = 0$. Also we note that (4.8.2) reduces to Legendre's equation if $m = 0$ and by (4.8.4) we have $P_n{}^0(x) = P_n(x)$. Finally, as a result of Rodrigues' formula we may write

$$P_n{}^m(x) = \frac{1}{2^n n!}(1 - x^2)^{m/2}D^{n+m}(x^2 - 1)^n, \quad m, n = 0, 1, 2, \ldots. \tag{4.8.5}$$

The second solution of (4.8.2),

$$Q_n{}^m(x) = (1 - x^2)^{m/2}Q_n^{(m)}(x),$$

the associated Legendre function of the *second kind*, like $Q_n(x)$, has the properties of being an infinite series and having infinite discontinuities at $x = \pm 1$. For this reason the $P_n{}^m(x)$ are far more important to us than the $Q_n{}^m(x)$ in solving boundary-value problems. We will therefore concentrate on the $P_n{}^m(x)$ and refer to them henceforth as the associated Legendre functions.

Thus far, in our definition of the associated Legendre function we have considered m to be a non-negative integer since it represents a number of differentiations. In (4.8.2), however, m appears only as m^2, and the equation is unchanged if we replace m by $-m$. Accordingly we *define* the associated Legendre function for m negative, by (4.8.5), as

$$P_n^{-m}(x) = \frac{1}{2^n n!}(1 - x^2)^{-m/2}D^{n-m}(x^2 - 1)^n, \quad n \geq m \geq 0. \tag{4.8.6}$$

We leave as an exercise (see Exercise 4.9.5) to show that (4.8.6) is a solution of Legendre's associated equation, and (see Exercise 4.9.6) that

$$P_n^{-m}(x) = (-1)^m \frac{(n - m)!}{(n + m)!} P_n{}^m(x), \quad n \geq m \geq 0. \tag{4.8.7}$$

4.9 RECURRENCE RELATIONS FOR $P_n{}^m(x)$

We may obtain a pure recurrence relation directly from (4.8.1) if we multiply it through by $(1 - x^2)^{m/2}$ and identify the various associated functions. The result after some simplification is

$$P_n^{m+2}(x) - \frac{2(m+1)x}{\sqrt{1-x^2}} P_n^{m+1}(x) + [n(n+1) - m(m+1)]P_n^{\ m}(x) = 0. \tag{4.9.1}$$

Other recurrence relations may be obtained by differentiating the relations for $P_n(x)$. Differentiating the relations in (4.6.2) and Exercise 4.6.3(a) m times and $m - 1$ times, respectively, yields

$$(n+1)P_{n+1}^{(m)} - (2n+1)xP_n^{(m)} - (2n+1)mP_n^{(m-1)} + nP_{n-1}^{(m)} = 0, \tag{4.9.2}$$

$$P_{n+1}^{(m)} - (2n+1)P_n^{(m-1)} - P_{n-1}^{(m)} = 0. \tag{4.9.3}$$

Eliminating $P_n^{(m-1)}$ from these two equations and multiplying by $(1 - x^2)^{m/2}$ gives the relation

$$(n - m + 1)P_{n+1}^m(x) - (2n+1)xP_n^{\ m}(x) + (n+m)P_{n-1}^m(x) = 0. \tag{4.9.4}$$

Multiplication of (4.9.3) by $(1 - x^2)^{m/2}$ and simplification yields the mixed relation

$$P_{n+1}^m(x) - (2n+1)\sqrt{1-x^2}\, P_n^{m-1}(x) - P_{n-1}^m(x) = 0. \tag{4.9.5}$$

Other recurrence relations are left to the exercises.

EXERCISES

4.9.1. Differentiate (4.6.4) $m - 1$ times and obtain the mixed relation

$$xP_n^{\ m} - P_{n-1}^m + (m - n - 1)\sqrt{1-x^2}\, P_n^{m-1} = 0.$$

4.9.2. Obtain the recurrence relations

(a) $\sqrt{1-x^2}\, P_n^{m+1} = (n + m + 1)xP_n^{\ m} - (n - m + 1)P_{n+1}^m.$

(b) $(1 - x^2)DP_n^{\ m} = (n+1)xP_n^{\ m} - (n - m + 1)P_{n+1}^m.$

(c) $(1 - x^2)DP_n^{\ m} = (n+m)P_{n-1}^m - nxP_n^{\ m}.$

(d) $P_{n+1}^m = xP_n^{\ m} + (n+m)\sqrt{1-x^2}\, P_n^{m-1}.$

4.9.3. Show that $P_n^{\ m}(\cos\theta) = (\sin^m\theta)P_n^{(m)}(\cos\theta)$, where the differentiation is with respect to the argument, $\cos\theta$.

4.9.4. Show that $P_n^{\ m}(\cos\theta)$ satisfies the equation

$$\frac{d^2y}{d\theta^2} + (\cot\theta)\frac{dy}{d\theta} + \left[n(n+1) - \frac{m^2}{\sin^2\theta}\right]y = 0,$$

or its equivalent,

$$\frac{1}{\sin\theta}\frac{d}{d\theta}\left(\sin\theta\frac{dy}{d\theta}\right) + \left[n(n+1) - \frac{m^2}{\sin^2\theta}\right]y = 0.$$

4.9.5. Show that if $y = (1 - x^2)^{-m/2}u$ in (4.8.2), then

$$(1 - x^2)u'' - 2x(1 - m)u' + [n(n + 1) + m(-m + 1)]u = 0,$$

which has, as we see by comparison with (4.8.1), a solution $u = v^{(-m)} = (1/D^m)v$. Show that this implies that (4.8.6) is a solution of (4.8.2).

4.9.6. Use Leibnitz's rule to prove (4.8.7).

4.9.7. Derive the results given in the accompanying table.

m, n	$P_n^m(x)$	$P_n^m(\cos\theta)$
1, 1	$(1 - x^2)^{1/2}$	$\sin\theta$
1, 2	$3x(1 - x^2)^{1/2}$	$\frac{3}{2}\sin 2\theta$
2, 2	$3(1 - x^2)$	$\frac{3}{2}(1 - \cos 2\theta)$
1, 3	$\frac{3}{2}(5x^2 - 1)(1 - x^2)^{1/2}$	$\frac{3}{8}(5\sin 3\theta + \sin\theta)$
2, 3	$15x(1 - x^2)$	$-\frac{15}{4}(\cos 3\theta - \cos\theta)$
3, 3	$15(1 - x^2)^{3/2}$	$-\frac{15}{4}(\sin 3\theta - 3\sin\theta)$
1, 4	$\frac{5}{2}(7x^3 - 3x)(1 - x^2)^{1/2}$	$\frac{5}{16}(7\sin 4\theta + 2\sin 2\theta)$
2, 4	$\frac{15}{2}(7x^2 - 1)(1 - x^2)$	$-\frac{15}{16}(7\cos 4\theta - 4\cos 2\theta - 3)$
3, 4	$105x(1 - x^2)^{3/2}$	$-\frac{105}{8}(\sin 4\theta - 2\sin 2\theta)$
4, 4	$105(1 - x^2)^2$	$\frac{105}{8}(\cos 4\theta - 4\cos 2\theta + 3)$

4.10 ORTHOGONALITY AND GENERATING FUNCTION OF $P_n^m(x)$

As the definition (4.8.4) of $P_n^m(x)$ indicates, we may obtain a generating function of the form

$$g(x, m, t) = \sum_{n=0}^{\infty} P_n^m(x)t^n \tag{4.10.1}$$

by considering the generating function of $P_n(x)$,

$$(1 - 2xt + t^2)^{-1/2} = \sum_{n=0}^{\infty} P_n(x)t^n.$$

Differentiating this latter equation m times with respect to x, we have

$$\frac{(2m)!}{2^m m!}(1 - 2xt + t^2)^{-1/2-m}t^m = \sum_{n=0}^{\infty} P_n^{(m)}(x)t^n.$$

Multiplying through by $(1 - x^2)^{m/2}$ and identifying $P_n^m(x)$, we have the generating function (4.10.1) as

$$\frac{(2m)!}{2^m m!}(1 - x^2)^{m/2}t^m(1 - 2xt + t^2)^{-(2m+1)/2} = \sum_{n=0}^{\infty} P_n^m(x)t^n,$$

$$m = 0, 1, 2, \ldots. \tag{4.10.2}$$

To obtain an orthogonality relation for $P_n^m(x)$, it is convenient to use the general procedure outlined in Sec. 1.9. The self-adjoint form of (4.8.2) is

$$D[(1 - x^2)DP_n^m(x)] + \left[n(n+1) - \frac{m^2}{1 - x^2}\right]P_n^m(x) = 0,$$

and comparison of this equation with (1.9.5) gives

$$r(x) = 1 - x^2, \quad p(x) = -\frac{m^2}{1 - x^2}, \quad \lambda_n = n(n+1), \quad w(x) = 1.$$

Equation (1.9.8) for this case (using k instead of m) is

$$[k(k+1) - n(n+1)](P_n^m, P_k^m) = (1 - x^2)(P_k^m DP_n^m - P_n^m DP_k^m)\Big|_a^b.$$

Choosing (a, b) to be $(-1, 1)$ we see that the right-hand member vanishes, giving the orthogonality relation

$$(P_n^m, P_k^m) = \int_{-1}^{1} P_n^m(x)P_k^m(x)\,dx = 0, \quad n \neq k. \tag{4.10.3}$$

An alternative orthogonality relation, which we leave as an exercise (see Exercise 4.10.4), is

$$\int_{-1}^{1} \frac{1}{1 - x^2} P_n^m(x)P_n^k(x)\,dx = 0, \quad m^2 \neq k^2. \tag{4.10.4}$$

The procedure which leads to the value of the norm

$$\|P_n^m\|^2 = \int_{-1}^{1} [P_n^m(x)]^2\,dx$$

gives a reduction formula for this integral. We begin by differentiating the associated Legendre function, obtaining

$$\begin{aligned} DP_n^m(x) &= D[(1 - x^2)^{m/2}P_n^{(m)}(x)] \\ &= (1 - x^2)^{-1/2}P_n^{m+1}(x) - mx(1 - x^2)^{-1}P_n^m(x), \end{aligned}$$

which is equivalent to

$$P_n^{m+1}(x) = (1 - x^2)^{1/2}DP_n^m(x) + mx(1 - x^2)^{-1/2}P_n^m(x).$$

Squaring both sides and integrating, we have

$$\begin{aligned} \int_{-1}^{1} [P_n^{m+1}(x)]^2\,dx \\ = \int_{-1}^{1} (1 - x^2)[DP_n^m(x)]^2\,dx + 2m\int_{-1}^{1} xP_n^m(x)DP_n^m(x)\,dx \\ + m^2\int_{-1}^{1} x^2(1 - x^2)^{-1}[P_n^m(x)]^2\,dx. \end{aligned} \tag{4.10.5}$$

Integrating the first and second terms on the right by parts gives, respectively,

$$\int_{-1}^{1} (1 - x^2)[DP_n^m(x)]^2\, dx = (1 - x^2)[DP_n^m(x)]P_n^m(x) \Big|_{-1}^{1} - \int_{-1}^{1} P_n^m(x) D[(1 - x^2)DP_n^m(x)]\, dx$$

and

$$2m\int_{-1}^{1} xP_n^m(x)DP_n^m(x)\, dx = mx(P_n^m)^2 \Big|_{-1}^{1} - \int_{-1}^{1} m(P_n^m)^2\, dx.$$

Therefore, since $P_n^m(\pm 1) = 0$, we have

$$\int_{-1}^{1} [P_n^{m+1}(x)]^2\, dx = \int_{-1}^{1} P_n^m(x)\left\{-D[(1 - x^2)DP_n^m] - mP_n^m + \frac{m^2x^2}{1 - x^2} P_n^m\right\} dx.$$

The expression in brackets can be simplified by noting that

$$\frac{m^2x^2}{1 - x^2} = \frac{m^2}{1 - x^2} - m^2,$$

and, from the differential equation for $P_n^m(x)$, that

$$D[(1 - x^2)DP_n^m(x)] - \frac{m^2}{1 - x^2} P_n^m(x) = -n(n + 1)P_n^m(x).$$

The result is

$$\int_{-1}^{1} [P_n^{m+1}(x)]^2\, dx = \int_{-1}^{1} [P_n^m(x)]^2[n(n + 1) - m(m + 1)]\, dx.$$

Finally, writing $n(n + 1) - m(m + 1) = (n - m)(n + m + 1)$, we have the reduction formula

$$\int_{-1}^{1} [P_n^{m+1}(x)]^2\, dx = (n - m)(n + m + 1)\int_{-1}^{1} [P_n^m(x)]^2\, dx. \qquad (4.10.6)$$

Applying this formula repeatedly leads to the desired formula for the norm $\|P_n^m\|$. The orthogonality relation is therefore

$$\int_{-1}^{1} P_n^m(x)P_k^m(x)\, dx = \frac{(n + m)!}{(n - m)!}\, \frac{2}{2n + 1}\, \delta_{nk}. \qquad (4.10.7)$$

EXERCISES

4.10.1. Derive a Rodrigues' formula for $P_n^m(x)$.

4.10.2. Show that

$$\int_{-1}^{1} P_n^{-k}(x)P_m^k(x)\, dx = \frac{2(-1)^k}{2n + 1}\, \delta_{mn}.$$

4.10.3. Prove (4.10.7) for $n = k$ by mathematical induction.

4.10.4. Prove (4.10.4).

4.10.5. Using (4.9.4), obtain the reduction formula

$$(n + m)(2n - 1)\int_{-1}^{1}(P_{n-1}^{m})^2\,dx = (n - m)(2n + 1)\int_{-1}^{1}(P_n{}^m)^2\,dx.$$

Use this formula to obtain (4.10.7).

4.10.6. Eliminate the term involving x in the equations in Exercise 4.9.2(b) and (c) and show that

$$\int_{-1}^{1}\frac{1}{1 - x^2}[P_n{}^m(x)]^2\,dx = \frac{1}{m}\frac{(n + m)!}{(n - m)!}$$

4.10.7. Determine a generating function of the form

$$g(x, t) = \sum_{n=0}^{\infty}\frac{P_n{}^m(x)}{n!}t^n.$$

4.10.8. Evaluate $\int_{-1}^{1}[DP_n(x)]^2\,dx$.

4.10.9. Show that

$$\int_0^{\pi}(\sin\theta)P_n{}^m(\cos\theta)P_k{}^m(\cos\theta)\,d\theta = \frac{2}{2n + 1}\frac{(n + m)!}{(n - m)!}\delta_{nk}.$$

4.11 SPHERICAL HARMONICS

A solution to Laplace's equation in three variables is often called a *harmonic* function. Such a solution, $\psi(x, y, z)$, is *homogeneous* of degree n if

$$\psi(tx, ty, tz) = t^n\psi(x, y, z),$$

and is called a *solid spherical harmonic*.

Laplace's equation in spherical coordinates (see Exercise 1.1.2), which we may write as

$$\frac{\partial}{\partial r}\left(r^2\frac{\partial\psi}{\partial r}\right) + \frac{1}{\sin\theta}\frac{\partial}{\partial\theta}\left(\sin\theta\frac{\partial\psi}{\partial\theta}\right) + \frac{1}{\sin^2\theta}\frac{\partial^2\psi}{\partial\phi^2} = 0, \tag{4.11.1}$$

may be solved by the method of separation of variables by first assuming the solution

$$\psi(r, \theta, \phi) = R(r)Y(\theta, \phi). \tag{4.11.2}$$

Substituting this expression in $\nabla^2\psi$ and dividing through by RY, we have

$$\frac{r^2R''}{R} + \frac{2rR'}{R} + \frac{1}{Y\sin\theta}\frac{\partial}{\partial\theta}\left(\sin\theta\frac{\partial Y}{\partial\theta}\right) + \frac{1}{Y\sin^2\theta}\frac{\partial^2 Y}{\partial\phi^2} = 0,$$

which, by equating the first two terms to a separation constant λ, may be written as two equations,

$$r^2R'' + 2rR' - \lambda R = 0 \tag{4.11.3}$$

and

$$\frac{1}{\sin\theta}\frac{\partial}{\partial\theta}\left(\sin\theta\,\frac{\partial Y}{\partial\theta}\right)+\frac{1}{\sin^2\theta}\frac{\partial^2 Y}{\partial\phi^2}+\lambda Y=0. \tag{4.11.4}$$

The latter equation may be separated by assuming

$$Y(\theta,\phi)=\Theta(\theta)\Phi(\phi), \tag{4.11.5}$$

a substitution which results in the two equations

$$\Theta''+(\cot\theta)\Theta'+\left[\lambda-\frac{\mu^2}{\sin^2\theta}\right]\Theta=0 \tag{4.11.6}$$

and

$$\Phi''+\mu^2\Phi=0, \tag{4.11.7}$$

where μ is another separation constant.

If it is required that $\Phi(\phi)$ be periodic of period 2π, then we must require that $\mu=m$, an integer, so that from (4.11.7) we have

$$\Phi=A\cos m\phi+B\sin m\phi.$$

The solution of (4.11.3) for $R(r)$ can be facilitated if we let $\lambda=n(n+1)$. Then following the procedure of Exercise 1.9.3 we have

$$R(r)=Cr^n+Dr^{-n-1}. \tag{4.11.8}$$

Substituting these values of λ and μ into (4.11.6) yields

$$\Theta''+(\cot\theta)\Theta'+\left[n(n+1)-\frac{m^2}{\sin^2\theta}\right]\Theta=0,$$

which is Legendre's associated equation (see Exercise 4.9.4). If the solution is required to be finite at $\theta=0$, we must have $n=0,1,2,\ldots,$ and

$$\Theta=EP_n^{\,m}(\cos\theta).$$

Also if ψ is to be defined for $r=0$, we must have $D=0$ in (4.11.8).

Collecting our results, we have

$$Y_{nm}(\theta,\phi)=(A_{nm}\cos m\phi+B_{nm}\sin m\phi)P_n^{\,m}(\cos\theta);$$
$$n=0,1,2,\ldots;\quad m=0,1,2,\ldots,n; \tag{4.11.9}$$

and a set of solutions,

$$\psi_{nm}=r^n(a_{nm}\cos m\phi+b_{nm}\sin m\phi)P_n^{\,m}(\cos\theta). \tag{4.11.10}$$

Negative values of m do not contribute to the set of solutions, since they change only the arbitrary constants a_{nm} and b_{nm}, as may be seen from the relationship between $P_n^{\,m}$ and P_n^{-m} in (4.8.7).

The functions in ψ_{nm},

$$\begin{aligned} u_{nm} &= (r^n \cos m\phi) P_n^{\,m}(\cos\theta), \\ v_{nm} &= (r^n \sin m\phi) P_n^{\,m}(\cos\theta), \end{aligned} \tag{4.11.11}$$

are solid spherical harmonics, and the functions u_{nm}/r^n and v_{nm}/r^n are sometimes referred to as *surface* spherical harmonics. The functions u_{n0}, u_{nm}, v_{nm}, $m = 1, 2, \ldots, n$, thus constitute a set of $2n + 1$ linearly independent spherical harmonics.

4.12 SERIES EXPANSIONS INVOLVING $P_n^{\,m}(x)$

If it is desired to expand a function $f(x)$ in a series of associated Legendre functions,

$$f(x) = \sum_{n=0}^{\infty} C_n P_n^{\,m}(x), \tag{4.12.1}$$

where m is known, we may use the procedure of Chapter 1 to obtain formally

$$C_n = \frac{(f, P_n^{\,m})}{\|P_n^{\,m}\|^2}. \tag{4.12.2}$$

By (4.10.7), this result is

$$C_n = \frac{(2n+1)}{2}\frac{(n-m)!}{(n+m)!}\int_{-1}^{1} f(x) P_n^{\,m}(x)\, dx. \tag{4.12.3}$$

As we might readily surmise from the preceding section, any series expansion involving $P_n^{\,m}(x)$ is very likely to be an expansion of the type

$$f(\theta, \phi) = \sum_{n=0}^{\infty} \psi_n(\theta, \phi), \tag{4.12.4}$$

where

$$\psi_n(\theta, \phi) = \sum_{m=0}^{n} (a_{nm} \cos m\phi + b_{nm} \sin m\phi) P_n^{\,m}(\cos\theta).$$

The coefficients may be determined as before, except that the process must be repeated.

First we note that

$$\begin{aligned} \int_0^{2\pi} f(\theta, \phi) \cos k\phi \, d\phi &= \sum_{n=0}^{\infty} \sum_{m=0}^{n} a_{nm} \pi \, \delta_{mk} P_n^{\,m}(\cos\theta), \quad k \neq 0, \\ &= \sum_{n=0}^{\infty} \pi a_{nk} P_n^{\,k}(\cos\theta). \end{aligned}$$

Finally, using the orthogonality relation for $P_n^{\,k}$ (see Exercise 4.10.9) and (4.12.2), we obtain, for $m \neq 0$,

$$a_{nm} = \frac{(2n+1)}{2\pi}\frac{(n-m)!}{(n+m)!}\int_0^{\pi} (\sin\theta) P_n^{\,m}(\cos\theta)\, d\theta \int_0^{2\pi} f(\theta, \phi) \cos m\phi \, d\phi.$$

We leave as an exercise the determination of a_{n0} and b_{nm}.

An alternative method of obtaining the spherical harmonics in Sec. 4.11 is to write the solution of (4.11.7) in the form

$$\Phi = Ke^{im\phi},$$

where m takes on both positive and negative values. In this case the solutions ψ_{nm} given in (4.11.10) become

$$\psi_{nm} = C_{nm} r^n e^{im\phi} P_n{}^m(\cos\theta). \tag{4.12.5}$$

where $m = 0, \pm 1, \pm 2, \ldots, \pm n$.

If, for ψ given by

$$\psi(r, \theta, \phi) = \sum_{n=0}^{\infty} \sum_{m=-n}^{n} \psi_{nm},$$

we have the condition $\psi(1, \theta, \phi) = f(\theta, \phi)$, it becomes necessary to obtain the expression

$$f(\theta, \phi) = \sum_{n=0}^{\infty} \sum_{m=-n}^{n} C_{nm} e^{im\phi} P_n{}^m(\cos\theta). \tag{4.12.6}$$

This may be accomplished by the use of (4.10.7) and the known formula

$$\int_0^{2\pi} e^{im\phi} e^{-ik\phi}\, d\phi = 2\pi\, \delta_{mk}.$$

The coefficients are thus

$$C_{nm} = \frac{(2n+1)}{4\pi} \frac{(n-m)!}{(n+m)!} \int_0^{\pi} (\sin\theta) P_n{}^m(\cos\theta)\, d\theta \int_0^{2\pi} e^{-im\phi} f(\theta, \phi)\, d\phi. \tag{4.12.7}$$

EXERCISES

4.12.1. Find the first three non-zero terms of the expansion

$$\sin\theta = \sum_{n=0}^{\infty} C_n P_n{}^3(\cos\theta).$$

4.12.2. Show that u_{nm} in (4.11.11) is homogeneous in x, y, and z.

4.12.3. Show that the set $\{\psi_{nm}\}$, where $\psi_{nm} = e^{im\phi} P_n{}^m(\cos\theta)$, $n = 0, 1, 2, \ldots,$ $m = 0, \pm 1, \pm 2, \ldots, \pm n$, has the orthogonality relation

$$(\psi_{nm}, \psi_{rs}) = \int_0^{\pi}\int_0^{2\pi} \psi_{nm}\, \bar{\psi}_{rs} \sin\theta\, d\phi\, d\theta = \frac{4\pi}{2n+1} \frac{(n+m)!}{(n-m)!} \delta_{nm}^{rs}.$$

4.12.4. Express $f(x, y, z) = ax + by + cz$ as a sum of solid spherical harmonics.

4.12.5. Express $f(x, y, z) = x^2 - 2y^2 + z^2$ as a sum of solid spherical harmonics.

4.13 ANOTHER EXPRESSION FOR $Q_n(x)$

In the last three sections of this chapter we develop some relations for the Legendre functions of the second kind which are of interest in a study of special functions. Because, as we have seen, the appearance of $Q_n(x)$ in a physical problem is extremely rare by comparison with $P_n(x)$, the reader may wish to omit these last three sections. For boundary-value problems which involve the $Q_n(x)$, the interested reader is referred to [Hn], Chapter 10, and [McR], Chapter 11.

We recall that $Q_n(x)$ is a second solution of Legendre's equation for the case n a non-negative integer. The expressions for $Q_n(x)$ for the various cases of n and x are given by (4.2.1), (4.2.2), and (4.2.3); and in (4.2.4) we have seen that $Q_n(x)$ may be written as

$$Q_n(x) = CP_n(x)\int \frac{dx}{[P_n(x)]^2(1 - x^2)} \,. \tag{4.13.1}$$

In this section we determine C for the case $|x| < 1$ and n even. In this case $Q_n(x)$ is given by (4.2.1).

For n even, we may factor out the constant term a_n in the expansion for $P_n(x)$ given by (4.1.9) and write

$$[P_n(x)]^2 = \left\{\frac{n!\,(-1)^{n/2}}{2^n[(n/2)!]^2}\right\}^2[1 + R(x)] = a_n{}^2[1 + R(x)], \tag{4.13.2}$$

where $R(x)$ is a polynomial which has no constant term. Substituting this expression into the integrand of (4.13.1) and performing a long division operation, we may write

$$Q_n(x) = CP_n(x)\int \frac{1 + S(x)}{a_n{}^2}\,dx,$$

where $S(x)$ is a series which has no constant term. Carrying out the integration, we have

$$Q_n(x) = \frac{CP_n(x)\left[x + \int S(x)\,dx\right]}{a_n{}^2} \,.$$

The constant of integration is absorbed in the term $c_2P_n(x)$ of (4.2.4). Comparing coefficients of x in this expression for $Q_n(x)$ and the one given by (4.2.1) we have

$$\frac{2^n[(n/2)!]^2(-1)^{n/2}C}{n!} = \frac{2^n[(n/2)!]^2(-1)^{n/2}}{n!} \,,$$

where we have again substituted from (4.1.9) for the constant term in $P_n(x)$.

Therefore we see that $C = 1$, and for n even and $|x| < 1$ we have

$$Q_n(x) = P_n(x)\int \frac{dx}{[P_n(x)]^2(1 - x^2)}. \tag{4.13.3}$$

We leave as exercises to show that this formula holds when n is any non-negative integer and $|x| \neq 1$ (see Exercises 4.15.1 and 4.15.2).

We may use this expression to compute $Q_n(x)$ in terms of $P_n(x)$. For example, when $n = 0$,

$$Q_0(x) = \int \frac{dx}{1 - x^2} = \tfrac{1}{2} \log \left| \frac{1 + x}{1 - x} \right|.$$

Similarly, when $n = 1$,

$$\begin{aligned} Q_1(x) &= x\int \frac{dx}{x^2(1 - x^2)} \\ &= \frac{x}{2} \log \left| \frac{1 + x}{1 - x} \right| - 1. \end{aligned}$$

We can extend the relation (4.13.3) by using partial fractions in the form

$$\frac{1}{[P_n(x)]^2(1 - x^2)} = \frac{T(x)}{[P_n(x)]^2} + \frac{\frac{1}{2}}{1 - x} + \frac{\frac{1}{2}}{1 + x},$$

where $T(x)$ is a polynomial of degree less than $2n$, which may be found if the value of n is given. Using this result in (4.13.3) we obtain

$$Q_n(x) = \tfrac{1}{2}P_n(x) \log \left| \frac{1 + x}{1 - x} \right| - f_{n-1}(x), \tag{4.13.4}$$

where

$$f_{n-1}(x) = -P_n(x) \int \frac{T(x)\, dx}{[P_n(x)]^2}. \tag{4.13.5}$$

4.14 RECURRENCE RELATIONS FOR $Q_n(x)$

In this section we will use *Sister Celine's technique* (see [SCF]) for determining a pure recurrence relation for the $Q_n(x)$. The technique is to find constants A, B, C, D, E such that

$$Q_n(x) + (A + Bx)Q_{n-1}(x) + (C + Dx)Q_{n-2}(x) + EQ_{n-3}(x) \equiv 0. \tag{4.14.1}$$

For the case $|x| > 1$, we have by (4.2.3)

$$Q_n = \sum_{k=0}^{\infty} \frac{2^n(n + k)!\,(n + 2k)!}{k!\,(2n + 2k + 1)!} x^{-(n+2k+1)}.$$

Substituting into (4.14.1) for the various Q_n's and simplifying somewhat, we have

$$\frac{2^{n-1}[(n-1)!]^2}{(2n-1)!}\left[B+\frac{2n-1}{n-1}C\right]x^{-n+1}$$
$$+\sum_{k=1}^{\infty}\frac{2^{n-1}(n+k-1)!\,(n+2k-2)!}{k!\,(2n+2k-1)!}[2k+(2n+2k-1)C+(n+2k-1)B]x^{-n-2k+1}$$
$$+\frac{2^{n-2}[(n-2)!]^2}{(2n-3)!}\left[D+\frac{2n-3}{n-2}E\right]x^{-n+2}$$
$$+\sum_{k=1}^{\infty}\frac{2^{n-2}(n+k-2)!\,(n+2k-3)!}{k!\,(2n+2k-3)!}[2kA+(n+2k-2)D+(2n+2k-3)E]x^{-n-2k+2}\equiv 0.$$

When the coefficient of each power of x is made zero, the values of the constants are

$$A=-\frac{(n-1)D}{2n-3},\quad E=-\frac{(n-2)D}{2n-3},$$
$$B=-\frac{2n-1}{n},\qquad C=\frac{n-1}{n}.$$

The recurrence relation (4.14.1) becomes, upon substitution of these values

$$\left(Q_n-\frac{2n-1}{n}xQ_{n-1}+\frac{n-1}{n}Q_{n-2}\right)$$
$$+D\left(-\frac{n-1}{2n-3}Q_{n-1}+xQ_{n-2}-\frac{n-2}{2n-3}Q_{n-3}\right)=0.$$

This relation is satisfied if

$$nQ_n-(2n-1)xQ_{n-1}+(n-1)Q_{n-2}=0,$$

since this relation makes the expressions in both parentheses zero. Writing this recurrence relation as

$$(n+1)Q_{n+1}-(2n+1)xQ_n+nQ_{n-1}=0, \tag{4.14.2}$$

we see that it is the same relation as (4.6.2) for $P_n(x)$.

It can be shown by direct substitution that $Q_n(x)$ satisfies the same relation as (4.6.4) for $P_n(x)$; that is,

$$xQ'_n-Q'_{n-1}-nQ_n=0. \tag{4.14.3}$$

Since $Q_n(x)$ and $P_n(x)$ satisfy the same two basic recurrence relations, it follows immediately that all the relations obtained for $P_n(x)$ from the two basic relations hold also for $Q_n(x)$. For example, we have

$$Q'_{n+1}-xQ'_n-(n+1)Q_n=0, \tag{4.14.4}$$
$$Q'_{n+1}-(2n+1)Q_n-Q'_{n-1}=0, \tag{4.14.5}$$
$$(x^2-1)Q'_n-nxQ_n+nQ_{n-1}=0. \tag{4.14.6}$$

These are the relations corresponding to those for $P_n(x)$ given in Exercises 4.6.2 and 4.6.3.

4.15 GENERATING FUNCTION FOR $Q_n(x)$

Numerous relations involving the Legendre functions can be derived by means of complex variable theory. One such relation (see [WW], page 319) is an integral relation for $Q_n(x)$,

$$Q_n(x) = \int_0^\infty [x + \sqrt{x^2 - 1} \cosh \theta]^{-n-1} \, d\theta, \quad |x| > 1. \tag{4.15.1}$$

Since the integrand is a power, we can determine a generating function of the form (see Sec. 1.8),

$$\sum_{n=0}^{\infty} Q_n(x)t^n = \sum_{n=0}^{\infty} \int_0^\infty [x + \sqrt{x^2 - 1} \cosh \theta]^{-n-1} t^n \, d\theta.$$

Assuming that the order of summation and integration can be interchanged, we have

$$\sum_{n=0}^{\infty} Q_n(x)t^n = \int_0^\infty \frac{d\theta}{(x - t) + \sqrt{x^2 - 1} \cosh \theta},$$

where we have used the property

$$\sum_{n=0}^{\infty} [x + \sqrt{x^2 - 1} \cosh \theta]^{-n} t^n = \frac{1}{1 - t[x + \sqrt{x^2 - 1} \cosh \theta]^{-1}}.$$

Substituting $\tanh \frac{1}{2}\theta = Z$, we obtain

$$\sum_{n=0}^{\infty} Q_n(x)t^n = -2\int_0^1 \frac{dZ}{[x - t - \sqrt{x^2 - 1}]\, Z^2 - [x - t + \sqrt{x^2 - 1}]},$$

and performing the integration, we find for the generating function,

$$(1 - 2xt + t^2)^{-1/2} \log \frac{t - x - \sqrt{1 - 2xt + t^2}}{\sqrt{x^2 - 1}} = \sum_{n=0}^{\infty} Q_n(x)t^n. \tag{4.15.2}$$

Performing the integration differently, we may write the generating function as

$$(1 - 2xt + t^2)^{-1/2} \cosh^{-1} \frac{t - x}{\sqrt{x^2 - 1}} = \sum_{n=0}^{\infty} Q_n(x)t^n. \tag{4.15.3}$$

EXERCISES

4.15.1. Show that $C = 1$ in (4.13.1) for n odd and $|x| < 1$.

4.15.2. Show that $C = 1$ in (4.13.1) for $|x| > 1$.

4.15.3. Compute $Q_2(x)$ in the form given for $Q_0(x)$ and $Q_1(x)$ in Sec. 4.13.

4.15.4. Show that $Q_0(x)$, $Q_1(x)$, and $Q_2(x)$ satisfy the relations given in
(a) (4.14.2).
(b) (4.14.3).
(c) (4.14.4).
(d) (4.14.5).
(e) (4.14.6).

4.15.5. Verify that $Q_n(x)$ satisfies (4.14.2) when $|x| < 1$.

4.15.6. Verify that $Q_n(x)$ satisfies (4.14.3) for $|x| \neq 1$.

4.15.7. Show that, for f_n defined in (4.13.4),

$$(n + 1)f_n - (2n + 1)xf_{n-1} + nf_{n-2} = 0.$$

4.15.8. Note that $f_{-1}(x) = 0$ and $f_0(x) = 1$ and determine $f_1(x)$. Use this to check the answer for $Q_2(x)$ in Exercise 4.15.3.

4.15.9. Develop a general formula for $f_{n-1}(x)$ in series form.

4.15.10. Using the relation

$$P_n(x) = \frac{1}{\pi}\int_0^\pi \frac{d\phi}{[x + \sqrt{x^2 - 1}\cos\phi]^{n+1}}$$

and the procedure used to derive (4.15.2), obtain a generating function for $P_n(x)$.

4.15.11. Consider the expansion of the generating function for $Q_n(x)$ as a Maclaurin series and show that

$$Q_n(x) = \tfrac{1}{2}P_n(x)\log\left|\frac{x - 1}{x + 1}\right| - \sum_{k=0}^{n-1}\frac{P_k(x)P_{n-k-1}(x)}{n - k};$$

and hence by (4.13.4), that

$$f_{n-1}(x) = \sum_{k=0}^{n-1}\frac{P_k(x)P_{n-k-1}(x)}{n - k}.$$

4.15.12. Defining

$$Q_n{}^m(x) = (1 - x^2)^{m/2}Q_n^{(m)}(x), \quad -1 < x < 1,$$
$$Q_n{}^m(x) = (-1)^m(x^2 - 1)^{m/2}Q_n^{(m)}(x), \quad x^2 > 1,$$

obtain $Q_n{}^m(x)$ for $0 \leq n, m \leq 2$, for $-1 < x < 1$ and for $x^2 > 1$.

5

The Gamma Function

5.1 INTEGRAL DEFINITION

In this chapter we will give a brief discussion of the *gamma function* defined as

$$\Gamma(n) = \int_0^\infty x^{n-1}e^{-x}\,dx, \quad n > 0. \tag{5.1.1}$$

The gamma function is not one of the special functions obtained as a solution of one of the separated differential equations in a boundary-value problem, but it is important for our purposes because of the role it plays as a *generalized factorial* function. This is especially true in the development of the Bessel functions, which we will undertake in Chapter 6.

To see that the gamma function behaves as a generalized factorial function we replace n by $n + 1$ in (5.1.1) and integrate by parts, to obtain

$$\Gamma(n + 1) = -x^n e^{-x}\Big|_0^\infty + n\int_0^\infty x^{n-1}e^{-x}\,dx,$$

which may be written as

$$\Gamma(n + 1) = n\Gamma(n). \tag{5.1.2}$$

From (5.1.1) if $n = 1$ we see that $\Gamma(1) = 1$. Using this result and letting $n = 1, 2, 3, \ldots$ successively in (5.1.2), we can obtain by induction

$$\Gamma(n + 1) = n!, \quad n = 0, 1, 2, \ldots. \tag{5.1.3}$$

It is convenient to extend our definition (5.1.1) to negative values of n by defining the gamma function in that case to be consistent with (5.1.2). We also define a generalized factorial function by means of (5.1.3) to hold for all real values of n. Writing (5.1.2) in the form

$$\Gamma(n) = \frac{\Gamma(n + 1)}{n},$$

and taking $n = 0, -1, -2, \ldots$ in that order, we see that the gamma function of zero and negative integers is undefined. However for $0 < n < 1$, (5.1.1) exists and hence by (5.1.2) the gamma function is defined for all negative non-integral values of n. For example, if $x = y^2$ in (5.1.1), we have

$$\Gamma(n) = 2\int_0^\infty y^{2n-1}e^{-y^2}\,dy,$$

which for $n = \frac{1}{2}$ becomes

$$\Gamma(\tfrac{1}{2}) = 2\int_0^\infty e^{-y^2}\,dy = \sqrt{\pi}. \tag{5.1.4}$$

Using (5.1.2) and (5.1.4) we may obtain

$$\Gamma(-\tfrac{1}{2}) = \frac{\Gamma(\frac{1}{2})}{-\frac{1}{2}} = -2\sqrt{\pi},$$

$$\Gamma(-\tfrac{3}{2}) = \tfrac{4}{3}\sqrt{\pi}, \quad \text{etc.}$$

From the information we have obtained we may now sketch the gamma function, as shown in Fig. 5–1.

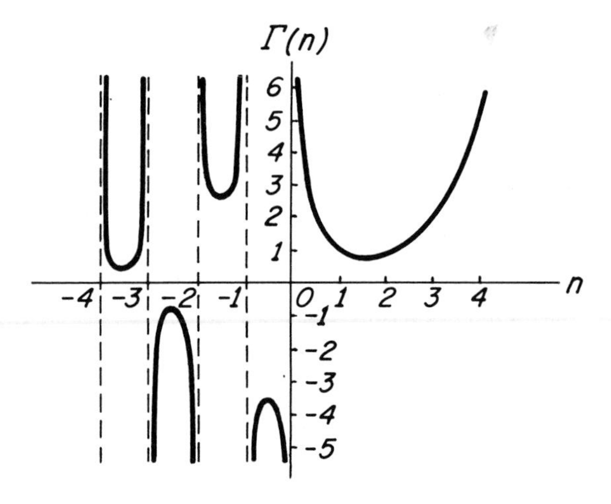

Fig. 5–1

EXERCISES

5.1.1. Prove Eq. (5.1.3).

5.1.2. Show that $\int_0^\infty x^n e^{-xz}\,dx = \Gamma(n+1)/z^{n+1}$.

5.1.3. Write $\Gamma(m)\Gamma(n)$ as a double integral,

$$\int_0^\infty u^{m-1}e^{-u}\,du \int_0^\infty v^{n-1}e^{-v}\,dv,$$

and let $u = x^2$, $v = y^2$, to show that

$$\int_0^{\pi/2} \cos^{2m-1}\theta\,\sin^{2n-1}\theta\,d\theta = \frac{\Gamma(m)\Gamma(n)}{2\Gamma(m+n)}, \quad m > 0, n > 0.$$

5.1.4. Using the results of Exercise 5.1.3, show that the *beta function* of m and n,

$$B(m, n) = \int_0^1 x^{m-1}(1 - x)^{n-1}\, dx,$$

is given by

$$B(m, n) = \frac{\Gamma(m)\Gamma(n)}{\Gamma(m + n)}, \quad m > 0, n > 0.$$

5.1.5. Derive Eq. (5.1.4). Use a technique similar to that used in Exercise 5.1.3.

5.1.6. Evaluate $\int_0^1 (x^n/\sqrt{1 - x^2})\, dx$ if n is an even positive integer; if n is an odd positive integer.

5.1.7. Show that $\int_{-1}^1 (1 + x)^{p-1}(1 - x)^{q-1}\, dx = 2^{p+q-1}B(p, q)$, $p, q > 0$, by replacing x by $\cos 2\theta$.

5.1.8. Show that $\int_0^\infty x^m \exp(-x^n)\, dx = (1/n)\Gamma[(m + 1)/n]$, $m > -1$, $n > 0$.

5.1.9. Show that

$$\int_0^{\pi/2} \sin^n \theta\, d\theta = \int_0^{\pi/2} \cos^n \theta\, d\theta = \frac{\sqrt{\pi}\, \Gamma[(n + 1)/2]}{2\Gamma[(n + 2)/2]}, \quad n > -1.$$

5.1.10. Show that

$$\int_0^{\pi/2} \tan^n \theta\, d\theta = \tfrac{1}{2}\Gamma\left(\frac{1 + n}{2}\right)\Gamma\left(\frac{1 - n}{2}\right), \quad -1 < n < 1.$$

5.1.11. By evaluating the integral

$$\int_0^{\pi/2} \sin^{2n-1} \theta \cos^{2n-1} \theta\, d\theta$$

two ways, show that $\Gamma(n)\Gamma(n + \frac{1}{2}) = 2^{1-2n}\sqrt{\pi}\, \Gamma(2n)$. This is *Legendre's duplication formula*.

5.2 EULER'S CONSTANT

If a function satisfies the conditions

$$\begin{aligned} f(x) &> 0, \quad \text{for all } x > 0, \\ f(x) &> f(x + \epsilon), \quad x, \epsilon > 0, \end{aligned} \tag{5.2.1}$$

then it can be shown that a finite constant C exists such that

$$\lim_{n\to\infty} \left\{ \sum_{k=1}^{n} f(k) - \int_1^n f(x)\, dx \right\} = C. \tag{5.2.2}$$

To establish (5.2.2) we first define

$$S_n = \sum_{k=1}^{n} f(k), \quad I_n = \int_1^n f(x)\, dx, \quad q_n = S_n - I_n.$$

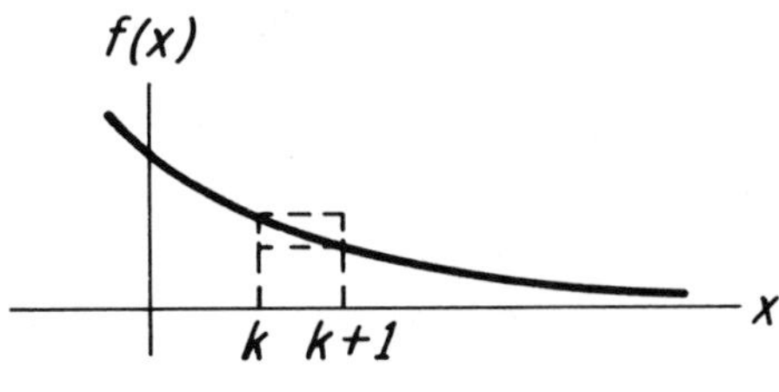

Fig. 5–2

From geometrical considerations of areas in Fig. 5–2, the inequality

$$f(k) > \int_k^{k+1} f(x)\,dx > f(k+1) > 0 \tag{5.2.3}$$

is evidently true.

Considering (5.2.3) successively for $k = 1, 2, 3, \ldots, n-1$, and adding the resulting inequalities, we have

$$\sum_{k=1}^{n-1} f(k) > \int_1^n f(x)\,dx > \sum_{k=2}^{n} f(k),$$

which we may write as

$$S_n - f(n) > I_n > S_n - f(1). \tag{5.2.4}$$

The first inequality in (5.2.4) is equivalent to

$$q_n = S_n - I_n > f(n),$$

and thus q_n is positive. We may also write

$$q_{n+1} - q_n = S_{n+1} - S_n - (I_{n+1} - I_n) = f(n+1) - \int_n^{n+1} f(x)\,dx.$$

Since by (5.2.3) the last member is negative, we have

$$q_{n+1} - q_n < 0, \quad n = 1, 2, 3, \ldots.$$

Hence $\{q_n\}$ is a sequence of decreasing positive numbers and must tend to a finite limit as $n \to \infty$. This is what is stated in (5.2.2).

As an example, the function $f(x) = 1/x$ satisfies the conditions in (5.2.1), and for this case (5.2.2) yields the limit

$$\gamma = \lim_{n\to\infty} \left\{\sum_{k=1}^{n} \frac{1}{k} - \log n\right\}, \tag{5.2.5}$$

which is known as *Euler's constant*. Its value to seven decimal places is

$$\gamma = .5772157. \tag{5.2.6}$$

As we shall see, Euler's constant figures prominently in alternative definitions of the gamma function, but it is of primary interest to us because of the role it plays in the treatment of the Bessel functions in the next chapter.

5.3 WEIERSTRASS' DEFINITION

An alternative expression for the gamma function given by *Weierstrass* is the infinite-product form

$$\Gamma(z) = z^{-1}e^{-\gamma z}\prod_{n=1}^{\infty}\left[\left(1+\frac{z}{n}\right)^{-1}e^{z/n}\right], \tag{5.3.1}$$

where γ is Euler's constant. It will be noted that $\Gamma(z)$ is undefined for z a negative integer or zero.

The formula $\Gamma(z+1) = z\Gamma(z)$ may be obtained from (5.3.1) by considering

$$\frac{\Gamma(z+1)}{\Gamma(z)} = \frac{ze^{\gamma z}\prod_{n=1}^{\infty}[(1+z/n)e^{-z/n}]}{(z+1)e^{\gamma(z+1)}\prod_{n=1}^{\infty}\{[1+(z+1)/n]e^{-(z+1)/n}\}},$$

which by (5.2.5) may be written as

$$\begin{aligned}\frac{\Gamma(z+1)}{\Gamma(z)} &= \frac{z}{z+1}\lim_{m\to\infty}\frac{(1+z)(1+z/2)\cdots(1+z/m)\exp\left[-z\sum_{n=1}^{m}1/n\right]}{\left\{\exp\left[\sum_{n=1}^{m}1/n-\log m\right][1+(z+1)][1+(z+1)/2]\cdots[1+(z+1)/m]\exp\left[-(z+1)\sum_{n=1}^{m}1/n\right]\right\}}\\ &= \frac{z}{z+1}\lim_{m\to\infty}\frac{m(1+z)(2+z)\cdots(m+z)}{(2+z)(3+z)\cdots(m+z+1)}\\ &= \frac{z}{z+1}\lim_{m\to\infty}\frac{m(z+1)}{m+z+1} = z.\end{aligned}$$

Using this result we may write (5.3.1) as

$$\Gamma(z+1) = e^{-\gamma z}\prod_{n=1}^{\infty}\left[\left(1+\frac{z}{n}\right)^{-1}e^{z/n}\right]. \tag{5.3.2}$$

EXERCISES

5.3.1. Show that $0 < a_n = \int_0^1[x/n(n+x)]\,dx < \int_0^1(1/n^2)\,dx = 1/n^2$, and hence that $\sum_{n=1}^{\infty} a_n$ converges. Use this result to show that Euler's constant exists.

5.3.2. Find $\Gamma(1)$, $\Gamma(2)$, $\Gamma(3)$, by (5.3.1); by (5.3.2).

5.3.3. It can be shown (see [WW], page 136) that if $f(z)$ is analytic for all z and has no zeros except simple zeros at $a_1, a_2, a_3, \ldots$, with $|a_n| \to \infty$ as $n \to \infty$, then if $a_n \neq 0$,

$$f(z) = f(0)e^{zf'(0)/f(0)} \prod_{n=1}^{\infty} \left[\left(1 - \frac{z}{a_n}\right)e^{z/a_n}\right].$$

Show that if $f(z) = (\sin z)/z$ for $z \neq 0$, $f(0) = \lim_{z\to 0} f(z)$, $f'(0) = \lim_{z\to 0} f'(z)$, then

$$\frac{\sin z}{z} = \prod_{n=1}^{\infty} \left[\left(1 - \frac{z}{n\pi}\right)e^{z/n\pi}\left(1 + \frac{z}{n\pi}\right)e^{-z/n\pi}\right].$$

5.3.4. Show that $\Gamma(z)\Gamma(-z) = -\pi/z \sin \pi z$ and hence that

$$\Gamma(z) = \frac{\pi}{(\sin \pi z)\Gamma(1 - z)}.$$

5.4 OTHER FORMS FOR THE GAMMA FUNCTION

To show that the Weierstrass definition of the gamma function is the same as the integral definition, we may derive one from the other. In this section we will obtain the integral form from that of Weierstrass, and in the process we will exhibit other forms of the gamma function.

We began by writing, from (5.3.1),

$$\begin{aligned}\frac{1}{\Gamma(z)} &= z \lim_{m\to\infty} \exp\left[\left(\sum_{n=1}^{m} \frac{1}{n} - \log m\right)z\right] \lim_{m\to\infty} \prod_{n=1}^{m} \left[\left(1 + \frac{z}{n}\right)e^{-z/n}\right] \\ &= z \lim_{m\to\infty} \left\{m^{-z} \prod_{n=1}^{m} \left(1 + \frac{z}{n}\right)\right\}\end{aligned}$$

Since $m^{-z} = \prod_{n=1}^{m-1} (1 + 1/n)^{-z}$, we may write

$$\frac{1}{\Gamma(z)} = z \lim_{m\to\infty} \left\{\left[\prod_{n=1}^{m} \left(1 + \frac{z}{n}\right)\left(1 + \frac{1}{n}\right)^{-z}\right]\left(1 + \frac{1}{m}\right)^{z}\right\}.$$

On taking the limit and inverting we have

$$\Gamma(z) = \frac{1}{z} \prod_{n=1}^{\infty} \left[\left(1 + \frac{1}{n}\right)^{z}\left(1 + \frac{z}{n}\right)^{-1}\right], \tag{5.4.1}$$

a form attributed to Euler.

On expanding (5.4.1), we have

$$\begin{aligned}\Gamma(z) &= \frac{1}{z} \lim_{m\to\infty} \frac{(1 + 1/1)^z(1 + 1/2)^z(1 + 1/3)^z \cdots (1 + 1/m)^z}{(1 + z/1)(1 + z/2)(1 + z/3) \cdots (1 + z/m)} \\ &= \lim_{m\to\infty} \left[\frac{m!\,(m + 1)^z}{z(z + 1)(z + 2) \cdots (z + m)}\right].\end{aligned}$$

Finally, replacing m by $m - 1$, we have another form of the gamma function:

$$\Gamma(z) = \lim_{m\to\infty} \frac{(m-1)!\, m^z}{z(z+1)(z+2)\cdots(z+m-1)}. \tag{5.4.2}$$

Now let us consider the integral

$$P(z, m) = \int_0^m \left(1 - \frac{x}{m}\right)^m x^{z-1}\, dx, \quad m = 1, 2, 3, \ldots, \tag{5.4.3}$$

which is equivalent to

$$P(z, m) = m^z \int_0^1 (1-t)^m t^{z-1}\, dt, \quad m = 1, 2, 3, \ldots.$$

Integrating by parts m times, we have

$$P(z, m) = \frac{[m(m-1)\cdots 1]m^z}{z(z+1)(z+2)\cdots(z+m-1)} \int_0^1 t^{z+m-1}\, dt,$$

which is equivalent to

$$P(z, m) = \frac{m!\, m^z}{z(z+1)\cdots(z+m)}.$$

Taking the limit as $m \to \infty$, we note that

$$\begin{aligned}\lim_{m\to\infty} P(z, m) &= \lim_{m\to\infty} \frac{(m-1)!\, m^z}{z(z+1)\cdots(z+m-1)} \lim_{m\to\infty} \frac{m}{z+m} \\ &= \lim_{m\to\infty} \frac{(m-1)!\, m^z}{z(z+1)\cdots(z+m-1)},\end{aligned}$$

and hence, from (5.4.2) we have

$$\lim_{m\to\infty} P(z, m) = \Gamma(z). \tag{5.4.4}$$

It may be shown that the operations of taking the limit and performing the integration in (5.4.3) may be interchanged (see [WW], Sec. 12.2); that is,

$$\Gamma(z) = \lim_{m\to\infty} P(z, m) = \int_0^\infty \left[\lim_{m\to\infty} \left(1 - \frac{x}{m}\right)^m\right] x^{z-1}\, dx.$$

Noting that

$$\lim_{m\to\infty} \left(1 - \frac{x}{m}\right)^m = e^{-x}$$

we obtain finally

$$\Gamma(z) = \int_0^\infty e^{-x} x^{z-1}\, dx,$$

which is in agreement with (5.1.1).

5.5 LOGARITHMIC DERIVATIVE

To obtain the derivative of the gamma function we begin by taking the logarithm of both members of (5.3.2), obtaining

$$\log \Gamma(z+1) = -\gamma z + \sum_{n=1}^{\infty}\left[\frac{z}{n} - \log\left(1+\frac{z}{n}\right)\right].$$

Differentiating, we have the *logarithmic derivative*,

$$\psi(z+1) = \frac{(d/dz)\Gamma(z+1)}{\Gamma(z+1)} = -\gamma + \sum_{n=1}^{\infty}\left[\frac{1}{n} - \frac{1}{z+n}\right], \tag{5.5.1}$$

which, for z a positive integer, is equivalent to

$$\psi(z+1) = -\gamma + \left[1 + \tfrac{1}{2} + \cdots + \frac{1}{z} + \left(\frac{1}{z+1} - \frac{1}{z+1}\right) + \left(\frac{1}{z+2} - \frac{1}{z+2}\right) + \cdots\right],$$

or

$$\psi(z+1) = -\gamma + \sum_{n=1}^{z}\frac{1}{n}, \quad z = 1, 2, 3, \ldots. \tag{5.5.2}$$

The case $z = 0$ is an exception and may be obtained from (5.5.1) as

$$\psi(1) = -\gamma. \tag{5.5.3}$$

We combine (5.5.2) and (5.5.3) in one equation, valid for z a non-negative integer, as

$$\psi(z+1) = -\gamma + \phi(z), \tag{5.5.4}$$

where

$$\begin{aligned}\phi(z) &= \sum_{n=1}^{z}\frac{1}{n}, \quad z = 1, 2, 3, \ldots, \\ \phi(0) &= 0.\end{aligned} \tag{5.5.5}$$

EXERCISES

5.5.1. Obtain $\Gamma(1)$, $\Gamma(2)$, $\Gamma(3)$ from (5.4.1); from (5.4.2).

5.5.2. Show that $\psi(z+1) = \psi(z) + 1/z$.

5.5.3. Show that $\psi(-z+1) = -\dfrac{(d/dz)\Gamma(-z+1)}{\Gamma(-z+1)}$.

5.5.4. Show that $\gamma = -\Gamma'(1) = -\int_0^{\infty}(\log t)e^{-t}\,dt$.

5.5.5. Show that $(d^2/dz^2)\log\Gamma(z) = \sum_{n=0}^{\infty} 1/(z+n)^2$.

5.5.6. Show that $\Gamma(n+\frac{1}{2}) = (2n)!\sqrt{\pi}/4^n n!$.

5.5.7. Show that

$$(n + \tfrac{1}{2} - k)_k = \frac{\Gamma(2n + 1)\Gamma(n - k + 1)}{2^{2k}\Gamma(n + 1)\Gamma(2n - 2k + 1)}.$$

5.5.8. Show that $(k + \frac{1}{2})_{n-k} = (2n)!k!/2^{2n-2k}(2k)!n!$.

5.5.9. Show that

$$(n + \tfrac{1}{2})_k = \frac{\Gamma(n + 1)\Gamma(2n + 2k + 1)}{2^{2k}\Gamma(n + k + 1)\Gamma(2n + 1)}.$$

6

Bessel Functions

6.1 BESSEL'S DIFFERENTIAL EQUATION

The equation

$$x^2 \frac{d^2y}{dx^2} + x \frac{dy}{dx} + (x^2 - n^2)y = 0, \tag{6.1.1}$$

where n is a constant, is known as *Bessel's differential equation* and can be solved by the series method. We note that $x = 0$ is a regular singular point, and hence the solution is of the type

$$y = \sum_{m=0}^{\infty} a_m x^{m+k}.$$

The reader may verify that substitution of this value of y into (6.1.1) and equating coefficients of like powers of x yields

$$\begin{aligned} (k^2 - n^2)a_0 &= 0, \\ [(k+1)^2 - n^2]a_1 &= 0, \end{aligned} \tag{6.1.2}$$

and the recurrence relation

$$a_{m+2} = -\frac{a_m}{(m+k+2)^2 - n^2}. \tag{6.1.3}$$

If $k = \pm n$, then a_0 is arbitrary, $a_1 = 0$, and for $k = n$, (6.1.3) may be written

$$a_{m+2} = -\frac{a_m}{(m+2)(2n+m+2)}. \tag{6.1.4}$$

(We note that $n = -\frac{1}{2}$ is an exceptional case, making both a_0 and a_1 arbitrary. This case is considered in Exercise 6.1.1.)

Thus all the a's with odd subscripts vanish, and upon writing a few of the even-subscript coefficients we obtain in general

$$a_{2m} = \frac{(-1)^m a_0}{2^{2m}\, m!\, (n+1)(n+2)\cdots(n+m)}, \quad n \text{ not a negative integer.}$$

This gives as a solution

$$y = \sum_{m=0}^{\infty} \frac{(-1)^m a_0 x^{n+2m}}{2^{2m}\, m!\, (n+1)(n+2)\cdots(n+m)}. \tag{6.1.5}$$

If we choose $a_0 = 1/2^n\Gamma(n+1)$, we have a particular solution, the standard notation for which is

$$J_n(x) = \sum_{m=0}^{\infty} \frac{(-1)^m (x/2)^{n+2m}}{m!\, \Gamma(n+m+1)}. \tag{6.1.6}$$

If n is not an integer, the other independent solution of Bessel's equation is found by using the other root of the indicial equation, $k = -n$, which is equivalent to replacing n by $-n$ in (6.1.6). The general solution in this case is then

$$y = C_1 J_n(x) + C_2 J_{-n}(x). \tag{6.1.7}$$

Equation (6.1.6) for n a positive real number defines the *Bessel function* $J_n(x)$ of order n of the *first kind*.

If $n = 0$, $J_n(x)$ and $J_{-n}(x)$ are identical and (6.1.7) is not the general solution. We also fail to have a general solution if n is a positive integer, which may be seen by examining the expression

$$J_{-n}(x) = \sum_{m=0}^{\infty} \frac{(-1)^m (x/2)^{-n+2m}}{m!\, \Gamma(-n+m+1)}. \tag{6.1.8}$$

This formula is meaningless for n a positive integer because the denominators of the first n terms, $m = 0, 1, 2, \ldots, n-1$, are all undefined. In fact, the assumption that $a_0 \neq 0$, $a_1 = 0$, $k = -n$, leads to a contradiction, for in this case (6.1.4) may be written

$$a_{m+2}(m+2)(m+2-2n) + a_m = 0, \quad m \geq 0, \tag{6.1.9}$$

from which we observe that $a_{2n-2} = 0$. It then follows that

$$a_{2i} = 0, \quad i = 0, 1, 2, \ldots, n-1,$$

a result which contradicts $a_0 \neq 0$.

However, we may attempt to find a second solution for the root $k = -n$ of the form

$$y = \sum_{m=0}^{\infty} a_m x^{m-n},$$

placing no restriction on a_0. If we take $a_1 = 0$ we obtain (6.1.9) and, as before, have $a_{2i} = 0$ for $i < n$. In this case, taking $m = 2n - 2$, we observe from (6.1.9) that a_{2n} is arbitrary. Writing out a few of the coefficients for $m > 2n$, we see that

$$a_{2n+2r} = \frac{(-1)^r n!\, a_{2n}}{r!\, 2^{2r}(n+r)!}, \quad r = 1, 2, 3, \ldots,$$

with the second solution given by

$$y = \sum_{m=n}^{\infty} a_{2m} x^{2m-n} = \sum_{r=0}^{\infty} a_{2n+2r} x^{n+2r}.$$

This second solution, which we *call* J_{-n}, is then

$$J_{-n}(x) = 2^n n!\, a_{2n} \sum_{r=0}^{\infty} \frac{(-1)^r (x/2)^{2r+n}}{r!\,(n+r)!} = 2^n n!\, a_{2n} J_n(x).$$

Thus we see that J_{-n} and J_n are not independent for n an integer. As we shall see in Sec. 6.3, when we define $J_n(x)$ by its generating function, the exact relation will be

$$J_{-n}(x) = (-1)^n J_n(x). \tag{6.1.10}$$

If n is an integer, (6.1.7) therefore fails to give the general solution and we must look for another solution to combine with $J_n(x)$. This second solution will be found in the next section, where it will be shown that the general solution of Bessel's equation of order n is

$$y = C_1 J_n(x) + C_2 Y_n(x), \tag{6.1.11}$$

with $Y_n(x)$ given in (6.2.10).

EXERCISES

6.1.1. Note that Eq. (3.2.1), which has already been solved, is Bessel's equation with $n = \pm\frac{1}{2}$. Show directly from (6.1.6) that

$$J_{1/2}(x) = \sqrt{\frac{2}{\pi x}} \sin x, \quad J_{-1/2}(x) = \sqrt{\frac{2}{\pi x}} \cos x.$$

6.1.2. Show that the substitution $y = u/\sqrt{x}$ in Bessel's equation results in

$$u'' + \left[1 - \frac{n^2 - \frac{1}{4}}{x^2}\right] u = 0.$$

6.1.3. Verify that $y = \sqrt{x}\, J_n(ax)$ is a solution of

$$y'' + \left[a^2 - \frac{n^2 - \frac{1}{4}}{x^2}\right] y = 0.$$

6.1.4. Show that $y = x^n J_n(x)$ is a solution of

$$y'' + \frac{1 - 2n}{x} y' + y = 0.$$

6.1.5. Show that the series for $J_n(x)$ converges absolutely and uniformly for all x [see Appendix A], and hence $J_n(x)$ is a continuous function. Show also that $J_n(x)/x^n$ is continuous.

6.1.6. Show that for n an integer, $J_n(-x) = (-1)^n J_n(x)$, and hence that $J_n(x)$ is even or odd according as n is even or odd. Show also that $J_n(-x) = J_{-n}(x)$.

6.1.7. Show from (6.1.6) that if $n \geq 1$, $|J_n(x)| \leq |x|^n e^{x^2/4}/2^n \Gamma(n+1)$, and hence $\lim_{n\to\infty} J_n(x) = 0$. Note that this inequality holds also for $n = 0$:

$$|J_0(x)| \leq e^{x^2/4}.$$

6.1.8. Why can we not conclude by the method used in Exercise 6.1.7 that the first inequality there holds for $0 < n < 1$? It is known (see [Mc-1], page 178) that the minimum value of the gamma function for $n \geq 0$ is .8856..., occurring at $n = 1.4616$. Hence if we replace $\Gamma(n+1)$ by .8856 (terminating the decimal) in Exercise 6.1.7, we still have a valid inequality,

$$|J_n(x)| \leq \frac{|x|^n e^{x^2/4}}{2^n(.8856)}.$$

Show that this is true for $n \geq 0$.

6.1.9. Show that if we define the first n terms in (6.1.8) as zero, then (6.1.10) follows directly from (6.1.8).

6.2 BESSEL FUNCTION OF THE SECOND KIND

As we have seen, if n is an integer, (6.1.7) fails to yield the general solution of Bessel's equation because $J_n(x)$ is a multiple of $J_{-n}(x)$. To obtain a second linearly independent solution, we begin by noting that the combination first proposed by *Weber* and *Schlafli*,

$$Y_\nu(x) = \frac{J_\nu(x)\cos\nu\pi - J_{-\nu}(x)}{\sin\nu\pi}, \tag{6.2.1}$$

is obviously a solution of Bessel's equation of order ν if ν is not an integer. Also, for this case, it is linearly independent of $J_\nu(x)$. For $\nu = n$, an integer, $Y_\nu(x)$ assumes the indeterminate form 0/0, but the limit as $\nu \to n$ exists and can be shown to be a solution of Bessel's equation of order n (see, for example [W], pages 58–59). We define, then, the Bessel function of the *second kind* of order n as

$$Y_n(x) = \lim_{\nu\to n} \frac{J_\nu(x)\cos\nu\pi - J_{-\nu}(x)}{\sin\nu\pi}. \tag{6.2.2}$$

By l'Hospital's rule, (6.2.2) is equivalent to

$$\pi Y_n(x) = \lim_{\nu\to n}\left[\frac{\partial J_\nu(x)}{\partial\nu} - (-1)^n\frac{\partial J_{-\nu}(x)}{\partial\nu}\right]. \tag{6.2.3}$$

The first term in the brackets is found readily by differentiating $J_\nu(x)$ as defined by (6.1.6) and making use of the logarithmic derivative of the gamma function, given by (5.5.1). This result is

$$\frac{\partial}{\partial \nu} J_\nu(x) = \sum_{m=0}^{\infty} \frac{(-1)^m (x/2)^{2m+\nu}[\log (x/2) - \psi(m + \nu + 1)]}{m!\, \Gamma(m + \nu + 1)}$$

$$= J_\nu(x) \log (x/2) - \sum_{m=0}^{\infty} \frac{(-1)^m (x/2)^{2m+\nu} \psi(m + \nu + 1)}{m!\, \Gamma(m + \nu + 1)} . \quad (6.2.4)$$

To obtain the second term in brackets in (6.2.3), we first remove the gamma functions whose arguments are negative, since for these values the gamma function will be undefined in the limit. To accomplish this removal, we make use of Exercise 5.3.4, from which

$$\Gamma(m - \nu + 1) = \frac{\pi}{\sin [\pi(\nu - m)]\Gamma(\nu - m)} = \frac{(-1)^m \pi}{\sin \pi\nu\, \Gamma(\nu - m)} . \quad (6.2.5)$$

If $\nu > 0$ and n is an integer such that $0 < \nu - n < 1$, then the argument of $\Gamma(m - \nu + 1)$ is negative for $m = 0, 1, 2, \ldots, n - 1$. Therefore we will substitute (6.2.5) into $J_{-\nu}(x)$ for the first n terms, giving

$$J_{-\nu}(x) = \sum_{m=0}^{n-1} \frac{(x/2)^{2m-\nu}(\sin \pi\nu)\Gamma(\nu - m)}{m!\, \pi} + \sum_{m=n}^{\infty} \frac{(-1)^m (x/2)^{2m-\nu}}{m!\, \Gamma(m - \nu + 1)} . \quad (6.2.6)$$

Differentiating, we have

$$\frac{\partial}{\partial \nu} J_{-\nu}(x) = \sum_{m=0}^{n-1} \frac{(x/2)^{2m}}{m!\, \pi} \frac{\partial f}{\partial \nu} + \sum_{m=n}^{\infty} \frac{(-1)^m (x/2)^{2m}}{m!} \frac{\partial g}{\partial \nu} , \quad (6.2.7)$$

where

$$\frac{\partial f}{\partial \nu} = \frac{\partial}{\partial \nu}\left[\left(\frac{x}{2}\right)^{-\nu} (\sin \pi\nu)\Gamma(\nu - m)\right]$$

$$= \left(\frac{x}{2}\right)^{-\nu} \Gamma(\nu - m)\left[-\left(\log \frac{x}{2}\right) \sin \pi\nu + \pi \cos \pi\nu + (\sin \pi\nu)\psi(\nu - m)\right] \quad (6.2.8)$$

and

$$\frac{\partial g}{\partial \nu} = \frac{\partial}{\partial \nu}\left[\frac{(x/2)^{-\nu}}{\Gamma(m - \nu + 1)}\right] = \frac{(x/2)^{-\nu}[- \log (x/2) + \psi(m - \nu + 1)]}{\Gamma(m - \nu + 1)} . \quad (6.2.9)$$

In (6.2.9) we have made use of Exercise 5.5.3.

To obtain $Y_n(x)$, we now substitute (6.2.4) and (6.2.7) into (6.2.3), making the substitutions given by (6.2.8) and (6.2.9). The result, after taking the limit, is

$$\pi Y_n(x) = J_n(x) \log \frac{x}{2} - \sum_{m=0}^{\infty} \frac{(-1)^m (x/2)^{2m+n} \psi(m + n + 1)}{m!\, \Gamma(m + n + 1)}$$

$$- \sum_{m=0}^{n-1} \frac{(x/2)^{2m-n}\Gamma(n - m)}{m!}$$

$$- \sum_{m=n}^{\infty} \frac{(-1)^{m+n} (x/2)^{2m-n}[-\log (x/2) + \psi(m - n + 1)]}{m!\, \Gamma(m - n + 1)} .$$

The last summation may be written, replacing m by $m + n$, as

$$\sum_{m=0}^{\infty} \frac{(-1)^m (x/2)^{2m+n}[-\log (x/2) + \psi(m + 1)]}{(m + n)!\, m!}.$$

Using (5.5.4) and the definition of $J_n(x)$, we may finally write, for $n \neq 0$,

$$Y_n(x) = \frac{2}{\pi} J_n(x)\left[\log \frac{x}{2} + \gamma\right] - \frac{1}{\pi} \sum_{m=0}^{n-1} \frac{(x/2)^{2m-n}(n - m - 1)!}{m!}$$
$$- \frac{1}{\pi} \sum_{m=0}^{\infty} \frac{(-1)^m (x/2)^{2m+n}[\phi(m + n) + \phi(m)]}{m!\,(m + n)!}, \qquad (6.2.10)$$

where γ is Euler's constant and

$$\phi(m) = \sum_{k=1}^{m} \frac{1}{k}, \quad \phi(0) = 0. \qquad (6.2.11)$$

We note that if $n = 0$, the first summation in (6.2.6) will not appear, and thus

$$Y_0(x) = \frac{2}{\pi} J_0(x)\left[\log \frac{x}{2} + \gamma\right] - \frac{2}{\pi} \sum_{m=0}^{\infty} \frac{(-1)^m (x/2)^{2m} \phi(m)}{(m!)^2}.$$

Since $\phi(0) = 0$, we may conclude that

$$Y_0(x) = \frac{2}{\pi}\left\{J_0(x)\left[\log \frac{x}{2} + \gamma\right] - \sum_{m=1}^{\infty} \frac{(-1)^m (x/2)^{2m} \phi(m)}{(m!)^2}\right\}. \qquad (6.2.12)$$

EXERCISES

6.2.1. Use the method outlined in (3.4.2) to obtain a second solution $\overline{Y}_0(x)$ to Bessel's equation of order zero. Note that if $\overline{Y}_0(x)$ is a solution, then $Y_0(x) = (2/\pi)[\overline{Y}_0(x) - (\log 2 - \gamma)J_0(x)]$ is also a solution, and obtain (6.2.12).

6.2.2. Show that $x^2y'' + (1 - 2a)xy' + [(bcx^c)^2 + a^2 - n^2c^2]y = 0$ has solution

$$y = x^a[C_1J_n(bx^c) + C_2Y_n(bx^c)].$$

6.2.3. Use Exercise 6.2.2 to work Exercises 6.1.3 and 6.1.4. Also, verify that $x^2y'' + xy' + (k^2x^2 - n^2)y = 0$ has solution $y = C_1J_n(kx) + C_2Y_n(kx)$.

6.2.4. Solve $x^2y'' + xy' + m^2(x^{2m} - n^2)y = 0$ by letting $z = x^m$. Also solve by Exercise 6.2.2. Ans. $y = C_1J_n(x^m) + C_2Y_n(x^m)$.

6.2.5. From Exercise 6.2.2. find solutions to

(a) $y'' + xy = 0$. Ans. $\sqrt{x}[C_1J_{1/3}(2x^{3/2}/3) + C_2J_{-1/3}(2x^{3/2}/3)]$.
(b) $y'' + x^2y = 0$.
(c) $y'' + x^4y = 0$.
(d) $xy'' + y' + k^2y = 0$. Ans. $C_1J_0(2k\sqrt{x}) + C_2Y_0(2k\sqrt{x})$.

6.2.6. We define the *modified Bessel functions* of order n of the first and second kinds, respectively, as

$$I_n(x) = i^{-n}J_n(ix), \quad K_n(x) = \frac{\pi i}{2}[(-1)^n I_n(x) + i^{n+1}Y_n(ix)],$$

where $i = \sqrt{-1}$ and n is an integer. Show that the equation

$$y'' + \frac{1}{x}y' - \left(\lambda^2 + \frac{n^2}{x^2}\right)y = 0$$

has a solution

$$y = AI_n(\lambda x) + BK_n(\lambda x).$$

6.2.7. Show that $I_n(x)$ and $K_n(x)$ are real. [See Exercise 6.2.6].

Ans. $$I_n(x) = \sum_{m=0}^{\infty} \frac{(x/2)^{n+2m}}{m!\,\Gamma(n+m+1)},$$

$$K_n(x) = \frac{1}{2}\sum_{m=0}^{n-1} \frac{(-1)^m (x/2)^{2m-n}\Gamma(n-m)}{m!}$$
$$+ (-1)^{n+1}\sum_{m=0}^{\infty} \frac{(x/2)^{n+2m}}{m!\,\Gamma(n+m+1)}\left\{\log\frac{x}{2} + \gamma - \tfrac{1}{2}[\phi(m+n) + \phi(m)]\right\}.$$

6.2.8. Show that

$$K_0(x) = \frac{\pi i}{2}[I_0(x) + iY_0(ix)]$$
$$= -I_0(x)\left[\gamma + \log\frac{x}{2}\right] + \sum_{m=1}^{\infty}\frac{(x/2)^{2m}\phi(m)}{(m!)^2}.$$

[See Exercise 6.2.6.]

6.2.9. If n is not an integer, show that the general solution of the equation in Exercise 6.2.6 is $y = AI_n(\lambda x) + BI_{-n}(\lambda x)$. Show that for n an integer $I_n(x)$ and $I_{-n}(x)$ are not linearly independent, but in fact are equal.

6.2.10. Find the ber, bei, ker, and kei functions, defined by

$$I_0(i^{1/2}x) = \text{ber } x + i \text{ bei } x,$$
$$K_0(i^{1/2}x) = \text{ker } x + i \text{ kei } x.$$

[See Exercise 6.2.6.]

Ans. $$\text{ber } x = \sum_{k=0}^{\infty}\frac{(-1)^k(x/2)^{4k}}{[(2k)!]^2}, \quad \text{bei } x = \sum_{k=0}^{\infty}\frac{(-1)^k(x/2)^{4k+2}}{[(2k+1)!]^2}$$

6.2.11. Show that

$$I_0(i^{1/2}x) = J_0(i^{3/2}x) \text{ and } K_0(i^{1/2}x) = \pi i/2\,[J_0(i^{3/2}x) + iY_0(i^{3/2}x)].$$

[See Exercise 6.2.6.]

6.3 GENERATING FUNCTION FOR $J_n(x)$

Let us consider the function

$$g(x, t) = \exp\left[\frac{x}{2}\left(t - \frac{1}{t}\right)\right], \tag{6.3.1}$$

which, when expanded in powers of t, results in a series of the form

$$\exp\left[\frac{x}{2}\left(t - \frac{1}{t}\right)\right] = \sum_{n=-\infty}^{\infty} A_n t^n.$$

We may determine the A_n's by the relation

$$\exp\left[\frac{x}{2}\left(t - \frac{1}{t}\right)\right] = \sum_{r=0}^{\infty}\left(\frac{x}{2}\right)^r \frac{t^r}{r!}\sum_{s=0}^{\infty}\left(-\frac{x}{2}\right)^s \frac{t^{-s}}{s!},$$

which may be written

$$\exp\left[\frac{x}{2}\left(t - \frac{1}{t}\right)\right] = \sum_{r=0}^{\infty}\sum_{s=0}^{\infty}\frac{(-1)^s(x/2)^{r+s}}{r!\,s!}\,t^{r-s} = \sum_{n=-\infty}^{\infty} A_n t^n.$$

The coefficient of t^n occurs when $r = s + n$, or

$$A_n = \sum_{s=0}^{\infty}\frac{(-1)^s(x/2)^{2s+n}}{s!\,(s+n)!},$$

which will be recognized as $J_n(x)$ for n an integer. Hence we may write the generating function of $J_n(x)$ for n an integer as

$$g(x, t) = \exp\left[\frac{x}{2}\left(t - \frac{1}{t}\right)\right] = \sum_{n=-\infty}^{\infty} J_n(x)t^n. \tag{6.3.2}$$

Since $g(x, t)$ is identical with $g(x, -1/t)$ we may also consider the generating function to be expanded as

$$g(x, t) = \sum_{n=-\infty}^{\infty} J_n(x)\left(-\frac{1}{t}\right)^n,$$

which, with n replaced by $-n$, is equivalent to

$$g(x, t) = \sum_{n=-\infty}^{\infty} (-1)^n J_{-n}(x)t^n. \tag{6.3.3}$$

Equating powers of t^n in (6.3.2) and (6.3.3), we have the relation previously given in (6.1.10),

$$J_n(x) = (-1)^n J_{-n}(x).$$

This method of manipulating the generating function can be employed to obtain a number of interesting results, some of which will be considered in the exercises and in the next section.

EXERCISES

6.3.1. Let $t = e^{i\theta}$ in (6.3.2) and show that

$$e^{ix\sin\theta} = J_0(x) + 2\sum_{m=1}^{\infty} J_{2m}(x)\cos 2m\theta + 2i\sum_{m=1}^{\infty} J_{2m-1}(x)\sin(2m-1)\theta.$$

6.3.2. Equating real and imaginary parts in Exercise 6.3.1, show that

$$\cos(x \sin\theta) = J_0(x) + 2\sum_{m=1}^{\infty} J_{2m}(x)\cos 2m\theta$$

and

$$\sin(x \sin\theta) = 2\sum_{m=1}^{\infty} J_{2m-1}(x)\sin(2m-1)\theta.$$

6.3.3. Show that $J_n(x) = (1/\pi)\int_0^\pi \cos(n\theta - x\sin\theta)\,d\theta$.

6.3.4. Show that

$$\sin x = 2\sum_{n=0}^{\infty}(-1)^n J_{2n+1}(x),$$

$$\cos x = J_0(x) + 2\sum_{n=1}^{\infty}(-1)^n J_{2n}(x),$$

and

$$1 = J_0(x) + 2\sum_{n=1}^{\infty} J_{2n}(x).$$

6.3.5. Show that the generating function for the modified Bessel function $I_n(x)$ is $g(x, t) = \exp[(x/2)(t + 1/t)]$.

6.3.6. Show that for x real, $|J_n(x)| \leq 1$, for $n = 0, 1, 2, \ldots$.

6.3.7. Show that $e^{ix\cos\theta} = J_0(x) + 2\sum_{n=1}^{\infty} i^n J_n(x)\cos n\theta$.

6.3.8. Write $\exp[\frac{1}{2}(x + y)(t - 1/t)]$, using (6.3.3), in two ways—one as the product of generating functions of $J_n(x)$ and $J_n(y)$, and one as the generating function of $J_n(x + y)$. Equate coefficients of t^n to obtain the *addition formula*,

$$J_n(x + y) = \sum_{m=-\infty}^{\infty} J_m(x)J_{n-m}(y), \quad n = 0, 1, 2, \ldots.$$

6.3.9. For $n = 0$ in Exercise 6.3.8, show that

$$J_0(x + y) = J_0(x)J_0(y) + 2\sum_{m=1}^{\infty}(-1)^m J_m(x)J_m(y).$$

6.3.10. Show that

$$J_1(x + y) = \sum_{n=-\infty}^{\infty}(-1)^{n+1}J_n(x)J_{n-1}(y)$$

and that

$$J_1(2x) = 2J_0(x)J_1(x) + 2\sum_{n=1}^{\infty}(-1)^n J_n(x)J_{n+1}(x).$$

6.3.11. Let $y = -x$ and $y = x$ in Exercise 6.3.9 to obtain, respectively,

$$1 = J_0^2(x) + 2\sum_{n=1}^{\infty} J_n^2(x),$$

$$J_0(2x) = J_0^2(x) + 2\sum_{n=1}^{\infty}(-1)^n J_n^2(x).$$

6.3.12. Show by Exercise 6.3.11 that if x is real, $|J_0(x)| \leq 1$ and $|J_n(x)| < 1/\sqrt{2}$, $n = 1, 2, 3, \ldots$.

6.3.13. Take the conjugate of both sides of the equation in Exercise 6.3.7 and replace x by ix to obtain

$$e^{x \cos \theta} = I_0(x) + 2 \sum_{n=1}^{\infty} I_n(x) \cos n\theta.$$

6.3.14. Show by taking real and imaginary parts in Exercise 6.3.7 that

$$\cos (x \cos \theta) = J_0(x) + 2 \sum_{n=1}^{\infty} (-1)^n J_{2n}(x) \cos 2n\theta,$$

$$\sin (x \cos \theta) = 2 \sum_{n=1}^{\infty} (-1)^{n+1} J_{2n-1}(x) \cos (2n - 1)\theta.$$

6.3.15. Show from Exercises 6.3.2 and 6.3.14 that

$$J_{2n}(x) = \frac{1}{\pi}\int_0^{\pi} \cos 2n\theta \cos (x \sin \theta)\, d\theta = \frac{(-1)^n}{\pi}\int_0^{\pi} \cos 2n\theta \cos (x \cos \theta)\, d\theta,$$

$$J_{2n+1}(x) = \frac{1}{\pi}\int_0^{\pi} \sin (2n + 1)\theta \sin (x \sin \theta)\, d\theta$$

$$= \frac{(-1)^n}{\pi}\int_0^{\pi} \cos (2n + 1)\theta \sin (x \cos \theta)\, d\theta.$$

6.3.16. Note that for $n = 0$ in Exercises 6.3.3 and 6.3.15 we have shown that

$$\int_0^{\pi} \cos (x \cos \theta)\, d\theta = \int_0^{\pi} \cos (x \sin \theta)\, d\theta = \pi J_0(x).$$

Obtain one of these integrals from the other analytically.

6.3.17. Show that the series (6.3.2) converges absolutely and uniformly with respect to x, for $t \neq 0$.

6.3.18. Show from (6.3.2) that $\lim_{n\to\infty} J_n(x) = 0$.

6.4 RECURRENCE RELATIONS

The method of manipulating the generating function as illustrated in Sec. 6.3 may be used to obtain a variety of recurrence relations involving $J_n(x)$, where n is an integer. For example if (6.3.2) is differentiated with respect to x, we have

$$\tfrac{1}{2}\left(t - \frac{1}{t}\right) \exp\left[\frac{x}{2}\left(t - \frac{1}{t}\right)\right] = \sum_{n=-\infty}^{\infty} \tfrac{1}{2} J_n(x)[t^{n+1} - t^{n-1}]$$

$$= \sum_{n=-\infty}^{\infty} J'_n(x) t^n.$$

Equating coefficients of like powers of t, we have the recurrence relation

$$2J'_n(x) = J_{n-1}(x) - J_{n+1}(x). \tag{6.4.1}$$

(We use $J'_n(x)$ throughout to mean differentiation with respect to the argument.) Following the same procedure, except for differentiation with respect to t, we can obtain

$$\frac{2n}{x} J_n(x) = J_{n-1}(x) + J_{n+1}(x), \tag{6.4.2}$$

where we have replaced n by $n - 1$. If we add (6.4.1) and (6.4.2) and simplify, we have

$$xJ'_n(x) = xJ_{n-1}(x) - nJ_n(x), \tag{6.4.3}$$

and if we subtract and simplify, the result is

$$xJ'_n(x) = nJ_n(x) - xJ_{n+1}(x), \tag{6.4.4}$$

Recurrence relations may also be obtained directly from the expression for $J_n(x)$, Eq. (6.1.6). For example, the expression

$$\frac{d}{dx} [x^n J_n(x)] = x^n J_{n-1}(x) \tag{6.4.5}$$

may be found by differentiating the quantity in brackets using the definition of $J_n(x)$. If $n = 0$ in (6.4.5), we have the special case,

$$J'_0(x) = -J_1(x). \tag{6.4.6}$$

The generating function is defined for integral values of n, but all the recurrence relations obtained in this section could be obtained from the expression for $J_n(x)$ and are valid for non-integral values of n. We give other recurrence relations in the exercises.

EXERCISES

6.4.1. Obtain (6.4.4) by substituting for $xJ'_n(x)$ directly from (6.1.6).

6.4.2. Derive (6.4.5).

6.4.3. Show that $2J''_0(x) = J_2(x) - J_0(x)$.

6.4.4. Show from (6.1.6) that $xJ'_n(x) = nJ_n(x) - xJ_{n+1}(x)$.

6.4.5. Multiply the result in Exercise 6.4.4 through by an appropriate factor to obtain

$$\frac{d}{dx} [x^{-n} J_n(x)] = -x^{-n} J_{n+1}(x).$$

6.4.6. Show from (6.4.3) and (6.4.4) that $J_n(x)$ satisfies Bessel's equation of order n.

6.4.7. Show that $J_{n+1}(x)J_{n-1}(x) = (n^2/x^2)J_n^2(x) - [J'_n(x)]^2$.

6.4.8. Show that $xI'_n(x) = nI_n(x) + xI_{n+1}(x)$ and $2nI_n(x) = xI_{n-1}(x) - xI_{n+1}(x)$. Use these to obtain other recurrence relations corresponding to those for $J_n(x)$.

6.4.9. Multiply Bessel's equation (6.1.1) through by $2y'$ and integrate, to obtain

$$2\int xy^2\,dx = x^2(y')^2 + x^2y^2 - n^2y^2,$$

or for $y = J_n(x)$, obtain

$$2\int xJ_n^2(x)\,dx = x^2[J'_n(x)]^2 + x^2J_n^2(x) - n^2J_n^2(x).$$

6.4.10. Using Exercise 6.4.7, write the result of Exercise 6.4.9 as

$$\int xJ_n^2(x)\,dx = \frac{x^2}{2}[J_n^2(x) - J_{n+1}(x)J_{n-1}(x)],$$

and replace x by kx to obtain a *Lommel integral*,

$$\int xJ_n^2(kx)\,dx = \frac{x^2}{2}[J_n^2(kx) - J_{n+1}(kx)J_{n-1}(kx)].$$

6.4.11. Show from Exercise 6.4.9 that

$$\int_0^b xJ_n^2(kx)\,dx = \frac{b^2}{2}\left\{[J'_n(kb)]^2 + J_n^2(kb) - \frac{n^2}{k^2b^2}J_n^2(kb)\right\}.$$

6.4.12. Show by repeated applications of (6.4.2) that

$$xJ_1(x) = 4\sum_{n=1}^{\infty}(-1)^{n+1}nJ_{2n}(x).$$

6.4.13. If $E^kJ_n(x) = J_{n+k}(x)$, show that

$$2^k\frac{d^k}{dx^k}J_n(x) = (E^{-1} - E)^kJ_n(x), \quad k = 0, 1, 2, \ldots.$$

Give some results for $k = 0, 1, 2, \ldots$.

6.4.14. Show that $|J'_n(x)| \le 1$, $n = 0, 1, 2, \ldots$, by using Exercise 6.3.6 and (6.4.1). Show also that $|J_n^{(k)}(x)| \le 1$, $k = 0, 1, 2, \ldots$, $n = 0, 1, 2, \ldots$. (Use Exercise 6.3.3.)

6.5 SPHERICAL BESSEL FUNCTIONS

We have seen in Exercise 6.1.1 that the Bessel functions of the first kind of orders $\frac{1}{2}$ and $-\frac{1}{2}$ may be written in the closed forms,

$$J_{1/2}(x) = \sqrt{\frac{2}{\pi x}}\sin x \tag{6.5.1}$$

and

$$J_{-1/2}(x) = \sqrt{\frac{2}{\pi x}}\cos x. \tag{6.5.2}$$

We will now show that, in general, the *spherical Bessel functions*, $J_{n+1/2}(x)$

for n an integer, may be represented in a closed form involving the circular functions $\sin x$ and $\cos x$.

We begin by defining the operator $\mathscr{D} = (1/x)d/dx$, and observing from Exercise 6.4.5 that

$$\mathscr{D}[x^{-n}J_n(x)] = -x^{-n-1}J_{n+1}(x).$$

We note that this operation changes the sign, lowers the exponent on x by 1 and increases the order of $J_n(x)$ by 1. Generalizing this result, we have

$$\mathscr{D}^k[x^{-n}J_n(x)] = (-1)^k x^{-n-k}J_{n+k}(x). \tag{6.5.3}$$

Replacing n by $\frac{1}{2}$ and substituting for $\mathscr{D}$ and $J_{1/2}(x)$, we may write (6.5.3) as

$$J_{n+1/2}(x) = (-1)^n \sqrt{\frac{2}{\pi}}\, x^{n+1/2}\left(\frac{1}{x}\frac{d}{dx}\right)^n \left(\frac{\sin x}{x}\right), \tag{6.5.4}$$

where we have replaced k by n, a non-negative integer.

In a similar manner we may start with (6.4.5) to obtain

$$\mathscr{D}^k[x^n J_n(x)] = x^{n-k}J_{n-k}(x),$$

which for $n = \frac{1}{2}$ may be written as

$$J_{1/2-n}(x) = \sqrt{\frac{2}{\pi}}\, x^{n-1/2}\left(\frac{1}{x}\frac{d}{dx}\right)^n (\sin x), \quad n = 0, 1, 2, \ldots. \tag{6.5.5}$$

Using (6.5.4) and (6.5.5), we may compile the accompanying table. The functions $f_n(x)$ are defined by the relation

$$\sqrt{\frac{\pi x}{2}}\, J_{n+1/2}(x) = f_n(x)\sin x - (-1)^n f_{-n-1}(x)\cos x. \tag{6.5.6}$$

n	$f_n(x)$	n	$f_n(x)$
0	1	-1	0
1	$\frac{1}{x}$	-2	-1
2	$\frac{3}{x^2} - 1$	-3	$\frac{3}{x}$
3	$\frac{15}{x^3} - \frac{6}{x}$	-4	$-\frac{15}{x^2} + 1$
4	$\frac{105}{x^4} - \frac{45}{x^2} + 1$	-5	$\frac{105}{x^3} - \frac{10}{x}$

EXERCISES

6.5.1. Derive $J_{5/2}(x)$.

6.5.2. Prove by induction on k that (6.5.3) holds.

6.5.3. Show that the recurrence relations (6.4.1) and (6.4.2) hold for $n = \frac{1}{2}$.

6.5.4. Show that

$$Y_{n+1/2}(x) = (-1)^{n+1}J_{-n-1/2}(x)$$

and

$$Y_{-n-1/2}(x) = (-1)^n J_{n+1/2}(x).$$

6.5.5. Show that

$$Y_{n+1/2}(x) = (-1)^{n+1}\sqrt{\frac{2}{\pi}}\, x^{n+1/2}\left(\frac{1}{x}\frac{d}{dx}\right)^{n+1} (\sin x).$$

6.5.6. Show that $I_{1/2}(x) = \sqrt{2/\pi x}\sinh x$ and $I_{-1/2}(x) = \sqrt{2/\pi x}\cosh x$.

6.6 ZEROS OF $J_n(x)$

Before proceeding to the orthogonality relation of the $J_n(x)$, we shall need some information about the roots of the equation $J_n(x) = 0$, for n and x real. We know from Exercise 6.1.2 that

$$u = \sqrt{x}\, J_n(x) \tag{6.6.1}$$

satisfies the differential equation

$$u'' + \left[1 - \frac{n^2 - \frac{1}{4}}{x^2}\right]u = 0.$$

If x is sufficiently large, the term $(n^2 - \frac{1}{4})/x^2$ can be neglected in comparison to 1, so that approximately

$$u'' + u = 0,$$

from which we have

$$u = C\cos(x - \phi),$$

where C and ϕ are constants. By (6.6.1) an approximation to $J_n(x)$ for large x is thus

$$J_n(x) = \frac{C}{\sqrt{x}}\cos(x - \phi), \tag{6.6.2}$$

or alternatively

$$J_n(x) = \frac{C}{\sqrt{x}}[\cos(x - \phi) + p(x)], \tag{6.6.3}$$

where

$$\lim_{x\to\infty} p(x) = 0.$$

The precise statement of (6.6.3) is known (see [W], Chapter 7) to be

$$J_n(x) \sim \sqrt{\frac{2}{\pi x}}\cos\left(x - \frac{\pi}{4} - \frac{n\pi}{2}\right). \tag{6.6.4}$$

(The symbol $\sim$ is read "is asymptotic to.") We note as a consequence of (6.6.3) that

$$\lim_{x \to \infty} J_n(x) = 0. \tag{6.6.5}$$

We may infer from (6.6.2) that $J_n(x)$ has an infinite number of real zeros and that consecutive large zeros differ approximately by π, the difference between consecutive zeros of $\cos (x - \phi)$ (for proof see [F], pages 168–170). We note from (6.1.6) that if λ is a zero of $J_n(x)$ then $-\lambda$ is also a zero, and hence we have an infinite number of zeros on both the positive and negative x-axis. Since each zero λ has a corresponding zero $-\lambda$ we may study their characteristics by considering only the positive zeros.

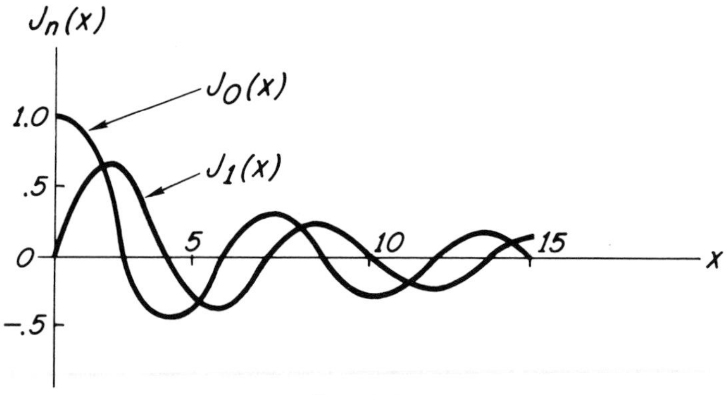

Fig. 6–1

All the zeros of $J_n(x)$, $x > 0$, are simple because if $x = \lambda$ is a double zero, then $J_n(\lambda)$ and $J'_n(\lambda)$ must both be zero. But by the uniqueness theorem of second-order ordinary differential equations (see for example [I], page 73), there is only one solution of Bessel's equation satisfying the two conditions $J_n(\lambda) = J'_n(\lambda) = 0$. Since $y \equiv 0$ is one such solution, it must be the only one. Thus $J_n(x)$ cannot have $x = \lambda$ as a double zero.

It may be shown also that the zeros of $J_n(x)$ are all real for $n > -1$ and that those of $J_{n+1}(x)$ and $J_n(x)$ interlace on the x-axis (see Exercises 6.7.1 and 6.7.2). The graph illustrating the interlacing of the zeros of $J_0(x)$ and $J_1(x)$ is given by Fig. 6–1 for $0 \le x \le 15$.

6.7 ORTHOGONALITY OF $J_n(x)$

We proceed now to show that the set $\{\phi_n(x)\}$, where $\phi_n(x) = J_k(\lambda_n x)$, is orthogonal over $(0, b)$ with weight function $w(x) = x$. The λ_n, with $0 < \lambda_1 < \lambda_2 < \lambda_3 < \cdots$, are the zeros of $J_k(\lambda)$, for k real. It will be noted that this set is different from the preceding examples in Chapter 1 in that it is enumerated by changing the index n of the roots rather than the index k of the function.

We begin by replacing x by $\lambda_n x$ in Bessel's equation with index k to obtain, for

$$y_n = J_k(\lambda_n x), \tag{6.7.1}$$

the self-adjoint equation

$$(xy'_n)' + \left(-\frac{k^2}{x} + \lambda_n^{\,2} x\right)y_n = 0. \tag{6.7.2}$$

We observe by comparison with Eq. (1.9.5) that $w(x) = x$, $r(x) = x$, $p(x) = -k^2/x$, with eigenvalue $\lambda_n^{\,2}$. We note from Eqs. (1.9.10) and (1.9.12) that if we choose the interval $(0, b)$, we have the orthogonality relation

$$\int_0^b xJ_k(\lambda_n x)J_k(\lambda_m x)\, dx = 0, \quad n \neq m, \tag{6.7.3}$$

where λ_n and λ_m satisfy

$$C_1\lambda J'_k(\lambda b) + C_2 J_k(\lambda b) = 0. \tag{6.7.4}$$

We may use Eq. (6.4.4) to write (6.7.4) in the form

$$(C_1 k + C_2 b)J_k(\lambda b) - C_1 b\lambda J_{k+1}(\lambda b) = 0,$$

and if $C_1 \neq 0$, we may write it finally as

$$(k + C)J_k(\lambda b) - \lambda b J_{k+1}(\lambda b) = 0, \tag{6.7.5}$$

where $C = C_2 b/C_1$. For this value of C, (6.7.4) becomes

$$b\lambda J'_k(\lambda b) + CJ_k(\lambda b) = 0. \tag{6.7.6}$$

If $C_1 = 0$, then λ_n and λ_m satisfy

$$J_k(\lambda b) = 0, \tag{6.7.7}$$

which, as we have seen, has an infinite number of positive, real, and simple roots. This is also the case with (6.7.5) for $C \geq 0$. (See [C-1], pages 178–184, and also Exercise 6.7.8.)

The norm $\|J_k(\lambda_n x)\|$ has been obtained in Exercise 6.4.11 as

$$\int_0^b xJ_k^{\,2}(\lambda x)\, dx = \frac{b^2}{2}\left\{[J'_k(\lambda b)]^2 + J_k^{\,2}(\lambda b) - \frac{k^2}{\lambda^2 b^2} J_k^{\,2}(\lambda b)\right\}. \tag{6.7.8}$$

Thus if $\lambda_n b$ and $\lambda_m b$ are zeros of $J_k(x)$, the case given by (6.7.7), the orthogonality relation is

$$\int_0^b xJ_k(\lambda_n x)J_k(\lambda_m x)\, dx = \frac{b^2}{2}\,[J'_k(\lambda_n b)]^2\, \delta_{mn}, \tag{6.7.9}$$

which by (6.4.4) and (6.4.3) may be written as

$$\int_0^b xJ_k(\lambda_n x)J_k(\lambda_m x)\,dx = \frac{b^2}{2}J_{k+1}^2(\lambda_n b)\,\delta_{mn} = \frac{b^2}{2}J_{k-1}^2(\lambda_n b)\,\delta_{mn}. \tag{6.7.10}$$

If λ_n and λ_m satisfy (6.7.6), we may eliminate the derivative term in (6.7.8) to obtain a simplified orthogonality relation,

$$\int_0^b xJ_k(\lambda_n x)J_k(\lambda_m x)\,dx = \left(\frac{C^2 + \lambda_n^{\,2}b^2 - k^2}{2\lambda_n^{\,2}}\right)J_k^{\,2}(\lambda_n b)\,\delta_{nm}. \tag{6.7.11}$$

We have tacitly assumed in (6.7.8) that $\lambda \neq 0$. This is generally the case since $\lambda = 0$ must correspond to the eigenfunction $J_k(0)$, and for $k \neq 0$ this is a trivial case, $J_k(0) = 0$. There is, however, one situation in which $\lambda = 0$ is an eigenvalue of the system given by (6.7.2) and (6.7.5). This is the case $k = C = 0$, in which $\lambda = 0$ is an eigenvalue corresponding to the eigenfunction $J_0(0) = 1$. For this case the norm is

$$\int_0^b x\,dx = \frac{b^2}{2}. \tag{6.7.12}$$

There is no case in which $\lambda = 0$ is an eigenvalue of (6.7.2) and (6.7.7).

If we seek an expansion, valid on $(0, b)$, of a function $f(x)$ of the type

$$f(x) = \sum_{n=1}^{\infty} C_n J_k(\lambda_n x),$$

then by (1.3.2) we will have

$$C_n = \frac{1}{\|J_k(\lambda_n x)\|^2}\int_0^b xf(x)J_k(\lambda_n x)\,dx, \tag{6.7.13}$$

where the norm $\|J_k(\lambda x)\|$ is given by (6.7.9) or by (6.7.11).

EXERCISES

6.7.1. If $a + bi$ is a complex root of $J_n(x) = 0$, then so is $a - bi$. Show by (6.7.10) with $n > -1$ to assure convergence, that there can be no roots of this type with $b \neq 0$.

6.7.2. *Rolle's theorem* says that if $f(x)$ is continuous for $a \leq x \leq b$ and differentiable for $a < x < b$, then if $f(a) = f(b)$, there is a point ξ, $a < \xi < b$, such that $f'(\xi) = 0$. Using Exercises 6.1.5 and 6.4.5, show that by Rolle's theorem there is a zero of J_{n+1} between any two zeros of J_n. Next use (6.4.5) and show that there is a zero of J_n between any two zeros of J_{n+1}, and hence the zeros of J_n and J_{n+1} interlace.

6.7.3. It is known (see for example [F], pages 201–202) that if λ_{ni} and $\lambda_{n,i+1}$ are consecutive zeros of $J_n(x)$, then $\lambda_{n,i+1} > \lambda_{ni} + \pi$, and the first positive zero λ_{n1} exceeds n. Show by induction on m that if λ_{nm}, $m = 1, 2, 3, \ldots$, are the positive zeros of $J_n(x)$, then $\lambda_{nm} > n + (m - 1)\pi$.

6.7.4. Show that if $g(x) = \sum_{n=1}^{\infty} C_n x^k J_r(\alpha_n x^m)$, where α_n are the zeros of $J_r(\alpha)$, then the Fourier coefficients are given by

$$C_n = \frac{2}{J_{r+1}^2(\alpha_n)} \int_0^1 z^{1-k/m} g(z^{1/m}) J_r(\alpha_n z)\, dz.$$

6.7.5. Show from Bessel's equation and Eq. (1.9.8) that

$$(\beta^2 - \alpha^2)\int_a^b x J_n(\alpha x) J_n(\beta x)\, dx = x[\alpha J'_n(\alpha x) J_n(\beta x) - \beta J'_n(\beta x) J_n(\alpha x)]\,\Big|_a^b.$$

6.7.6. If λ_n, λ_m are zeros of $J'_k(\lambda)$, obtain the orthogonality relation

$$\int_0^1 x J_k(\lambda_n x) J_k(\lambda_m x)\, dx = \frac{1}{2\lambda_n{}^2} J_k{}^2(\lambda_n)[\lambda_n{}^2 - k^2]\delta_{mn}.$$

6.7.7. Show, for the modified Bessel functions, that

$$2\lambda^2 \int x I_n{}^2(\lambda x)\, dx = (\lambda^2 x^2 + n^2) I_n{}^2(\lambda x) - \lambda^2 x^2 [I'_n(\lambda x)]^2$$

and that

$$2\int_0^1 x I_n{}^2(\lambda x)\, dx = \left(1 + \frac{n^2}{\lambda^2}\right) I_n{}^2(\lambda) - [I'_n(\lambda)]^2.$$

6.7.8. Show that if $C \geq 0$, $k = 0, 1, 2, \ldots$, then (6.7.5) has no purely imaginary roots.

6.8 INTEGRAL RELATIONS

In computing the Fourier coefficients given by (6.7.13) we need to evaluate integrals of the type

$$I(\alpha, n) = \int_0^1 x f(x) J_n(\alpha x)\, dx, \tag{6.8.1}$$

which can be expressed in terms of tabulated functions for certain cases of $f(x)$. In this section we will consider some of these cases and we will also consider some more general integrals involving J_n. For convenience we shall write them as indefinite integrals, omitting constants of integration. For example, the equation,

$$\int J_{n+1}(x)\, dx = n \int \frac{J_n(x)}{x}\, dx - J_n(x), \tag{6.8.2}$$

will represent the result of integrating both sides of Eq. (6.4.4).

We begin by collecting a few integral relations that we have already in different forms. Equation (6.4.5) may be written as

$$\int x^n J_{n-1}(x)\, dx = x^n J_n(x), \tag{6.8.3}$$

the result of Exercise 6.4.5 as

$$\int x^{-n} J_{n+1}(x)\,dx = -x^{-n} J_n(x), \tag{6.8.4}$$

Eq. (6.4.1) as

$$\int J_{n+1}(x)\,dx = \int J_{n-1}(x)\,dx - 2J_n(x), \tag{6.8.5}$$

and the result of Exercise 6.7.5 as

$$(\beta^2 - \alpha^2)\int xJ_n(\alpha x)J_n(\beta x)\,dx = x[\alpha J'_n(\alpha x)J_n(\beta x) - \beta J'_n(\beta x)J_n(\alpha x)]. \tag{6.8.6}$$

We note that (6.8.5) may be used as a *reduction formula* to evaluate $\int J_n(x)\,dx$ when n is an integer. If n is even, the problem is reduced to that of integrating $J_0(x)$. A solution to the latter problem cannot be given in a closed form, but approximate values of the integral

$$I = \int_0^x J_0(t)\,dt \tag{6.8.7}$$

for given values of x are known and have been tabulated (see for example [W], page 752). For the cases when n is an odd integer, we ultimately obtain $\int J_1(x)\,dx$, which is $-J_0(x)$ [see (6.4.6)]. Hence we see that the integral $\int_0^x J_n(t)\,dt$ can be found from tabulated values of the Bessel functions and the integral I in (6.8.7).

The next type of integral which we will consider is of the form

$$I_{mn} = \int x^m J_n(x)\,dx, \quad m \geq -n. \tag{6.8.8}$$

(We note that Eqs. (6.8.3) and (6.8.4) are trivial cases.)

If we insert the term x^{n-1}/x^{n-1} in the integrand, we may use (6.8.4) and integrate by parts to obtain

$$\int x^m J_n(x)\,dx = -x^m J_{n-1} + (m + n - 1)\int x^{m-1} J_{n-1}(x)\,dx, \tag{6.8.9}$$

which in the notation of (6.8.8) is the reduction formula

$$I_{mn} = -x^m J_{n-1}(x) + (m + n - 1)I_{m-1,n-1}. \tag{6.8.10}$$

After k applications of this reduction formula, we have I_{mn} expressed in terms of $I_{m-k,n-k}$. The process terminates when either $m - k$ or $n - k$ becomes zero, or in the case when either term becomes a positive fraction.

When m and n are non-negative integers the problem of evaluating I_{mn} reduces to that of evaluating I_{0p} or I_{q0}. The case of

$$I_{0p} = \int J_p(x)\,dx$$

has already been discussed. We now proceed to evaluate I_{q0}, which we write as

$$I_{n0} = \int x^n J_0(x)\, dx. \tag{6.8.11}$$

First we note by (6.8.3) that

$$I_{10} = \int x J_0(x)\, dx = x J_1(x). \tag{6.8.12}$$

When $n > 1$, the reduction formula,

$$I_{n0} = x^n J_1(x) + (n - 1)x^{n-1} J_0(x) - (n - 1)^2 I_{n-2,0}, \tag{6.8.13}$$

can be obtained by integrating (6.8.11) by parts twice (see Exercise 6.8.1). Repeated applications of (6.8.13) reduce the problem of finding I_{n0} to that of finding $I_{10} = xJ_1(x)$ if n is odd, or $I_{00} = \int J_0(x)\, dx$ if n is even.

Hence the integral I_{mn} can be expressed as a sum of Bessel functions of orders less than n and $\int J_0(x)\, dx$. It follows then that when $f(x)$ is a polynomial, (6.8.1) can be expressed in terms of tabulated functions. In view of Eq. (6.8.6), the integration of (6.8.1) can also be carried out when $f(x)$ is a Bessel function, $J_n(kx)$.

As an example of the application of (6.8.10), let us consider the integral

$$I_{23} = \int x^2 J_3(x)\, dx.$$

Applying (6.8.10) repeatedly, we have

$$\begin{aligned} I_{23} &= -x^2 J_2(x) + 4I_{12} \\ &= -x^2 J_2(x) + 4[-xJ_1(x) + 2I_{01}]. \end{aligned}$$

This may be written as

$$\int x^2 J_3(x)\, dx = -x^2 J_2(x) - 4xJ_1(x) - 8J_0(x),$$

since

$$I_{01} = \int J_1(x)\, dx = -J_0(x).$$

EXERCISES

6.8.1. Verify (6.8.13). (*Hint*: Use both (6.8.3) and (6.8.4).)

6.8.2. Evaluate the following integrals:

(a) $\int x^2 J_0(x)\, dx$.

(b) $\int x^2 J_0(\alpha x)\, dx$.

(c) $\int_0^1 x^3 J_0(\alpha x)\, dx$. Ans. $[(\alpha^2 - 4)/\alpha^3]J_1(\alpha) + (2/\alpha^2)J_0(\alpha)$.

(d) $\int_0^1 x(1 - x^2)J_0(\alpha x)\,dx$. Ans. $(4/\alpha^3)J_1(\alpha) - (2/\alpha^2)J_0(\alpha)$.

(e) $\int J_0(x)J_1(x)\,dx$.

(f) $\int xJ_0(x)J_1(x)\,dx$. Ans. $-\frac{1}{2}xJ_0^2(x) + \frac{1}{2}\int J_0^2(x)\,dx$.

(g) $\int x^2J_0(x)J_1(x)\,dx$. Ans. $\frac{1}{2}x^2J_1^2(x)$.

(h) $\int x(\log x)J_0(x)\,dx$. Ans. $J_0(x) + (x \log x)J_1(x)$.

(i) Show that $\int_0^\infty J_{n+1}(x)\,dx = \int_0^\infty J_{n-1}(x)\,dx,\ n > 0$.

6.8.3. Evaluate

(a) $\int J_4(x)\,dx$.

(b) $\int J_3(x)\,dx$.

(c) $\int x^3J_1(x)\,dx$.

(d) $\int xJ_2(x)\,dx$.

(e) $\int x^5J_4(x)\,dx$.

(f) $\int x^6J_4(x)\,dx$.

(g) $\int xJ_6^2(2x)\,dx$.

6.8.4. If $I_n = \int J_n(x)\,dx$, show from (6.4.1) that $I_{n-1} = 2J_n(x) + I_{n+1}$. Use this to show that

$$\int J_0(x)\,dx = 2\sum_{n=1}^{\infty} J_{2n-1}(x).$$

6.8.5. Determine the coefficients in the following expansions on $(0, 1)$. In each case λ_n, $n = 1, 2, 3, \ldots$, are the positive zeros of the Bessel functions in the series.

(a) $1 = \sum_{n=1}^{\infty} C_nJ_0(\lambda_n x)$.

(b) $x^2 = \sum_{n=1}^{\infty} C_nJ_0(\lambda_n x)$.

(c) $x^2 = \sum_{n=1}^{\infty} C_nJ_1(\lambda_n x)$.

(d) $\log x = \sum_{n=1}^{\infty} C_nJ_0(\lambda_n x)$.

(e) $J_0(\alpha x) = \sum_{n=1}^{\infty} C_nJ_0(\lambda_n x),\ J_0(\alpha) \neq 0$.

6.9 SOME PROPERTIES OF $Y_n(x)$

Many of the results obtained thus far for $J_n(x)$ remain valid when $J_n(x)$ is replaced by $Y_n(x)$ or, for that matter, by any other solution of Bessel's equation. For example, all of the results obtained for y, considered as a general solution of (6.1.1), are valid for $Y_n(x)$. As a result we see that $u = \sqrt{x}\, Y_n(x)$ is a solution of

$$u'' + \left[1 - \frac{n^2 - \frac{1}{4}}{x^2}\right]u = 0,$$

and by proceeding as in Sec. 6.6 it follows that $Y_n(x)$, like $J_n(x)$, has an infinite number of real zeros, all of which are simple. As a matter of fact, it can be shown that, over an interval not including $x = 0$, the zeros of any two linearly independent solutions are distinct and interlace each other (see for example [F], page 69). Therefore the positive zeros of $J_n(x)$ and $Y_n(x)$ are interlaced.

The integral relations derived for $J_n(x)$ directly from Bessel's differential equation must be valid for any solution, the most general being, for $x \neq 0$,

$$\begin{aligned} U_n(x) &= AJ_n(x) + BY_n(x), \qquad n \text{ an integer}, \\ &= A_1J_n(x) + B_1J_{-n}(x), \quad n \text{ not an integer}. \end{aligned} \tag{6.9.1}$$

Two of these relations, which we shall need later, were obtained in Exercises 6.4.9 and 6.7.5 and may now be given more generally as

$$2\int xU_n{}^2(kx)\,dx = x^2[U'_n(kx)]^2 + x^2U_n{}^2(kx) - \frac{n^2}{k^2}U_n{}^2(kx) \tag{6.9.2}$$

and

$$(\beta^2 - \alpha^2)\int_a^b xU_n(\alpha x)U_n(\beta x)\,dx = x[\alpha U'_n(\alpha x)U_n(\beta x) - \beta U'_n(\beta x)U_n(\alpha x)]\,\Big|_a^b. \tag{6.9.3}$$

These, of course, are also true for $J_n(x)$ and $Y_n(x)$ separately.

We note also that Eq. (6.6.2) has an analogy in the $Y_n(x)$ case, its precise statement (see [W], Chapter 7) being

$$Y_n(x) \sim \sqrt{\frac{2}{\pi x}}\sin\left(x - \frac{\pi}{4} - \frac{n\pi}{2}\right), \tag{6.9.4}$$

which implies that

$$\lim_{x\to\infty} Y_n(x) = 0. \tag{6.9.5}$$

In this section we will obtain other results pertaining to $Y_n(x)$, which will require the use of our definition given in (6.2.1). For convenience, we repeat this definition here:

$$\begin{aligned} Y_\nu(x) &= \frac{J_\nu(x)\cos \nu\pi - J_{-\nu}(x)}{\sin \nu\pi}, \\ Y_n(x) &= \lim_{\nu\to n} Y_\nu(x). \end{aligned} \tag{6.9.6}$$

We will first obtain a recurrence relation for $Y_n(x)$ analogous to (6.4.2), which we also repeat as

$$\frac{2n}{x} J_n(x) = J_{n-1}(x) + J_{n+1}(x). \tag{6.9.7}$$

Replacing ν by $\nu - 1$ in the first of (6.9.6) and simplifying, we have

$$Y_{\nu-1}(x) = \frac{J_{\nu-1}(x) \cos \nu\pi + J_{-\nu+1}(x)}{\sin \nu\pi}. \tag{6.9.8}$$

Similarly we may obtain

$$Y_{\nu+1}(x) = \frac{J_{\nu+1}(x) \cos \nu\pi + J_{-\nu-1}(x)}{\sin \nu\pi}, \tag{6.9.9}$$

and adding these two results we have

$$Y_{\nu-1}(x) + Y_{\nu+1}(x) = \frac{[J_{\nu-1}(x) + J_{\nu+1}(x)] \cos \nu\pi + [J_{-\nu+1}(x) + J_{-\nu-1}(x)]}{\sin \nu\pi},$$

which by virtue of (6.9.7) (which holds for non-integral n) is

$$Y_{\nu-1}(x) + Y_{\nu+1}(x) = \frac{(2\nu/x)J_\nu(x) \cos \nu\pi - (2\nu/x)J_{-\nu}(x)}{\sin \nu\pi}.$$

Factoring the right member and using (6.9.6), we have

$$Y_{\nu-1}(x) + Y_{\nu+1}(x) = \frac{2\nu}{x} Y_\nu(x),$$

which for $\nu \to n$ becomes the recurrence relation

$$\frac{2n}{x} Y_n(x) = Y_{n-1}(x) + Y_{n+1}(x). \tag{6.9.10}$$

By this method we may also obtain the analogue of (6.4.1), which for convenience we repeat:

$$2J'_n(x) = J_{n-1}(x) - J_{n+1}(x). \tag{6.9.11}$$

Differentiating the first of (6.9.6) with respect to x and eliminating the derivatives of J_ν and $J_{-\nu}$ by (6.9.11), we have upon rearrangement

$$2Y'_\nu(x) = \frac{J_{\nu-1}(x) \cos \nu\pi + J_{-\nu+1}(x)}{\sin \nu\pi} - \frac{J_{\nu+1}(x) \cos \nu\pi + J_{-\nu-1}(x)}{\sin \nu\pi}.$$

The two terms on the right may be identified by (6.9.8) and (6.9.9), to give

$$2Y'_\nu(x) = Y_{\nu-1}(x) - Y_{\nu+1}(x),$$

and passing to the limit we have

$$2Y'_n(x) = Y_{n-1}(x) - Y_{n+1}(x). \tag{6.9.12}$$

All the relations for $J_n(x)$ which were obtained from (6.9.7) and (6.9.11) now follow for $Y_n(x)$ from (6.9.10) and (6.9.12). Some of these are as follows:

$$xY'_n(x) = xY_{n-1}(x) - nY_n(x), \tag{6.9.13}$$

$$xY'_n(x) = nY_n(x) - xY_{n+1}(x), \tag{6.9.14}$$

$$Y_{n+1}(x)Y_{n-1}(x) = \frac{n^2}{x^2}Y_n^2(x) - [Y'_n(x)]^2, \tag{6.9.15}$$

$$\frac{d}{dx}[x^{-n}Y_n(x)] = -x^{-n}Y_{n+1}(x), \tag{6.9.16}$$

$$\frac{d}{dx}[x^nY_n(x)] = x^nY_{n-1}(x). \tag{6.9.17}$$

Since $Y_n(x)$ and $J_n(x)$ both satisfy these relations, then so does the function $U_n(x)$ in (6.9.1).

6.10 AN ORTHOGONALITY RELATION INVOLVING $Y_n(x)$

As a final example of a relation involving $Y_n(x)$, we shall obtain an orthogonality relation of the set $\{U_k(\lambda_n x)\}$ defined in (6.9.1). The λ_n, $n = 1, 2, 3, \ldots$, are the positive zeros of $U_k(\lambda a)$, which are known to be infinite in number, all simple, and such that if α is a zero then $-\alpha$ is also a zero (see [GM], page 82).

The case we consider is that in which A and B are such that

$$U_k(\lambda x) = Y_k(\lambda b)J_k(\lambda x) - J_k(\lambda b)Y_k(\lambda x). \tag{6.10.1}$$

The orthogonality relation will be obtained from (6.9.2) and (6.9.3) by a proper choice of a and b. We note from (6.10.1) that $U_k(\lambda b) \equiv 0$, and that if α and β are taken as λ_n and λ_m, two roots of $U_k(\lambda a) = 0$, the right member of (6.9.3) vanishes, and we have

$$(\lambda_m^2 - \lambda_n^2)\int_a^b xU_k(\lambda_n x)U_k(\lambda_m x)\,dx = 0.$$

Since $\lambda_m \neq \lambda_n$ for $m \neq n$, the integral must be zero and thus the set $\{U_k(\lambda_n x)\}$ is orthogonal over $[a, b]$ with weight function x. Also, for this choice of a and b in (6.9.2), we have the orthogonality relation,

$$\int_a^b xU_k(\lambda_n x)U_k(\lambda_m x)\,dx = \tfrac{1}{2}\{b^2[U'_k(\lambda_n b)]^2 - a^2[U'_k(\lambda_n a)]^2\}\,\delta_{mn}. \tag{6.10.2}$$

We leave as an exercise (see Exercise 6.10.7) to show that the orthogonality relation (6.10.2) may be put in the simpler form,

$$\int_a^b xU_k(\lambda_n x)U_k(\lambda_m x)\,dx = \frac{2\{J_k^2(\lambda_n a) - J_k^2(\lambda_n b)\}}{\pi^2\lambda_n^2 J_k^2(\lambda_n a)}\,\delta_{mn} \tag{6.10.3}$$

The denominator is never zero, because $\lambda_n > 0$ and $J_k(x)$ and $U_k(x)$ are linearly independent.

EXERCISES

6.10.1. Show from (6.9.10) that

$$xY_1(x) = 4 \sum_{n=1}^{\infty} (-1)^{n+1} n Y_{2n}(x).$$

6.10.2. Obtain (6.9.16) and (6.9.17).

6.10.3. Show by a consideration of the dominant terms of the Bessel functions that for small values of x the following asymptotic relations hold:

$$J_\nu(x) \sim \frac{(x/2)^\nu}{\Gamma(\nu+1)}; \qquad J_{-\nu}(x) \sim \frac{(2/x)^\nu}{\Gamma(-\nu+1)};$$

$$Y_n(x) \sim -\left(\frac{2}{x}\right)^n \frac{(n-1)!}{\pi}, \quad n \geq 1; \qquad Y_0(x) \sim \frac{2 \log x}{\pi}.$$

6.10.4. Show that the Wronskian of any two solutions u, v of Bessel's equation is given by (see Exercise 3.4.11)

$$W(x) = u'v - v'u = \frac{k}{x}.$$

6.10.5. Show from Exercise 6.10.4 that if u and v are linearly independent solutions of the Bessel differential equation, they can have no common zeros except possibly $x = 0$.

6.10.6. Using Exercise 6.10.3 and $\lim_{x\to 0} xW(x)$ in Exercise 6.10.4, show that

$$J_n(x)Y'_n(x) - Y_n(x)J'_n(x) = \frac{2}{\pi x}, \quad x \neq 0, n = 0, 1, 2, \ldots$$

and

$$J_\nu(x)J'_{-\nu}(x) - J_{-\nu}(x)J'_\nu(x) = -\frac{2}{\pi x}\sin \nu\pi, \quad \nu \text{ real}.$$

6.10.7. Use the result of Exercise 6.10.6 to verify (6.10.3).

6.10.8. Evaluate for $U_0(\alpha a) = U_0(\alpha b) = 0$:

(a) $\int_a^b xU_0(\alpha x)\,dx.$

(b) $\int_a^b x(\log x)U_0(\alpha x)\,dx.$

6.10.9. Show that an orthogonality relation for $\{U_\nu(\lambda_n x)\}$, where

$$U_\nu(\lambda x) = J_\nu(\lambda x)J_{-\nu}(\lambda b) - J_\nu(\lambda b)J_{-\nu}(\lambda x),$$

is

$$\int_a^b xU_\nu(\lambda_n x)U_\nu(\lambda_m x)\,dx = \frac{2\{J_\nu^2(\lambda_n a) - J_\nu^2(\lambda_n b)\}(\sin^2 \nu\pi)\delta_{mn}}{\pi^2\lambda_n^2 J_\nu^2(\lambda_n a)}.$$

7

Boundary-Value Problems

7.1 LINEAR OPERATORS AND BOUNDARY-VALUE PROBLEMS

Before proceeding to the solution of boundary-value problems, we need to consider some basic concepts and definitions. For example, we shall want to classify the various boundary-value problems according to their differential equations and boundary conditions, and consider what constitutes a solution and under what conditions our methods are valid.

Toward this end we need to state our boundary-value problem in more precise mathematical terms. We begin by defining a *linear manifold* M as a set of functions or a function class having the property that for every pair of functions ψ_1 and ψ_2 belonging to the set the linear combination $C_1\psi_1 + C_2\psi_2$ belongs to the set, where C_1 and C_2 are real constants. An *operator* L is a transformation which associates with every function ψ of M a function $L(\psi)$. The latter function may or may not belong to M. Unless otherwise specified, the manifold under consideration will be the class of sectionally continuous functions.

The operator L is *linear* if for each pair of functions ψ_1 and ψ_2 of M and for any constants C_1 and C_2 it is true that

$$L(C_1\psi_1 + C_2\psi_2) = C_1L(\psi_1) + C_2L(\psi_2). \tag{7.1.1}$$

By induction we may extend (7.1.1) to the case of n functions, resulting in

$$L\left(\sum_{k=1}^{n} C_k\psi_k\right) = \sum_{k=1}^{n} C_kL(\psi_k). \tag{7.1.2}$$

As an example, $L = \nabla^2$ in Laplace's equation is a linear operator. A simpler example is $L = d/dx$ applied to differentiable functions of a single variable x. Unless otherwise specified, all the operators we consider henceforth will be linear operators.

A differential equation in ψ is *linear* if it may be expressed as

$$L(\psi) = f, \tag{7.1.3}$$

where L is linear and f is a function of the independent variables of ψ. The case $f = 0$,

$$L(\psi) = 0, \tag{7.1.4}$$

is a *linear, homogeneous* differential equation, and the case $f \neq 0$ is linear and non-homogeneous. Laplace's equation, for example, is a linear homogeneous equation, and Poisson's equation is linear and non-homogeneous.

The terms *linear, homogeneous, and non-homogeneous* also characterize the boundary conditions. For example, the condition $N_1[\psi(x, t)] = \psi(0, t) + \psi_x(1, t) = 0$ is a linear homogeneous boundary condition while $N_2[\psi(x)] = \psi(a) = 1$ is linear and non-homogeneous. A homogeneous condition or equation is thus one in which each term is of the first degree in ψ and its derivatives.

Finally, we define a *linear boundary-value problem* as one whose differential equation and boundary conditions are all linear. The terms *homogeneous* and *non-homogeneous* have the same meaning as before. For example we may state our boundary-value problem as follows:

$$\begin{aligned} L(\psi) &= f, \\ N_i(\psi) &= g_i, \quad i = 1, 2, \ldots, n. \end{aligned} \tag{7.1.5}$$

The first formula represents the differential equation and the second the n boundary conditions. If $f = g_i = 0$, $i = 1, 2, \ldots, n$, then (7.1.5) is a homogeneous, linear boundary-value problem. If at least one of the functions $f, g_1, g_2, \ldots, g_n$ is not zero, then the system is non-homogeneous.

We may further classify the boundary conditions according to the operators N_i. If $N_i(\psi) = g_i$ specifies the values of ψ on the boundary, it is known as a *Dirichlet* boundary condition. A boundary condition prescribing the normal derivative of ψ at the boundary is a *Neumann* condition. The example N_2 given previously is of Dirichlet type, and the example N_1 is a mixture of the two.

EXERCISES

7.1.1. Show that the set of real sectionally continuous functions is a linear manifold.

7.1.2. Show that if ψ_1 and ψ_2 are solutions of $L(\psi) = 0$, then $C_1\psi_1 + C_2\psi_2$ is also a solution.

7.1.3. Extend Exercise 7.1.2 by induction to show that if ψ_n, $n = 1, 2, \ldots, m$, is a solution of $L(\psi) = 0$, then $\sum_{n=1}^{m} C_n\psi_n$ is also a solution.

7.1.4. (a) Show that if ψ_1 satisfies $L(\psi) = f$ and ψ_2 satisfies $L(\psi) = g$, then $C_1\psi_1 + C_2\psi_2$ satisfies $L(\psi) = C_1 f + C_2 g$.

(b) Show that if ψ_n, $n = 1, 2, \ldots, m$, satisfies $L(\psi) = 0$ and ψ_{m+1} satisfies $L(\psi) = f$, then $\sum_{n=1}^{m+1}\psi_n$ satisfies $L(\psi) = f$.

7.2 PRINCIPLE OF SUPERPOSITION

As an example of a linear boundary-value problem, let us consider the system of equations

$$\psi_{xx}(x, y) + \psi_{yy}(x, y) = 0, \quad 0 < x < 1, 0 < y < \infty, \tag{7.2.1}$$

$$\psi(0, y) = \psi(1, y) = \lim_{y \to \infty} \psi(x, y) = 0, \tag{7.2.2}$$

$$\psi(x, 0) = f(x). \tag{7.2.3}$$

The reader may verify that each of the functions

$$\psi_n(x, y) = C_n e^{-n\pi y} \sin n\pi x, \quad n = 1, 2, 3, \ldots, \tag{7.2.4}$$

is a solution of the linear homogeneous system formed by the four equations of (7.2.1) and (7.2.2). Since each of these equations is of the form $L(\psi) = 0$, we may apply Exercise 7.1.3 to obtain the more general solution

$$\psi_m(x, t) = \sum_{n=1}^{m} C_n e^{-n\pi y} \sin n\pi x. \tag{7.2.5}$$

To show that ψ_m is a solution of the system we may substitute $\psi = \psi_m$ directly in (7.2.1) and (7.2.2). Because the series involved is finite, termwise differentiation is valid, as is a termwise limit operation indicated in the third member of (7.2.2).

The principle embodied in Exercise 7.1.2 and the foregoing example is known as the *principle of superposition*, and because it is of fundamental importance in solving boundary-value problems we state it formally as follows:

> **Principle of Superposition.** If each member of a set $\{\psi_n\}$, $n = 1, 2, \ldots, m$, satisfies the condition $L(\psi) = 0$, then $\Sigma_{n=1}^{m} C_n\psi_n$, with C_n arbitrary constants, also satisfies the condition.

Superposition holds if $L(\psi) = 0$ is a differential equation or a boundary condition, as long as L is linear.

EXERCISES

7.2.1. Show that if ψ_n, $n = 1, 2, \ldots, m$, satisfies the system

(1) $$L(\psi) = 0, \qquad N_i(\psi) = 0, \quad i = 1, 2, \ldots, k,$$

and if ψ_0 satisfies the system

(2) $$L(\psi) = f, \qquad N_i(\psi) = 0,$$

then $\psi_m = \psi_0 + \Sigma_{n=1}^{m} C_n\psi_n$ satisfies (2).

7.2.2. As an example of Exercise 7.2.1 let $L[\psi(x)] = 12x$, where

$$L = \frac{d^2}{dx^2} + 4\frac{d}{dx} + 3,$$

and find ψ_0, ψ_1, ψ_2.

7.3 INFINITE SERIES OF SOLUTIONS

In the previous section we saw that the solution (7.2.5) of the system formed by (7.2.1) and (7.2.2) could be checked by direct substitution because the series involved was finite. However, if we attempt to extend our solution to satisfy also the boundary condition (7.2.3), we will see that the finite series is, in general, insufficient; for this condition requires us to seek a set of C_n such that

$$f(x) = \sum_{n=1}^{m} C_n \sin n\pi x. \tag{7.3.1}$$

We know from Chapter 2 that for a general $f(x)$ no such finite set exists, but if m is allowed to become infinite, the series on the right may converge to $f(x)$ under certain conditions. Our series is then the half-range sine series, and the Fourier coefficients are given by (2.4.3) as

$$C_n = 2\int_0^1 f(x) \sin n\pi x \, dx. \tag{7.3.2}$$

If $f(x)$ satisfies the Dirichlet conditions, the half-range sine series converges to $\frac{1}{2}[f(x+0) + f(x-0)]$ for every x on $(0, 1)$; and at points of continuity of $f(x)$ the series

$$\psi(x, y) = \sum_{n=1}^{\infty} C_n e^{-n\pi y} \sin n\pi x, \tag{7.3.3}$$

with C_n given by (7.3.2), satisfies the boundary condition (7.2.3).

The question now remaining is whether (7.3.3) satisfies (7.2.1) and (7.2.2). It no longer may be possible to check by direct substitution, as was done in Sec. 7.2, because termwise differentiation may not be valid, nor may the operation indicated in the last of (7.2.2). If these operations are valid, then (7.3.3) is a solution of the system made up of (7.2.1), (7.2.2), and (7.2.3). It may even be possible that we have a solution if these operations are not valid, in which case the solution cannot be confirmed by direct substitution of the series form of $\psi(x, t)$. In this case we must be able to work with a closed form of $\psi(x, t)$; that is, we must be able to sum the series. An example of this type is given in Exercise 7.5.2.

Because of the difficulty involved in summing the series, it is important to know under what conditions the operations on the series itself are valid. These conditions are given in detail in Appendix A, but because of their importance to us we summarize them here. For the series $\psi = \Sigma_{n=1}^{\infty} \psi_n$, the following properties hold:

1. If ψ and each ψ_n are integrable and the series is uniformly convergent with respect to x over $a \le x \le b$, then

$$\int_a^b \psi\, dx = \sum_{n=1}^{\infty} \int_a^b \psi_n\, dx.$$

2. If each ψ_n is continuous and the series is uniformly convergent over $a \le x \le b$, then ψ is continuous on $a < x < b$ and

$$\lim_{x\to x_0} \psi = \sum_{n=1}^{\infty} \lim_{x\to x_0} \psi_n, \quad \text{for } a < x_0 < b.$$

3. If the series converges, the derivative of each ψ_n is continuous, and $\sum_{n=1}^{\infty} \psi'_n$ converges uniformly for $a \le x \le b$, then $\psi'(x) = \sum_{n=1}^{\infty} \psi'_n(x)$.
4. Weierstrass M-test: If a convergent series of positive constants M_n exists such that $|\psi_n| \le M_n$, $n = 1, 2, 3, \ldots$, then $\sum_{n=1}^{\infty} \psi_n$ converges uniformly.

These results may be extended to hold for series of functions of more than one independent variable.

The foregoing properties of uniformly convergent series may now be used to answer the question raised earlier in this section, whether the function

$$\psi(x, y) = \sum_{n=1}^{\infty} C_n e^{-n\pi y} \sin n\pi x = \sum_{n=1}^{\infty} C_n \psi_n, \tag{7.3.4}$$

with

$$C_n = 2\int_0^1 f(x) \sin n\pi x\, dx,$$

satisfies the conditions

$$\nabla^2 \psi(x, y) = 0, \quad 0 < x < 1, 0 < y < \infty \tag{7.3.5}$$

$$\psi(0, y) = \psi(1, y) = \lim_{y\to\infty} \psi(x, y) = 0. \tag{7.3.6}$$

From Chapter 1 we know that the Fourier coefficients satisfy Bessel's inequality (1.4.6), which for this case is

$$\sum_{n=1}^{m} C_n^{\,2} \le 2\, \|f\|^2,$$

for any integer $m \ge 1$. Thus if the norm of f exists [and it does if f is sectionally continuous on $(0, 1)$], then the C_n are bounded; that is, a number N exists such that

$$|C_n| \le N, \quad n = 1, 2, 3, \ldots.$$

For $y \ge y_0 > 0$, we have

$$|C_n e^{-n\pi y} \sin n\pi x| \le N e^{-n\pi y_0} = M_n. \tag{7.3.7}$$

For M_n defined this way, we have

$$\sum_{n=1}^{\infty} M_n = \sum_{n=1}^{\infty} N e^{-n\pi y_0} = \frac{N e^{-\pi y_0}}{1 - e^{-\pi y_0}}.$$

Since the series ΣM_n converges, by the Weierstrass M-test the series $\Sigma C_n\psi_n$ converges uniformly with respect to both x and y, for $0 \le x \le 1$, $y \ge y_0$. By Abel's test [see Appendix A] the series converges uniformly with respect to y for $y \ge 0$. The ψ_n are continuous functions and therefore, by property (2), $\psi(x, y)$ satisfies conditions (7.3.6). For example,

$$\lim_{y\to\infty} \psi(x, y) = \lim_{y\to\infty} \sum_{n=1}^{\infty} C_n\psi_n = \sum_{n=1}^{\infty} C_n \lim_{y\to\infty} \psi_n = 0.$$

In general, for $y \to \infty$ rather than $y \to a$, this interchange of operations is valid if the limiting process is uniform; that is $\lim_{y\to\infty} \psi_n = 0$ for all x on $(0, 1)$. This is true in our case since $|\psi_n| < |C_n e^{-n\pi y}|$.

To show that condition (7.3.5) is satisfied we need to establish the validity of termwise differentiation. Considering differentiation with respect to x, by property (3) we need to show that the series $\Sigma C_n(\partial\psi_n/\partial x)$ and $\Sigma C_n(\partial^2\psi_n/\partial x^2)$ are uniformly convergent. Following the same procedure as in obtaining (7.3.7), we may write, for $y \ge y_0 > 0$,

$$\left| C_n \frac{\partial \psi_n}{\partial x} \right| = |C_n e^{-n\pi y} n\pi \cos n\pi x| \le Nn\pi e^{-n\pi y_0} = M_n.$$

Using the ratio test, we have

$$\frac{M_{n+1}}{M_n} = \frac{n+1}{n} e^{-\pi y_0} \sim e^{-\pi y_0} < 1.$$

Hence, the series ΣM_n converges, and the series $\Sigma C_n(\partial\psi_n/\partial x)$ converges uniformly. Similarly we may show that the series

$$\sum C_n \frac{\partial^2 \psi_n}{\partial x^2}, \quad \sum C_n \frac{\partial \psi_n}{\partial y}, \quad \sum C_n \frac{\partial^2 \psi_n}{\partial y^2}$$

converge uniformly and hence condition (7.3.5) is satisfied; that is

$$\nabla^2\psi(x, y) = \nabla^2 \sum_{n=1}^{\infty} C_n\psi_n = \sum_{n=1}^{\infty} C_n\nabla^2\psi_n = 0.$$

The function $\psi(x, y)$ defined in (7.3.3) is therefore a solution to our boundary-value problem given by (7.2.1), (7.2.2), and (7.2.3).

We have extended the principle of superposition of solutions of $L\psi = 0$ to an infinite set, obtaining $\psi = \Sigma_{n=1}^{\infty} C_n\psi_n$ as a solution provided $L\psi_n = 0$ for each ψ_n, $n = 1, 2, 3, \ldots$, and the order of the operations L and Σ is interchangeable. In general, we may determine the validity of our formal operations by investigating the convergence properties of the infinite series. However, since the proof, as we have seen, is rather involved, henceforth for the most part we will assume that the formal steps taken in obtaining the solution are valid; that is, we will conclude our solution with obtaining a form such as (7.3.4). In practice, of course, one should validate these steps.

7.4 SEPARATION-OF-VARIABLES METHOD

An obvious question which we have left unanswered in the previous two sections is how we can obtain the set of solutions in (7.2.4),

$$\psi_n(x, y) = C_n e^{-n\pi y} \sin n\pi x, \quad n = 1, 2, 3, \ldots, \tag{7.4.1}$$

each of which is a solution of the system given by (7.2.1) and (7.2.2):

$$\begin{aligned} \nabla^2\psi(x, y) &= 0, \quad 0 < x < 1,\ 0 < y < \infty, \\ \psi(0, y) &= \psi(1, y) = \lim_{y \to \infty} \psi(x, y) = 0. \end{aligned} \tag{7.4.2}$$

To answer this question, we shall use the *separation-of-variables method* outlined in Sec. 1.1 to seek a solution of the form

$$\psi(x, y) = X(x)Y(y). \tag{7.4.3}$$

Substituting this expression into the first of (7.4.2) and rearranging, we have

$$\frac{X''(x)}{X(x)} = -\frac{Y''(y)}{Y(y)} = -\lambda,$$

where $-\lambda$ is the separation constant. This results in two ordinary differential equations, and (7.4.3) substituted into the second of (7.4.2) results in a new set of boundary conditions. Collecting the results, we have the systems

$$\begin{aligned} X'' + \lambda X &= 0, \\ X(0) = X(1) &= 0, \end{aligned} \tag{7.4.4}$$

and

$$\begin{aligned} Y'' - \lambda Y &= 0, \\ \lim_{y \to \infty} Y(y) &= 0. \end{aligned} \tag{7.4.5}$$

As an example of how the boundary conditions are obtained, consider

$$\psi(0, y) = 0,$$

which, using (7.4.3), becomes

$$\psi(0, y) = X(0)Y(y) = 0.$$

Since this must hold for all y, it follows that $X(0) = 0$.

We note that Eqs. (7.4.4) constitute a Sturm-Liouville system, and hence by Exercise 1.9.7 the eigenvalue λ is real. Hence we need to check only the cases $\lambda = 0$, λ real and positive, and λ real and negative. Considering $\lambda = 0$, we have from (7.4.4) $X'' = 0$, or

$$X(x) = ax + b.$$

Applying the boundary conditions $X(0) = X(1) = 0$, we have the trivial case $X \equiv 0$.

Next we consider the possibility that λ is negative. That is, $\lambda = -\mu$ for $\mu > 0$. For this case, the solution of (7.4.4) is

$$X = k_1 e^{\sqrt{\mu}\,x} + k_2 e^{-\sqrt{\mu}\,x},$$

and the two boundary conditions in (7.4.4) again require $X \equiv 0$.

Finally, the last possibility, $\lambda > 0$, leads to

$$X(x) = C_1 \cos \sqrt{\lambda}\, x + C_2 \sin \sqrt{\lambda}\, x, \tag{7.4.6}$$

and the boundary conditions require that

$$C_1 = 0, \quad C_2 \sin \sqrt{\lambda} = 0. \tag{7.4.7}$$

For a non-trivial solution, $C_2 \neq 0$, and hence

$$\lambda = n^2\pi^2, \quad n = \pm 1, \pm 2, \ldots. \tag{7.4.8}$$

($n = 0$ has already been ruled out.) Therefore we have a set of solutions for the system (7.4.4), each member of which is of the form, for n an integer,

$$X_n(x) = C_2(n) \sin n\pi x. \tag{7.4.9}$$

The negative values of n will not contribute any new functions to the set since these values will simply alter the sign of the arbitrary constants.

Substituting for λ into (7.4.5) and solving that system for $Y(y)$ we have

$$Y(y) = C_3 e^{n\pi y} + C_4 e^{-n\pi y}, \tag{7.4.10}$$

and

$$\lim_{y \to \infty} (C_3 e^{n\pi y} + C_4 e^{-n\pi y}) = 0. \tag{7.4.11}$$

Again we see that the negative values of n contribute nothing new, since if n is positive, (7.4.11) requires that $C_3 = 0$, and if n is negative, we must have $C_4 = 0$. In either case $Y(y)$ is of the form

$$Y_n(y) = C_4(n) e^{-n\pi y}, \quad n = 1, 2, \ldots. \tag{7.4.12}$$

Combining the values of $X_n(x)$ in (7.4.9) and of $Y_n(y)$ in (7.4.12) as indicated in (7.4.3), and letting $C_n = C_2(n)C_4(n)$, we have the set of solutions given in (7.4.1).

7.5 SUMMARY OF THE METHOD

We will now collect our results of the previous sections and consider the method of separation of variables in some detail. The method, while quite powerful, is not applicable to the general case, as we shall see, but a wide variety of boundary-value problems can be successfully attacked in this way.

First we attempt to find a set of solutions to the differential equation,

$$L(\psi) = 0, \tag{7.5.1}$$

by assuming a solution of the type

$$\psi(q_i) = \prod_{i=1}^{k} Q_i(q_i), \tag{7.5.2}$$

where k is the number of independent variables q_i. If it is possible to separate the variables in $L(\psi)$, (7.5.1) may be split into k ordinary differential equations, whose solutions are the $Q_i(q_i)$, $i = 1, 2, \ldots, k$.

Substitution of (7.5.2) into the boundary conditions

$$M_j\psi = f_j, \quad j = 1, 2, \ldots, m, \tag{7.5.3}$$

results in a new set of boundary conditions, which together with the k ordinary differential equations constitute k new systems. One or more of the boundary conditions, say

$$M_r\psi = f_r, \tag{7.5.4}$$

may not be included in any of the new systems and must be considered separately.

If it is possible to solve the k systems for the Q_i, then substitution into (7.5.2) will yield a set of solutions $\{\psi_n\}$, each member of which will satisfy every condition except (7.5.4). By the superposition principle a linear combination of the ψ_n will also satisfy every condition except (7.5.4). In general, to satisfy (7.5.4) the linear combination will have to be an infinite series,

$$\psi(q_i) = \sum_{n=1}^{\infty} \psi_n(q_i). \tag{7.5.5}$$

If (7.5.4) includes more than one condition, (7.5.5) may be a multiple Fourier series; that is, it may be a double sum, triple sum, etc. To show that (7.5.5) holds, we must finally show that, from its convergence properties, the operations (7.5.1) and (7.5.3) are valid.

EXERCISES

7.5.1. Solve for $\psi(x, t)$ if $\partial^2\psi/\partial x^2 = (1/c^2)\,\partial^2\psi/\partial t^2$, $\psi(0, t) = \psi(a, t) = 0$, $\psi_t(x, 0) = 0$, $\psi(x, 0) = f(x)$. This is the problem of the vibrating string, which will be considered in detail in Chapter 8.

7.5.2. For a string plucked at its midpoint and released from rest, the conditions are the same as in Exercise 7.5.1 except that

$$\begin{aligned} f(x) &= \frac{2bx}{a}, & 0 \le x \le \frac{a}{2} \\ &= -\frac{2b}{a}(x - a), & \frac{a}{2} \le x \le a. \end{aligned}$$

Solve for $\psi(x, t)$ for this case.

Ans. $\psi(x, t) = \sum_{n=1}^{\infty} b_n (\sin n\pi x/a) \cos n\pi ct/a$, $b_n = (8b/n^2\pi^2) \sin n\pi/2$.

7.5.3. Note that the series in Exercise 7.5.2 cannot be differentiated twice termwise, and hence the solution cannot be checked directly. Show that for the general case in Exercise 7.5.1

$$\psi(x, t) = \frac{f_1(x - ct) + f_1(x + ct)}{2},$$

where f_1 is the odd periodic extension of f. Show that the second derivatives exist, and that the solution obtained in Exercise 7.5.1 is valid if $f''(x)$ exists on $0 \leq x \leq a$.

7.5.4. Solve for $\psi(x, y)$ if $\nabla^2\psi(x, y) = 0$, $\psi(x, 0) = \psi(x, b) = 0$, $\lim_{x\to\infty} \psi(x, y) = 0$, $\psi(0, y) = \psi_0$ (constant).

Ans. $\psi(x, y) = (4\psi_0/\pi) \sum_{n=1}^{\infty} [1/(2n - 1)][\sin (2n - 1)\pi y/b]e^{-n\pi x/b}$.

7.5.5. Solve for y if $\partial^2 y/\partial x^2 = \partial^2 y/\partial t^2$, $y(0, t) = y(1, t) = 0$, $y_t(x, 0) = 0$, $y(x, 0) = \sin \pi x$.

7.5.6. Solve for $\psi(r, \theta)$ where r and θ are spherical coordinates, ψ is independent of ϕ, and $\psi(r, \theta)$ is bounded, if $\nabla^2\psi = 0$, $0 < \theta < \pi$, $0 < r < 1$, $\psi(1, \theta) = f(\theta)$, $0 \leq \theta \leq \pi$.

7.5.7. Find $\psi(r, z)$ if $\nabla^2\psi = 0$ inside the cylinder $r = 1$, $z = 0$, and $z = 1$, and $\psi(1, z) = \psi(r, 1) = 0$, $\psi(r, 0) = f(r)$.

7.5.8. Find $\psi(r, z)$ if $\nabla^2\psi = 0$, $\psi(r, 0) = \psi(r, 1) = 0$, $\psi(1, z) = f(z)$, $|\psi(r, z)| < M$. (See Exercise 6.2.6.)

Ans. $\psi(r, z) = \sum_{n=1}^{\infty} a_n I_0(n\pi r) \sin n\pi z$, where $a_n = [2/I_0(n\pi)]\int_0^1 f(z) \sin n\pi z\, dz$.

7.5.9. Solve for $\psi(r, \theta)$ inside and outside a sphere $r = 1$, if $\nabla^2\psi = 0$, $\psi(1, \theta) = f(\theta)$ and $\psi(r, \theta)$ is bounded.

Ans. Inside sphere: $\psi = \sum_{n=0}^{\infty} A_n r^n P_n (\cos \theta)$,
Outside sphere: $\psi = \sum_{n=0}^{\infty} (B_n/r^{n+1}) P_n (\cos \theta)$,
$A_n = B_n = [(2n + 1)/2]\int_0^{\pi} f(\theta) P_n (\cos \theta) \sin \phi\, d\theta$.

7.6 AN EXAMPLE

To illustrate a boundary-value problem requiring a doubly infinite series solution, we consider the temperature distribution $\psi(r, \theta, t)$ of a thin circular plate, which, as we shall see in Chapter 8, may satisfy the following conditions:

$$\nabla^2\psi = \frac{\partial^2\psi}{\partial r^2} + \frac{1}{r}\frac{\partial\psi}{\partial r} + \frac{1}{r^2}\frac{\partial^2\psi}{\partial\theta^2} = \frac{\partial\psi}{\partial t}, \tag{7.6.1}$$

$$\psi(1, \theta, t) = 0, \tag{7.6.2}$$

$$\psi(r, \theta, 0) = f(r, \theta). \tag{7.6.3}$$

Assuming a solution

$$\psi(r, \theta, t) = R(r)\Theta(\theta)T(t), \tag{7.6.4}$$

substituting into (7.6.1) and rearranging, we have

$$\frac{R''}{R} + \frac{R'}{rR} + \frac{\Theta''}{r^2\Theta} = \frac{T'}{T} = -\lambda^2. \tag{7.6.5}$$

We have separated the expression involving t only from that involving r and θ, and have chosen $-\lambda^2$ as the separation constant. Equating the first and third members of (7.6.5) and rearranging, we obtain a separation of the variables r and θ, with a second separation constant μ^2,

$$\frac{r^2R'' + rR'}{R} + r^2\lambda^2 = -\frac{\Theta''}{\Theta} = \mu^2. \tag{7.6.6}$$

We will consider first the cases $\lambda^2 > 0$ and $\mu^2 > 0$. The three separated equations are then

$$T' + \lambda^2 T = 0, \tag{7.6.7}$$

$$\Theta'' + \mu^2\Theta = 0, \tag{7.6.8}$$

$$r^2R'' + rR' + (r^2\lambda^2 - \mu^2)R = 0. \tag{7.6.9}$$

The solution of (7.6.8) is

$$\Theta(\theta) = C_1 \cos \mu\theta + C_2 \sin \mu\theta, \tag{7.6.10}$$

and because (r, θ) and $(r, \theta + 2\pi)$ represent the same point on the plate, we must have $\psi(r, \theta, t) = \psi(r, \theta + 2\pi, t)$ and consequently $\Theta(\theta) = \Theta(\theta + 2\pi)$. Therefore μ must be an integer. Replacing $\mu = n$ by $-n$ in (7.6.10) will merely replace C_2 by $-C_2$ and will not alter the form of the general solution. Since μ appears only as μ^2 in (7.6.9), replacing μ by its negative there will not change $R(r)$. Hence we need only the positive values

$$\mu = n, \quad n = 1, 2, 3, \ldots .$$

(Note that $n = 0$ was excluded at the beginning.)

Equation (7.6.9) is a form of Bessel's equation and has as its solution (see Exercise 6.2.3)

$$R(r) = C_3J_n(\lambda r) + C_4Y_n(\lambda r). \tag{7.6.11}$$

Because the temperature is defined at $r = 0$, we must suppress the term involving Y_n by making $C_4 = 0$.

Solving (7.6.7) and substituting for R, Θ, and T in (7.6.4), we have

$$\psi(r, \theta, t) = e^{-\lambda^2 t}J_n(\lambda r)[C_5 \cos n\theta + C_6 \sin n\theta]. \tag{7.6.12}$$

Condition (7.6.2) requires that $J_n(\lambda) = 0$ and hence that $\lambda = \lambda_{nm}$, $m = 1, 2. 3, \ldots$, are the ordered positive zeros of J_n, which are known to be real and distinct [see Sec. 6.6]. The negative zeros contribute nothing new to the solution (7.6.12) because they are the negatives of the positive zeros, as seen by the equation

$$J_n(-x) = (-1)^nJ_n(x).$$

Therefore the only change in (7.6.12) due to negative zeros would be a possible change in sign of the constants.

We now have found a set of solutions $\{\psi_{nm}\}$ satisfying all the conditions except (7.6.3). Using (7.6.12), we write a typical member of the set as

$$\psi_{nm} = e^{-\lambda_{nm}^2 t} J_n(\lambda_{nm} r)[a_{nm} \cos n\theta + b_{nm} \sin n\theta]. \tag{7.6.13}$$

Before proceeding further with the solution let us check the possibility that there are other solutions of the type (7.6.4) which are not included in the set $\{\psi_{nm}\}$. We have established that λ and μ are real, but we have not considered the cases where one or both may be zero. If $\lambda = 0$ and $\mu \neq 0$, the separated equations become

$$T'(t) = 0,$$

$$\Theta'' + \mu^2\Theta = 0,$$

$$r^2R'' + rR' - \mu^2 R = 0.$$

The last equation is Euler's equation (see Exercise 1.9.3) and has as a solution

$$R(r) = K_1 r^{\mu} + K_2 r^{-\mu}.$$

The condition that $\psi(0, \theta, t)$ is finite requires that $K_2 = 0$, and condition (7.6.2) then requires that $K_1 = 0$. Hence for this case $R \equiv 0$ and we have the trivial solution.

Of the remaining two cases, $\lambda = 0$, $\mu = 0$, and $\lambda \neq 0$, $\mu = 0$ only the latter yields a non-trivial solution, as the reader may verify, and this solution is included in the set $\{\psi_{nm}\}$ if $n = 0$.

By the superposition theorem we know that if ψ_{nm} is a solution of (7.6.1) and (7.6.2), then

$$\begin{aligned} \psi_{NM}^* &= \sum_{n=0}^{N} \sum_{m=1}^{M} \psi_{nm} \\ &= \sum_{n=0}^{N} \sum_{m=1}^{M} e^{-\lambda_{nm}^2 t} J_n(\lambda_{nm} r)[a_{nm} \cos n\theta + b_{nm} \sin n\theta] \end{aligned} \tag{7.6.14}$$

is also a solution.

To require that ψ_{NM}^* satisfy also (7.6.3) is equivalent to

$$f(r, \theta) = \sum_{n=0}^{N} [A_{nM}(r) \cos n\theta + B_{nM}(r) \sin n\theta], \tag{7.6.15}$$

where

$$\begin{aligned} A_{nM}(r) &= \sum_{m=1}^{M} a_{nm} J_n(\lambda_{nm} r), \\ B_{nM}(r) &= \sum_{m=1}^{M} b_{nm} J_n(\lambda_{nm} r). \end{aligned} \tag{7.6.16}$$

For r fixed in (7.6.15) we see, as in Sec. 7.3, that in general an infinite series is required, in which case the coefficients are the Fourier coefficients,

$$A_{nM}(r) = \frac{1}{(1 + \delta_{n0})\pi} \int_0^{2\pi} f(r, \theta) \cos n\theta \, d\theta,$$
$$B_{nM}(r) = \frac{1}{\pi} \int_0^{2\pi} f(r, \theta) \sin n\theta \, d\theta. \tag{7.6.17}$$

Substituting these values into (7.6.16), we see, by the same argument, that we must have $M \to \infty$. The coefficients a_{nm} and b_{nm} are therefore the Fourier coefficients in the series of Bessel functions, given in Sec. 6.7. These equations together with (7.6.17) enable us finally to write

$$a_{nm} = \frac{2(1 + \delta_{n0})^{-1}}{\pi J^2_{n+1}(\lambda_{nm})} \int_0^1 \int_0^{2\pi} r f(r, \theta) J_n(\lambda_{nm} r) \cos n\theta \, d\theta \, dr,$$
$$b_{nm} = \frac{2}{\pi J^2_{n+1}(\lambda_{nm})} \int_0^1 \int_0^{2\pi} r f(r, \theta) J_n(\lambda_{nm} r) \sin n\theta \, d\theta \, dr, \tag{7.6.18}$$
$$n = 0, 1, 2, \ldots, \qquad m = 1, 2, 3, \ldots,$$

and from (7.6.14),

$$\psi(r, \theta, t) = \sum \sum \psi_{nm}(r, \theta, t)$$
$$= \sum_{n=0}^{\infty} \sum_{m=1}^{\infty} e^{-\lambda^2_{nm} t} J_n(\lambda_{nm} r)[a_{nm} \cos n\theta + b_{nm} \sin n\theta]. \tag{7.6.19}$$

7.7 LIMITATIONS OF THE METHOD

We have already discussed one obvious limitation of the method of separation of variables, that the variables are not always separable. For example, the equation

$$f_1 \frac{\partial^2 \psi}{\partial x^2} + f_2 \frac{\partial^2 \psi}{\partial x \, \partial y} + f_3 \frac{\partial^2 \psi}{\partial y^2} + f_4 \frac{\partial \psi}{\partial x} + f_5 \frac{\partial \psi}{\partial y} + f_6 \psi = 0, \tag{7.7.1}$$

with

$$\psi(x, y) = X(x) Y(y),$$

results in

$$\frac{1}{X}(f_1 X'' + f_4 X') + \frac{1}{Y}(f_3 Y'' + f_5 Y') + f_2 \frac{X' Y'}{XY} + f_6 = 0.$$

The separation of variables can be accomplished if f_1, f_4 are functions of x alone, f_3, f_5 are functions of y alone, $f_2 \equiv 0$, and f_6 is a sum of a function of x alone and a function of y alone. The class of second-order equations in which separation of variables can be effected can be represented by the equation

$$f_1(x)\psi_{xx} + g_1(y)\psi_{yy} + f_2(x)\psi_x + g_2(y)\psi_y + [f_3(x) + g_3(x)]\psi = 0. \tag{7.7.2}$$

A second limitation of the method is that, although separation of the variables may be achieved, the resulting system of equations may fail to yield a solution. After all, some solutions cannot be put in factored form. For example, $\psi = x + y$ is a solution of $\nabla^2\psi = 0$ and is not of the form $\psi = X(x)Y(y)$. In a given boundary-value problem, we seek a solution of the differential equation that satisfies the boundary conditions, and these conditions may be such that no member of the set of factored solutions is acceptable.

An example of a boundary-value problem in which this second limitation applies is the following:

$$\begin{aligned} \psi_t(x, t) &= \psi_{xx}(x, t), \quad x > 0, t > 0, \\ \psi(x, 0) &= 0, \quad x > 0, \\ \psi(0, t) &= 1, \lim_{x \to \infty} \psi(x, t) = 0, \quad t > 0. \end{aligned} \tag{7.7.3}$$

Assuming a solution of the type $\psi = X(x)T(t)$, we obtain, for $\lambda \neq 0$,

$$\begin{aligned} T(t) &= C_1 e^{-\lambda t}, \\ X(x) &= C_2 \cos \sqrt{\lambda}\, x + C_3 \sin \sqrt{\lambda}\, x. \end{aligned}$$

The second of equations (7.7.3) requires that $T(0) = 0$, which results in $C_1 = 0$ and $T \equiv 0$. When $\lambda = 0$, $T(t) = a$, and again $T \equiv 0$. Obviously this trivial solution, $\psi \equiv 0$, cannot satisfy $\psi(0, t) = 1$, and the method fails.

As we shall see in Exercise 14.4.7(c), (7.7.3) has the non-trivial solution

$$\psi(x, t) = \operatorname{erfc}\left(\frac{x}{2\sqrt{t}}\right), \tag{7.7.4}$$

where erfc (z) is the *complementary error function* defined as

$$\operatorname{erfc}(z) = \frac{2}{\sqrt{\pi}} \int_z^\infty e^{-r^2}\, dr. \tag{7.7.5}$$

It is related to the *error function*,

$$\operatorname{erf}(z) = \frac{2}{\sqrt{\pi}} \int_0^z e^{-r^2}\, dr, \tag{7.7.6}$$

by the expression

$$\operatorname{erfc}(z) = 1 - \operatorname{erf}(z). \tag{7.7.7}$$

The situation in the above example arises when T satisfies a first-order differential equation with the condition that $T(0) = 0$, or when T satisfies a second-order differential equation with conditions $T(0) = T'(0) = 0$, etc. For these situations the uniqueness theorem for ordinary differential equations requires that $T(t) \equiv 0$. We could have foreseen the result from the partial differential equation, since the order of the highest partial derivative with respect to t is also the order of the ordinary differential equation in t. These remarks about t apply, of course, to all the other variables involved.

There are many problems in which the above difficulty is encountered, but which can still be solved by separation of variables by modifying the system of equations. An example of such a problem is given in Exercise 7.7.2.

EXERCISES

7.7.1. Solve for $\psi(x, y, z)$ if

$$\begin{aligned}\nabla^2\psi &= 0, \quad 0 < x < 1,\ 0 < y < 1,\ 0 < z < 1,\\ \psi &= f(x, y) \text{ on surface } z = 0,\\ \psi &= 0 \text{ on other five faces of the cube.}\end{aligned}$$

7.7.2. Solve for $\psi(x, t)$ if $k\psi_{xx} = \psi_t$, $0 < x < 1$, $t > 0$, $\psi(x, 0) = \psi(0, t) = 0$, $\psi(1, t) = 1$. (*Hint*: Let $\psi(x, t) = U(x, t) + v(x)$ and determine $v(x)$ so that $U_t = kU_{xx}$, $U(0, t) = U(1, t) = 0$.)

7.7.3. Solve for $\psi(x, y, t)$ if $k\nabla^2\psi = \psi_t$, $t > 0$, $0 < x < a$, $0 < y < b$, $\psi(0, y, t) = \psi(a, y, t) = \psi(x, 0, t) = \psi(x, b, t) = 0$, $\psi(x, y, 0) = f(x, y)$.

7.7.4. (a) Show that if $y_\alpha = f(\alpha, x, t)$, $\alpha > 0$, is a solution of the system, $Ly = 0$, $M_i y = 0$, $i = 1, 2, 3, \ldots$, then

$$y(x, t) = \int_0^\infty g(\alpha) y_\alpha \, d\alpha$$

is also a solution, provided the operations L and M_i are interchangeable with the integration.

(b) Solve for $\psi(x, t)$ if $\psi(x, t)$ is bounded for $0 < x < \infty$, $t > 0$, and $k\psi_{xx}(x, t) = \psi_t(x, t)$, $\psi(0, t) = 0$, $\psi(x, 0) = f(x)$. (*Hint*: See Sec. 2.7.)

Ans. $\psi = (2/\pi)\int_0^\infty \int_0^\infty f(\xi)\, e^{-\alpha^2 kt} \sin \alpha\xi \sin \alpha x \, d\alpha \, d\xi$.

7.7.5. Use the known result,

$$\int_0^\infty e^{-a^2x^2} \cos bx \, dx = \frac{\sqrt{\pi}}{2a} e^{-b^2/4a^2}, \quad a > 0,$$

and show that the answer given in Exercise 7.7.4 is

$$\psi = \operatorname{erf}\left(\frac{x}{2\sqrt{kt}}\right).$$

7.7.6. Solve for $\psi(r, \theta, \phi)$ if $\nabla^2\psi = 0$, $\psi(1, \theta, \phi) = f(\theta, \phi)$, for $0 \le r \le 1$, $0 \le \theta \le \pi$, $0 \le \phi \le 2\pi$. The function $\psi(r, \theta, \phi)$ is the potential inside a unit sphere.

7.7.7. Show that the substitution

$$U(\phi, \theta, \psi) = e^{i(m\phi + n\psi)} F(\theta)$$

into

$$\csc^2 \theta \frac{\partial^2 U}{\partial \phi^2} + \frac{\partial^2 U}{\partial \theta^2} + \csc^2 \theta \frac{\partial^2 U}{\partial \psi^2} - 2 \cos \theta \csc^2 \theta \frac{\partial^2 U}{\partial \phi \, \partial \psi} + \cot \theta \frac{\partial U}{\partial \theta} = \lambda U$$

results in the separated equation

$$F''(\theta) + \cot \theta \, F'(\theta) - \left[\lambda + \frac{(m - n \cos \theta)^2}{\sin^2 \theta} + n^2 \right] F(\theta) = 0.$$

Note that this equation does not belong to the class of equations given by (7.7.2) but that it is separable by this method.

8

Partial Differential Equations of Mathematical Physics

8.1 HELMHOLTZ EQUATION

In the case of the wave equation,

$$\nabla^2 \psi = \frac{1}{c^2} \frac{\partial^2 \psi}{\partial t^2}, \tag{8.1.1}$$

and the diffusion equation,

$$\nabla^2 \psi = \frac{1}{k} \frac{\partial \psi}{\partial t}, \tag{8.1.2}$$

our separation-of-variables method may be begun by separating the space variables x,y,z from the time variable t. This could be carried out in each case by seeking a solution of the form

$$\psi(x, y, z, t) = W(x, y, z)T(t). \tag{8.1.3}$$

Substitution of this expression into (8.1.1) yields

$$\frac{\nabla^2 W}{W} = \frac{T''}{c^2 T} = -\lambda^2,$$

from which we have

$$\nabla^2 W + \lambda^2 W = 0, \tag{8.1.4}$$

and

$$T(t) = C_1 \cos c\lambda t + C_2 \sin c\lambda t. \tag{8.1.5}$$

Applying this method to Eq. (8.1.2) results in (8.1.4) again, and in

$$T(t) = C_3 e^{-\lambda^2 kt}. \tag{8.1.6}$$

Equation (8.1.4), known as the *Helmholtz* equation, is important to us since it is the differential equation which the space coordinates must satisfy in both the wave equation and the diffusion equation. When λ^2 is a real constant, not zero, (8.1.4) represents the wave equation for sinusoidal time dependence (8.1.5), and it also represents the diffusion equation for exponential time dependence, (8.1.6). If $\lambda = 0$, the Helmholtz equation becomes Laplace's equation. More generally, λ^2 may be a function of the space coordinates, and (8.1.4) would not then represent a separated part of another differential equation. Our primary interest in the Helmholtz equation will be in solving the diffusion and wave equations, in which cases λ^2 is a real constant.

If, in (8.1.4), $W = W(x)$ is a function of only one space coordinate, we have

$$W(x) = C_4 \cos \lambda x + C_5 \sin \lambda x. \tag{8.1.7}$$

When we multiply this expression by $T(t)$ as given in either (8.1.5) or (8.1.6), we have a solution $\psi(x, t)$ of the wave or diffusion equation.

If $W = W(x, y)$, the variables in (8.1.4) will separate upon letting $W = X(x)Y(y)$. For this case, (8.1.4) becomes

$$\frac{X''}{X} + \frac{Y''}{Y} = -\lambda^2,$$

and letting

$$\frac{Y''}{Y} = \mu^2,$$

where μ^2 is a second separation constant, we have as solutions

$$\begin{aligned} X(x) &= C_6 \cos \sqrt{\lambda^2 + \mu^2}\, x + C_7 \sin \sqrt{\lambda^2 + \mu^2}\, x, \\ Y(y) &= C_8 e^{\mu y} + C_9 e^{-\mu y}. \end{aligned} \tag{8.1.8}$$

Employing the same technique for the case $W(x, y, z) = X(x)Y(y)Z(z)$, we have the separated equations

$$\frac{Z''}{Z} = \nu^2, \quad \frac{Y''}{Y} = \mu^2, \quad \frac{X''}{X} = -(\lambda^2 + \mu^2 + \nu^2),$$

and the solutions

$$\begin{aligned} X(x) &= C_{10} \cos \sqrt{\lambda^2 + \mu^2 + \nu^2}\, x + C_{11} \sin \sqrt{\lambda^2 + \mu^2 + \nu^2}\, x, \\ Y(y) &= C_{12} e^{\mu y} + C_{13} e^{-\mu y}, \\ Z(z) &= C_{14} e^{\nu z} + C_{15} e^{-\nu z}. \end{aligned} \tag{8.1.9}$$

Thus we see that the Helmholtz equation with λ^2 a constant is separable in the cartesian coordinate system. It is also separable in cylindrical and spherical coordinates, as well as many others. For a more complete list of coordinate systems in which the Helmholtz equation is separable, see [MF], Chapter 5.

EXERCISES

8.1.1. Show that if $\psi = X(x)Y(y)Z(z)$ in the Helmholtz equation in rectangular coordinates, then for α^2, β^2 separation constants the separated equations are

$$Z'' + \alpha^2 Z = 0,\ Y'' + (\beta^2 - \alpha^2)Y = 0, \quad X'' + (\lambda^2 - \beta^2)X = 0.$$

8.1.2. Work Exercise 8.1.1 for cylindrical coordinates, using $\psi = R(r)\Theta(\theta)Z(z)$.
Ans. $Z'' - \alpha^2 Z = 0,\ \Theta'' + \beta^2\Theta = 0,\ R'' + (1/r)R' + (\alpha^2 + \lambda^2 - \beta^2/r^2)R = 0.$

8.1.3. Work Exercise 8.1.1 for spherical coordinates.
Ans. $(r^2R')' + \lambda^2 r^2 R - \alpha^2 R = 0, \quad [(\sin\theta)\Theta']' + \alpha^2(\sin\theta)\Theta - (\beta^2/\sin\theta)\Theta = 0,$
$\Phi'' + \beta^2\Phi = 0.$

8.2 WAVE EQUATION

One of the most important equations in mathematical physics is the equation

$$\nabla^2\psi = \frac{1}{c^2}\frac{\partial^2\psi}{\partial t^2}, \tag{8.2.1}$$

which is known as the *wave equation*. It is of fundamental importance because it governs the phenomenon of wave motion, which is so common in the world about us. The propagation of waves in space, sound waves, electromagnetic waves, tidal waves in an incompressible fluid, elastic waves in solids, vibrations of tightly stretched strings and membranes, longitudinal vibrations in a bar, and propagation of electric currents and voltages along a transmission line are all phenomena governed by this important equation.

In the next two sections we will show that the wave equation holds for some of these cases, and we will consider its solution, subject to given boundary conditions, by the method of separation of variables.

8.3 VIBRATING STRING

To introduce the wave equation, we consider the vibrations of a tightly stretched uniform string of density δ (mass per unit length) whose equilibrium position is along the x-axis and whose transverse displacements are given by $\psi(x, t)$. The conditions we assume are

(1) The motion is in the x-ψ plane in the ψ direction, and the deflections ψ and slopes $\partial\psi/\partial x$ are small.
(2) The tension T is constant and is so great that the weight of the string and other external forces can be neglected by comparison.
(3) The string is perfectly flexible.

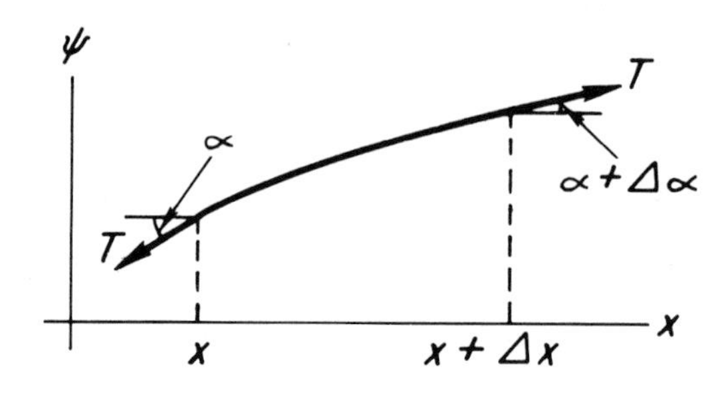

Fig. 8–1

At a given time t consider the vertical forces acting on the increment of length shown in Fig. 8–1. The net force in the vertical direction is

$$F = T \sin (\alpha + \Delta\alpha) - T \sin \alpha,$$

which, because the displacements are small, may be approximated by

$$F = T\psi_x(x + \Delta x, t) - T\psi_x(x, t).$$

By use of the mean-value theorem we may write

$$F = T\psi_{xx}(x + \theta\, \Delta x, t)\, \Delta x, \quad 0 < \theta < 1.$$

Application of Newton's second law of motion to the increment, its length considered to be approximately Δx, results in the approximation,

$$\delta\, \Delta x\, \psi_{tt}(x, t) \doteq T\, \psi_{xx}(x + \theta\, \Delta x, t)\, \Delta x. \tag{8.3.1}$$

The approximation improves as Δx becomes smaller, so that upon dividing by $T\, \Delta x$ and passing to the limit as $\Delta x \to 0$, we have the equation

$$\psi_{xx}(x, t) = \frac{1}{c^2}\, \psi_{tt}(x, t), \tag{8.3.2}$$

which is the wave equation (8.2.1) in one space coordinate. The constant $1/c^2$ is equal to δ/T.

If the tension is not constant, but is given by $T = T(x)$, then (8.3.1) would be replaced by

$$\delta\, \Delta x \psi_{tt}(x, t) \doteq \left\{ \frac{\partial}{\partial x}\, (T\psi_x) \bigg|_{(x+\theta\Delta x, t)} \right\} \Delta x$$

and (8.3.2) would be replaced by

$$\delta\psi_{tt}(x, t) = \frac{\partial}{\partial x}\, (T\psi_x). \tag{8.3.3}$$

EXERCISES

8.3.1. Find the displacement of a vibrating string with ends fixed at $x = 0$ and $x = L$, initially at rest, and with initial displacement given by $f(x)$.

8.3.2. Find the displacement of the string in Exercise 8.3.1 if it has an initial velocity $g(x)$ and is initially in its equilibrium position along the x-axis.

8.3.3. Solve the problem of the vibrating string with ends fixed at 0 and L, with initial displacement $f(x)$ and with initial velocity $g(x)$.

8.3.4. Show that for L, M_1, M_2 linear operators, the problem $L\psi = 0$, $M_1\psi = f$, $M_2\psi = g$ can be solved as $\psi = \psi_1 + \psi_2$, where $L\psi_1 = 0$, $M_1\psi_1 = f$, $M_2\psi_1 = 0$, and $L\psi_2 = 0$, $M_1\psi_2 = 0$, $M_2\psi_2 = g$. Use this idea of superposition to solve Exercise 8.3.3, using the results of Exercises 8.3.1 and 8.3.2.

8.3.5. Solve Exercise 8.3.2 if $g(x) = v_0 \sin(\pi x/L) + v_1 \sin(3\pi x/L)$, where v_0, v_1 are constants.

8.3.6. A tightly stretched string fixed at $x = 0$ and $x = 3$ is plucked at $x = 1$ and is initially in the shape of a triangle of height h. If it is released from rest in this position, find its displacement.

Ans. $\psi(x, t) = \sum_{n=1}^{\infty} b_n \cos(n\pi ct/3) \sin(n\pi x/3)$, $b_n = (9h/n^2\pi^2) \sin(n\pi/3)$.

8.3.7. Modify Eq. (8.3.1) to consider the case where the weight of the string is not negligible, and show that Eq. (8.3.2) becomes

$$\psi_{tt} = c^2\psi_{xx} - g,$$

where g is the acceleration of gravity. Solve this equation subject to the conditions $\psi(0, t) = \psi(L, t) = \psi(x, 0) = \psi_t(x, 0) = 0$, by assuming a solution of the form $\psi(x, t) = u(x, t) + v(x)$, where $u(x, t)$ satisfies the homogeneous equation $\psi_{tt} = c^2\psi_{xx}$ and $v(x)$ satisfies the non-homogeneous equation. We are at liberty to impose conditions on the two extra arbitrary constants occurring in $v(x)$, so long as the given conditions are satisfied. It will be seen that taking the conditions $v(0) = v(L) = 0$ simplifies the procedure.

8.3.8. Derive the equation for the tightly stretched string, with negligible weight, vibrating in a viscous medium. (An external damping force, F per unit length, is present, proportional to the velocity. Take $F = -2k\delta\psi_t$.) Solve for the displacement if the string is fixed at $x = 0$ and $x = L$, the initial displacement is $f(x)$, and initial velocity is $g(x)$. Ans. $\psi_{tt} = c^2\psi_{xx} - 2k\psi_t$.

8.3.9. Consider a cable hanging vertically with end $x = L$ fixed and end $x = 0$ free, so that the tension is not constant but varies with x. Derive its equation of motion, $x\psi_{xx} + \psi_x = (1/g)\psi_{tt}$, where $\psi(x, t)$ is the horizontal displacement from the vertical equilibrium position. Find $\psi(x, t)$ if the cable is initially at rest in a position $\psi(x, 0) = f(x)$.

8.3.10. Show that if the density is not constant, but that $\delta(x) = T\rho(x)$, with the other conditions remaining the same, the equation corresponding to (8.3.2) is

$$\rho(x)\psi_{tt} = \psi_{xx}.$$

Find the displacement of a string fixed at $x = 0$ and $x = L$ satisfying this equation for $\rho(x) = x$. The string is initially in the form $\psi(x, 0) = f(x)$, and released from rest. (*Hint*: See Exercise 6.2.2.)

Ans. $\psi(x, t) = \sum_{n=1}^{\infty} C_n \sqrt{x}\, J_{1/3}(\frac{2}{3}\lambda_n x^{3/2}) \cos \lambda_n t$, where $\frac{2}{3}\lambda_n L^{3/2} = \alpha_n$, $n = 1, 2, \ldots,$ are roots of $J_{1/3}(\alpha) = 0$, and $C_n = [2/\sqrt{L}\, J_{4/3}^2(\alpha_n)] \int_0^1 z^{2/3} f(z^{2/3}L) J_{1/3}(\alpha_n z)\, dz$.

8.3.11. Let the density and tension both vary in Eq. (8.3.3), say $\delta = \rho(x)$ and $T = f(x)$, and show that the differential equation for $X(x)$, where $\psi = X(x)T(t)$, satisfies the self-adjoint Sturm-Liouville equation,

$$[f(x)X'(x)]' + \lambda\rho(x)X(x) = 0.$$

Find values of $f(x)$ and $\rho(x)$ so that $X(x)$ is a solution of (a) Legendre's differential equation, (b) Bessel's, (c) Laguerre's, (d) Hermite's, and (e) Chebyshev's. See Exercises 1.9.4, 1.9.5, and 3.1.5.

8.3.12. The *transmission-line equations* are known to be (see for example [Ka], page 5)

$$-\frac{\partial e}{\partial x} = iR + L\frac{\partial i}{\partial t}, \qquad -\frac{\partial i}{\partial x} = eG + C\frac{\partial e}{\partial t},$$

where $e(x, t)$ is the line voltage at a point x (measured in miles) and time t, $i(x, t)$ is the line current, R = series resistance per loop mile, L = series inductance per loop mile, C = shunt capacitance per mile, and G = shunt conductance per mile. Show that both i and e satisfy the equation

$$LC\psi_{tt} + (LG + RC)\psi_t + RG\psi = \psi_{xx}.$$

For high frequencies we may assume that R and G are negligible. Note that for this case both i and e satisfy the wave equation.

8.3.13. Find the line voltage for a high-frequency transmission line of length c if there is initially no leakage current $[e_t(x, 0) = 0]$, the voltages at both sending and receiving ends are zero, and the initial line voltage is $e(x, 0) = f(x)$.

8.3.14. Solve the system

$$\nabla^2\psi(r, \theta, \phi, t) = \frac{1}{c^2}\frac{\partial^2\psi}{\partial t^2},$$

$$\psi(r, \theta, \phi, 0) = f(r, \theta, \phi),$$

$$\psi_t(r, \theta, \phi, 0) = 0,$$

$$\psi(1, \theta, \phi, t) = 0.$$

The function ψ is to be bounded, single-valued, and continuous inside the unit sphere. Suggestion: Let $R = r^{-1/2}y$ in the equation involving r.

8.3.15. Solve Exercise 8.3.14 if the region is to be that between two concentric spheres, $a < r < b$. The condition $\psi(1, \theta, \phi, t) = 0$ is to be replaced by $\psi = 0$ when $r = a$ and $r = b$.

8.3.16. The longitudinal displacements $\psi(x, t)$ in the x direction of an elastic bar satisfy the wave equation (see [C-1], page 7) with $c^2 = E/\delta$, E being the modulus of elasticity and δ being the mass per unit volume. The displacements must be small with no external longitudinal forces acting other than at the ends. The internal force exerted at the point x by the portion of the bar to the left of x is $F(x, t) = -AE\psi_x(x, t)$, where A is the cross-sectional area of the bar. Find $\psi(x, t)$ if the bar of length L is free at the two ends, is initially at rest, and $\psi(x, 0) = kx$.

8.4 VIBRATING MEMBRANE

As a second illustration of the wave equation we consider the transverse motion of a vibrating membrane, an increment of which is shown in Fig. 8–2.

We assume that the force T per unit length is constant, normal to the boundary, and so great that the weight of the membrane and other external forces can be neglected by comparison. We also assume that the membrane is of uniform density δ per unit area and that the displacements ψ (perpendicular to the x-y plane) and the slopes are small.

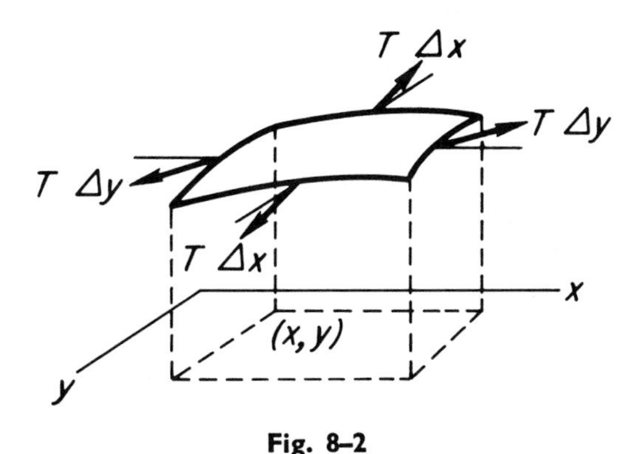

Fig. 8–2

The net vertical force acting on the increment is then approximately

$$F \doteq T[\Delta y\, \psi_x(x + \Delta x, y, t) - \Delta y\, \psi_x(x, y, t) + \Delta x\, \psi_y(x, y + \Delta y, t) - \Delta x\, \psi_y(x, y, t)].$$

Equating this force to

$$\delta\, \Delta x\, \Delta y\, \psi_{tt}(x, y, t),$$

dividing by $T\, \Delta x\, \Delta y$, and taking the limit as $\Delta x \to 0$ and $\Delta y \to 0$, we have the wave equation for two space coordinates,

$$\psi_{xx}(x, t) + \psi_{yy}(x, t) = \frac{1}{c^2}\, \psi_{tt}(x, t), \tag{8.4.1}$$

where again $1/c^2 = \delta/T$.

As an example of a boundary-value problem involving the wave equation, let us find the deflections $\psi(x, y, t)$ of a membrane fastened to a rectangular frame, $x = 0$, $x = a$, $y = 0$, $y = b$, initially displaced in a position $f(x, y)$ and released from rest. The mathematical statement of the problem is then

$$\begin{aligned}
\nabla^2\psi &= \frac{1}{c^2}\, \psi_{tt}, && 0 < x < a,\ 0 < y < b,\ t > 0,\\
\psi(0, y, t) &= \psi(a, y, t) = 0, && 0 < y < b,\ t > 0,\\
\psi(x, 0, t) &= \psi(x, b, t) = 0, && 0 < x < a,\ t > 0,\\
\psi_t(x, y, 0) &= 0, && 0 < x < a,\ 0 < y < b,\\
\psi(x, y, 0) &= f(x, y), && 0 < x < a,\ 0 < y < b.
\end{aligned} \tag{8.4.2}$$

The solution of the problem is left as an exercise.

EXERCISES

8.4.1. Solve the system given in (8.4.2).

Ans. $\psi(x, y, t) = \sum_n \sum_m A_{nm} \cos \omega_{nm} t \sin (n\pi x/a) \sin (m\pi y/b)$,
where $A_{nm} = (4/ab) \int_0^a \int_0^b f(x, y) \sin (n\pi x/a) \sin (m\pi y/b)\, dy\, dx$;
and $\omega_{nm} = c\pi\sqrt{n^2/a^2 + m^2/b^2}$; $n, m = 1, 2, 3, \ldots$.

8.4.2. Find the transverse displacements of a tightly stretched membrane of negligible weight fastened to a circular frame of radius a if the initial displacement is $f(r)$, a function only of the distance from the center, and the membrane is initially at rest. For this case the displacement is a function of r and t only.

8.4.3. Solve Exercise 8.4.2 if the initial displacement is $g(r, \theta)$, where r and θ are the polar coordinates in the plane of the frame.

8.4.4. Find the displacements of the membrane in Exercise 8.4.2 if the initial velocity is $v(r)$ and it is originally in its equilibrium position.

8.4.5. Solve Exercise 8.4.1 if $a = b = c = 1$ and $f(x, y) = \sin 3\pi x \sin 7\pi y$.

8.4.6. If the membrane in Exercise 8.4.2 is viscously damped, show that the equation of motion is

$$c^2\nabla^2\psi(r, t) = \frac{\partial^2\psi}{\partial t^2} + k\frac{\partial\psi}{\partial t},$$

where k is a positive constant. Solve for $\psi(r, t)$ under the conditions given in Exercise 8.4.2.

8.5 DIFFUSION EQUATION

Perhaps the best example of this important equation is in its application to the conduction of heat. In this section we will give a heuristic derivation to show that the conduction of heat in an isotropic medium (one in which the heat flows equally well in all directions) is governed by the diffusion equation. A more rigorous treatment may be found in [C-1], pages 10–14.

Let $U(x, y, z, t)$ be the temperature at a point (x, y, z) on a surface S in the medium at time t, and $Q(x, y, z, t)$ be the quantity of heat flowing through the surface per unit time. The quantity Q may be expressed in Btu per second or calories per second. Through a small slab of face area ΔA, and thickness Δn (measured normal to ΔA), it is known that the heat flow per second is directly proportional to ΔA and to the difference in temperature, ΔU, of the two faces, and inversely proportional to the thickness Δn. Thus if ΔQ is the rate of flow, then we have

$$\Delta Q = -K\,\Delta A\,\frac{\Delta U}{\Delta n}. \tag{8.5.1}$$

The constant K is known as the thermal conductivity, dependent on the composition of the medium, and the minus sign indicates that the flow is from the higher temperature to the lower. In other words the flow is considered positive if $\Delta U < 0$ for $\Delta n > 0$.

The ratio $\Delta Q/\Delta A$ is the rate of flow per unit area, and if we shrink the slab to a point by letting $\Delta A \to 0$ and $\Delta n \to 0$, this ratio becomes the *flux* $\phi(x, y, z, t)$ conducted through the surface at the point (x, y, z) at time t. The flux is then, by (8.5.1),

$$\phi(x, y, z, t) = \frac{\partial Q}{\partial A} = -K\frac{\partial U}{\partial n}. \tag{8.5.2}$$

This result is known as Fourier's conduction law and shows that the flux is positive when the directional derivative of U, or the *temperature gradient*, is negative. Thus we have a positive flow from a high to a low temperature.

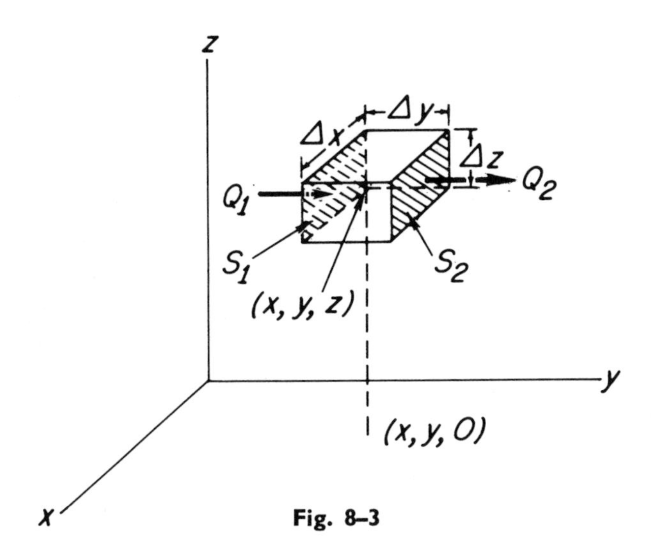

Fig. 8–3

From (8.5.2) we may obtain an expression for the total flow per second through the surface S in the form

$$Q = -\int_S K\frac{\partial U}{\partial n}\,dA. \tag{8.5.3}$$

Let us now consider a heat-conducting solid referred to the x,y,z-coordinate system and develop the equation describing its distribution of temperature. At an interior point (x, y, z) of the solid we will construct an element of volume, $\Delta V = \Delta x\,\Delta y\,\Delta z$, as shown in Fig. 8–3. The net heat flow, $Q_1 - Q_2$, in the y direction may be obtained approximately by the use of Eq. (8.5.3). If $\Delta x\,\Delta z$ is small, we may consider $\partial U/\partial n$, which for this case is $\partial U/\partial y$, to be virtually constant over each of the two faces S_1 and S_2.

Thus the net flow in the y direction is approximately, by (8.5.3),

$$Q_1 - Q_2 \doteq K\,\Delta x\,\Delta z\,[U_y(x, y + \Delta y, z, t) - U_y(x, y, z, t)]$$
$$= KU_{yy}(x, y + \theta_2\,\Delta y, z, t)\,\Delta V.$$

Similar expressions may be obtained in the x and z directions, and the net heat flow in all directions is the sum of the three expressions:

$$\text{Net flow} \doteq K[U_{xx}(x + \theta_1\,\Delta x, y, z, t) + U_{yy}(x, y + \theta_2\,\Delta y, z, t) + U_{zz}(x, y, z + \theta_3\,\Delta z, t)]\,\Delta V, \quad (8.5.4)$$
$$0 < \theta_i < 1, \quad i = 1, 2, 3.$$

Another expression for the net flow (say Btu/sec), as the units indicate, is

$$\text{Net flow} \doteq \delta C_p\,\Delta V\,\frac{\partial U}{\partial t}, \quad (8.5.5)$$

where δ is the density of the solid in mass per unit volume, C_p is the specific heat at constant pressure in Btu per unit mass per degree, and $\partial U/\partial t$ is the rate of change of temperature (assumed constant over the small volume) in degrees per second.

Equating the two values of net flow in (8.5.4) and (8.5.5) and dividing by $K\,\Delta V$, we have

$$U_{xx}(x + \theta_1\,\Delta x, y, z, t) + U_{yy}(x, y + \theta_2\,\Delta y, z, t) + U_{zz}(x, y, z + \theta_3\,\Delta z, t) \doteq \frac{\delta C_p}{K}\frac{\partial U}{\partial t}. \quad (8.5.6)$$

The approximation in (8.5.6) improves as the element of volume shrinks. Taking the limit as $\Delta V \to 0$, we have the equation for the temperature distribution,

$$\nabla^2 U(x, y, z, t) = \frac{1}{k}\frac{\partial U}{\partial t}, \quad (8.5.7)$$

which is the diffusion equation. We have assumed that the solid is homogeneous (δ constant), so that $k = K/\delta C_p$ is constant.

As an example of a heat-flow problem, we will consider the temperature distribution in a cylindrical rod of length L whose lateral surface is impervious to heat; that is, the heat flow is in one direction, along L, which we take as the x-axis. Let the two ends of the rod be kept at $0°$, and the initial temperature distribution be a given function $f(x)$. Then if the temperature $U(x, t)$ is bounded, it will be a solution of the boundary-value problem,

$$U_t(x, t) = kU_{xx}(x, t), \quad 0 < x < L, t > 0,$$
$$U(0, t) = U(L, t) = 0, \quad (8.5.8)$$
$$U(x, 0) = f(x), |U(x, t)| < \infty.$$

EXERCISES

8.5.1. Solve the system (8.5.8).

Ans. $U(x, t) = \sum_{n=1}^{\infty} C_n e^{-n^2\pi^2 kt/L^2} \sin (n\pi x/L)$,
where $C_n = (2/L)\int_0^L f(x) \sin (n\pi x/L)\, dx$, $n = 1, 2, 3, \ldots$.

8.5.2. For the rod in Exercise 8.5.1, instead of the temperatures being fixed at the two ends we have the following conditions. Find $U(x, t)$ in each case.

(a) The two ends are insulated; that is, the flux is zero at each end.

(b) The end $x = 0$ is kept at $0°$, and the end $x = L$ is insulated.

(c) The end $x = 0$ is insulated, and the end $x = L$ is kept at $0°$.

8.5.3. The function (see [WW], Chapter 21)

$$\vartheta_3(z, q) = 1 + 2 \sum_{n=1}^{\infty} q^{n^2} \cos 2nz$$

is known as a *Jacobi theta function*. Using the identity

$$2 \sin A \sin B = \cos (A - B) - \cos (A + B),$$

show that the solution in Exercise 8.5.1 may be written

$$U(x, t) = \int_0^L K(x, \xi) f(\xi)\, d\xi,$$

where

$$K(x, \xi) = \frac{1}{2L}\left\{\vartheta_3\left(\frac{\pi(x - \xi)}{2L}, p^4\right) - \vartheta_3\left(\frac{\pi(x + \xi)}{2L}, p^4\right)\right\}$$

and

$$p^4 = \exp \frac{-\pi^2 kt}{L^2}.$$

8.5.4. Give the solutions in Exercise 8.5.2 in terms of theta functions.

Ans. (b) $$K(x, \xi) = \frac{1}{2L}\left\{\vartheta_3\left(\frac{\pi(x - \xi)}{4L}, p\right) - \vartheta_3\left(\frac{\pi(x + \xi)}{4L}, p\right) + \vartheta_3\left(\frac{\pi(x + \xi - 2L)}{4L}, p\right) - \vartheta_3\left(\frac{\pi(x - \xi - 2L)}{4L}, p\right)\right\},$$

with $U(x, t)$ and p given in Exercise 8.5.3.

8.5.5. Find the temperature distribution of the rod in Exercise 8.5.1 if the temperature is initially zero for $0 < x < L$ and the ends are maintained at constant temperatures, $U(0, t) = a$, $U(L, t) = b$.

8.5.6. Solve Exercise 8.5.5 if $U(x, 0) = f(x)$ instead of zero.

8.5.7. Newton's law of cooling says (cf. [CJ], pages 18–19) that the flux across a surface is proportional to the temperature difference between the surface and the surrounding medium; that is, $\partial U/\partial n + h(U - U_0) = 0$. Find $U(x, t)$ for the rod in the previous exercises if heat is being radiated at the end $x = L$ into a medium of zero temperature. Let the end $x = 0$ be fixed at $0°$, and let the initial temperature be $f(x)$.

8.5.8. Solve Exercise 8.5.7 if instead of $U(0, t) = 0$ we have heat radiating into a medium of 0° at the end $x = 0$.

8.5.9. Find the temperature distribution of a thin circular plate of radius a if it depends only on the distance r from the center of the plate and the time, if the temperature on the circumference is zero, and the initial distribution is $f(r)$.

8.5.10. Solve Exercise 8.5.9 if the condition of zero temperature on the circumference is replaced by the condition that heat is radiating normal to the circumference into a medium of zero temperature. (See Exercise 6.4.11.)

8.5.11. Solve Exercise 8.5.9 if the condition of zero temperature on the circumference is replaced by the condition that the circumference is insulated.

8.5.12. Find the temperature distribution between two concentric infinite cylinders whose axes are along the z-axis and whose base is the region in the x-y plane $a < r < b$. The boundary conditions are

$$\psi(a, t) = \psi(b, t) = 0, \quad \psi(r, 0) = f(r).$$

$$\text{Ans.} \quad \psi = \frac{\pi^2}{2} \sum_{n=1}^{\infty} \left[\frac{\alpha_n^2 J_0^2(a\alpha_n) e^{-k\alpha_n^2 t}}{J_0^2(a\alpha_n) - J_0^2(b\alpha_n)} \int_a^b r f(r) U_0(r\alpha_n)\, dr \right] U_0(r\alpha_n),$$

$$\text{where} \quad U_0(\alpha r) = J_0(\alpha r) Y_0(\alpha b) - J_0(\alpha b) Y_0(\alpha r),$$
$$\text{and} \quad U_0(\alpha_n a) = 0, \quad n = 1, 2, 3, \ldots .$$

8.5.13. Find the temperature distribution of a thin rectangular plate $0 \le x \le a$, $0 \le y \le b$, with the conditions $U(0, y, t) = U(a, y, t) = U(x, 0, t) = U(x, b, t) = 0$, and an initial distribution $f(x, y)$.

8.5.14. Find the temperature distribution ψ of a thin wedge bounded by $\theta = 0$, $\theta = 2\alpha$, and $r = a$, subject to the conditions, $\psi(r, 0, t) = \psi(r, 2\alpha, t) = 0$, the curved boundary is insulated, and initially $\psi = f(r, \theta)$.

8.5.15. Find the temperature distribution ψ inside the unit sphere $r = 1$, if $\psi(r, \theta, \phi, 0) = f(r, \theta, \phi)$, $\psi(1, \theta, \phi, t) = 0$.

8.5.16. Solve Exercise 8.5.15 if ψ is independent of ϕ and the initial distribution is $g(r, \theta)$.

8.5.17. Solve for $u(x, t)$ if

$$u_t(x, t) = u_{xx}(x, t), \quad -\infty < x < \infty,\ t > 0,$$
$$u(x, 0) = f(x),$$
$$u(x, t) \text{ bounded.}$$

8.6 LAPLACE'S EQUATION

In this section we will consider the second-order equation

$$\nabla^2 \psi = 0, \tag{8.6.1}$$

which is perhaps the most important differential equation in mathematical physics, and the generalization of it,

$$\nabla^2 \psi = f. \tag{8.6.2}$$

These are known respectively as *Laplace's equation* and *Poisson's equation.* The first equation is a special case of the wave and diffusion equations when ψ is independent of time, and therefore it governs steady-state heat flow and static deflections of loaded strings and membranes, as well as many other time-independent phenomena.

To introduce Laplace's equation, we will consider its application to electrostatic potential. By Coulomb's law, the force exerted on a unit point charge at a point $P(x, y, z)$ due to a charge q located at a point $Q(\xi, \eta, \zeta)$ is given by

$$F = K\frac{q}{r^2},$$

where K is a constant of proportionality and r is the distance between P and Q. The *potential* V at P is defined as the negative of the work done in moving a unit charge from infinity to the point P. If there is no other charge except q present, the potential at P will be given by

$$V = -\int_{\infty}^{r} \frac{Kq}{R^2}\, dR,$$

where R is the distance between the position of the point charge and the point Q. Thus we have

$$V = \frac{Kq}{r}. \tag{8.6.3}$$

If the potential at P is due to a set of point charges q_i located at discrete points Q_i, then we have

$$V = \sum_i \frac{Kq_i}{r_i}. \tag{8.6.4}$$

By generalizing still further, if the potential is due to a continuous distribution of charge over a region R not including the point P, then

$$V = \iiint_R \frac{K\rho\, dV}{r} = \iiint_R \frac{K\rho(\xi, \eta, \zeta)\, d\xi\, d\eta\, d\zeta}{\sqrt{(x - \xi)^2 + (y - \eta)^2 + (z - \zeta)^2}}, \tag{8.6.5}$$

where the integration is taken over the region R, and ρ is the charge density (charge per unit volume).

By applying the operator ∇^2 to V given in each of the previous three equations, we have

$$\nabla^2 V = 0. \tag{8.6.6}$$

For example, in the case of (8.6.5), we have

$$\frac{\partial V}{\partial x} = -\iiint_R K\rho\frac{(x - \xi)\, dV}{r^3}$$

and

$$\frac{\partial^2 V}{\partial x^2} = -\iiint_R K\rho\left[\frac{1}{r^3} - \frac{3(x-\xi)^2}{r^5}\right] dV.$$

Computing the other second derivatives and adding, we have

$$\nabla^2 V = -\iiint_R K\rho\left[\frac{3}{r^3} - \frac{3}{r^3}\right] dV = 0.$$

The foregoing operations are valid if the first partial derivatives are continuous in the closed region R and the resulting integral, when improper, converges uniformly. (See [Bu], pages 71 and 153.)

If the point P is inside the region R, the potential is known (see, for example, [CL], pages 30–35) to satisfy Poisson's equation,

$$\nabla^2 V = -\frac{\rho}{\epsilon_0}, \tag{8.6.7}$$

where ϵ_0 is the *permittivity* of the medium. Laplace's equation in electrostatics is the special case of Poisson's where the density ρ is zero.

EXERCISES

8.6.1. Find the steady-state temperature in a rod, $0 < x < 1$, if the temperatures at $x = 0$ and $x = 1$ are held at 0° and 10° respectively.

8.6.2. Find the potential distribution inside the rectangle (no charges inside), $0 < x < a$, $0 < y < b$, if $V(x, 0) = V(x, b) = V(0, y) = 0$, $V(a, y) = V_1$, a constant.

8.6.3. Solve Exercise 8.6.2 if the potential is zero on all sides except that $V(0, y) = V_2$, a constant.

8.6.4. Solve Exercise 8.6.2 if the potential is zero on the two sides $y = 0$ and $y = b$, and $V(a, y) = V_1$, $V(0, y) = V_2$. (Use superposition.)

8.6.5. Find the potential inside the rectangle of Exercise 8.6.2 if $V(0, y) = V_3$, $V(x, b) = -V_4$, for $0 < x < \frac{1}{2}a$, $V(x, b) = V_4$, $\frac{1}{2}a < x < a$, $V(x, 0) = V(a, y) = 0$.

8.6.6. Find the potential inside the rectangle of Exercise 8.6.2 if $V = f(x)$ on face $y = b$, $\partial V/\partial y = g(x)$ on $y = 0$ and is zero on the other two faces.

8.6.7. Show that the static transverse displacements $\psi(x, y)$ of a membrane subject to a constant transverse force per unit area satisfies Poisson's equation,

$$\nabla^2 \psi = -K.$$

If the membrane is fixed on a square frame defined by $x = 0$, $x = \pi$, $y = 0$, $y = \pi$, find $\psi(x, y)$. (See Exercise 8.3.7.)

8.6.8. Determine the potential $V(r, \theta)$ inside a long cylinder (no variation with z) parallel to the z-axis with a circular base of radius a in the x-y plane. Consider that there is no charge inside and that $V(a, \theta) = 0$, $0 < \theta < \pi$, $V(a, \theta) = V_0$, $\pi < \theta < 2\pi$.

8.6.9. Find the potential inside the annular region between $r = a$ and $r = b > a$ if there are no charges present, $V(a, \theta)$ is the same as in Exercise 8.6.8, and $V(b, \theta) = V_1$.

8.6.10. Find the potential $V(r, z)$ inside the cylinder $0 < r < a$, $0 < z < b$ if there are no charges present, $V(a, z) = V_0$, and $V(r, 0) = V(r, b) = 0$.

8.6.11. Solve for $\psi(x, y)$ if $\nabla^2\psi = 0$, $\psi(x, 0) = \psi(x, b) = 0$, $\psi(0, y) = \psi_1$, $\psi(a, y) = \psi_2$; ψ_1, ψ_2 constants. Do not use superposition.

8.6.12. Find the steady-state temperature distribution inside a cube of edges π if one face is kept at a temperature ψ_0 and the other faces are at 0°.

8.6.13. Find the potential at (r, θ, ϕ) due to a thin ring of total charge Q, radius a, lying in the x-y plane with center at the origin. (*Hint*: Determine the coefficients by noting that the potential at a point on the z-axis is $Q/\sqrt{a^2 + r^2}$. Expand this by the binomial theorem for the two cases $r < a$ and $r > a$ and compare coefficients obtained with those in the Fourier series.

Ans. For $r > a$, $V = (Q/a)[(a/r)\, P_0 (\cos \theta) - (a^3/2r^3)\, P_2 (\cos \theta) + (3a^5/2 \cdot 4r^5)\, P_4 (\cos \theta) + \cdots]$.

8.6.14. Find the potential inside a sphere of radius a if there are no charges within and the potential on the sphere is given by $f(\theta, \phi)$.

8.6.15. Find the steady-state temperature in a solid hemisphere $0 \le r \le 1$, $0 \le \theta \le \frac{1}{2}\pi$, if the base is kept at 0° and the hemispherical surface is kept at ψ_0.

8.6.16. The faces of a parallelepiped, $0 \le x \le a$, $0 \le y \le b$, $0 \le z \le c$, are held at a potential $V = 0$, and the interior is filled with charge with a density given by

$$\rho = \sin \frac{\pi x}{a} \sin \frac{\pi z}{c} [y(y - b)].$$

Find a solution for the potential inside. Suggestion: Assume that V can be represented by a three-dimensional Fourier series,

$$V = \sum_{n=1}^{\infty} \sum_{m=1}^{\infty} \sum_{k=1}^{\infty} C_{nmk} \sin \frac{n\pi x}{a} \sin \frac{m\pi y}{b} \sin \frac{k\pi z}{c},$$

and equate this to the similar Fourier series for ρ.

8.6.17. Find the potential V inside the thin wedge bounded by $\theta = 0$, $\theta = 2\alpha$, $r = a$, if there are no charges inside, $V(r, 0) = V(r, 2\alpha) = 0$, $V(a, \theta) = f(\theta)$, where

$$f(\theta) = 0, \quad 0 < \theta < \alpha$$
$$= V_0, \quad \alpha < \theta < 2\alpha.$$

Ans. $V(r, \theta) = \sum_{n=1}^{\infty} A_n r^{\pi n/2\alpha} \sin(n\pi\theta/2\alpha)$,

$A_n = (2V_0/n\pi)[\cos (n\pi/2) - (-1)^n]a^{-n\pi/2\alpha}$.

8.6.18. (a) Find the potential inside an infinite cylinder with axis along the z-axis and base a unit circle in the x-y plane. There are no charges inside, and the lateral surface is kept at a potential $f(\theta) = \cos\theta$. Ans. $V = x$.
(b) Find the potential outside the cylinder. Ans. $V = (\cos\theta)/r$.

8.6.19. Find the potential $V(x, y)$ inside the semi-infinite strip, $0 < x < \infty$, $0 < y < 1$, if there are no charges inside, $V_x(0, y) = V_y(x, 0) = 0$, $V(x, y)$ is bounded, and

$$V(x, 1) = 1, \quad 0 < x < 1$$
$$= 0, \quad x > 1.$$

Ans. $V = (2/\pi)\int_0^\infty [(\sin\alpha \cos\alpha x \cosh\alpha y)/(\alpha\cosh\alpha)]\,d\alpha$.

9

Hermite Polynomials

9.1 DEFINITION

In our discussion of the Bessel and Legendre functions we began with their differential equations as a means of defining them. In this section, to illustrate a different approach, we define the *Hermite polynomials* by means of the Rodrigues' formula,

$$H_n(x) = (-1)^n e^{x^2} D^n e^{-x^2}. \tag{9.1.1}$$

In the notation of Sec. 1.11, for a differential equation of the form

$$f(x)y'' + g(x)y' + \left[\frac{n}{2}(1-n)f'' - ng'\right]y = 0, \tag{9.1.2}$$

f and g polynomials of degrees less than or equal to 2 and 1 respectively, the Rodrigues' formula for the polynomial solution is

$$y_n = C_n f \exp\left(-\int \frac{g\,dx}{f}\right) D^n\left[f^{n-1} \exp\left(\int \frac{g\,dx}{f}\right)\right]. \tag{9.1.3}$$

Clearly, $H_n(x)$ given by (9.1.1) is a polynomial, so that to obtain its differential equation we have only to find f and g by comparing (9.1.1) and (9.1.3). This results in

$$e^{x^2} = (-1)^n C_n f \exp\left(-\int \frac{g\,dx}{f}\right) = f^{1-n} \exp\left(-\int \frac{g\,dx}{f}\right), \tag{9.1.4}$$

and since the multiplier C_n in (9.1.3) will not affect the differential equation for y_n, we take it as $(-1)^n$. Then (9.1.4) has the solutions $f = 1$ and $g = -2x$ and the differential equation (9.1.2) for $H_n(x)$ becomes

$$y'' - 2xy' + 2ny = 0. \tag{9.1.5}$$

This is called *Hermite's differential equation.*

9.2 GENERATING FUNCTION

Recalling that the generating function $g(x, t)$ of a set $\{\phi_n(x)\}$ is merely a Maclaurin expansion with the coefficients given by

$$\phi_n(x) = \frac{1}{n!} \frac{\partial^n g(x, t)}{\partial t^n}\bigg|_{t=0}, \tag{9.2.1}$$

and noting the similarity between this expression and $H_n(x)$, we see that to obtain $g(x, t)$ we need to express $D^n e^{-x^2}$ in terms of $(\partial^n g/\partial t^n)|_{t=0}$. We note that one way of accomplishing this is to write

$$D^n e^{-x^2} = \frac{d^n}{dz^n} e^{-z^2}\bigg|_{z=x} = (-1)^n \frac{\partial^n}{\partial t^n} e^{-(x-t)^2}\bigg|_{t=0}, \tag{9.2.2}$$

where we have let $z = x - t$ in the last member.

Substituting (9.2.2) into the Rodrigues' formula, we have

$$\frac{H_n(x)}{n!} = \frac{1}{n!} \frac{\partial^n}{\partial t^n} e^{2xt-t^2}\bigg|_{t=0},$$

and by comparison with (9.2.1) we have the generating function

$$g(x, t) = \exp(2xt - t^2) = \sum_{n=0}^{\infty} \frac{H_n(x)}{n!} t^n. \tag{9.2.3}$$

We may use the generating function to exhibit $H_n(x)$ in summation form by expanding it in a Maclaurin series and equating coefficients of t^n. We begin by writing

$$\exp(2xt - t^2) = \sum_{n=0}^{\infty} \frac{(2x)^n t^n}{n!} \sum_{n=0}^{\infty} \frac{(-1)^n t^{2n}}{n!}.$$

Using the Cauchy product, defined in (3.3.3) as

$$\sum_{n=0}^{\infty} a_n t^n \sum_{n=0}^{\infty} b_n t^n = \sum_{n=0}^{\infty} \sum_{m=0}^{n} a_{n-m} b_m t^n,$$

and noting that

$$a_n = \frac{(2x)^n}{n!},$$

$$b_n = \frac{(-1)^{n/2}}{(n/2)!}, \qquad n \text{ even},$$

$$b_n = 0, \qquad n \text{ odd},$$

we have

$$\frac{H_n(x)}{n!} = \sum_{m=0}^{n} a_{n-m} b_m = \sum_{\substack{m=0 \\ m \text{ even}}}^{N} \frac{(-1)^{m/2} (2x)^{n-m}}{(m/2)!\,(n-m)!},$$

where N is the largest even integer $\leq n$. Letting $m = 2k$ we obtain the Hermite polynomial, of degree n,

$$H_n(x) = \sum_{k=0}^{[n/2]} \frac{(-1)^k n!\,(2x)^{n-2k}}{k!\,(n-2k)!}. \tag{9.2.4}$$

The following properties of $H_n(x)$ are quite evident from (9.2.4):

(1) $H_n(x)$ is a polynomial of degree n.
(2) $H_n(x)$ contains only even powers of x when n is even and odd powers of x when n is odd.
(3) The coefficient of x^n is 2^n.

9.3 RECURRENCE RELATIONS

Differentiating the generating function partially with respect to x and then t, we obtain, respectively,

$$\frac{\partial g}{\partial x} = 2tg(x, t) = \sum_{n=0}^{\infty} \frac{H'_n(x)t^n}{n!} \tag{9.3.1}$$

and

$$\frac{\partial g}{\partial t} = (2x - 2t)g(x, t) = \sum_{n=0}^{\infty} \frac{nH_n(x)t^{n-1}}{n!} \tag{9.3.2}$$

Upon replacing $g(x, t)$ by its series expansion and equating coefficients of powers of t, we obtain from (9.3.1) and (9.3.2), respectively,

$$H'_n(x) = 2nH_{n-1}(x), \quad n \geq 1, \tag{9.3.3}$$

and

$$2xH_n(x) - 2nH_{n-1}(x) = H_{n+1}(x), \quad n \geq 1. \tag{9.3.4}$$

We note that (9.3.4) is a pure three-term recurrence relation. It is also a difference equation for the Hermite polynomials.

Equations (9.3.3) and (9.3.4) are basic recurrence relations and many others, including the differential equation itself, may be obtained from these two. We leave other recurrence relations to the exercises.

EXERCISES

9.3.1. Using (9.2.4), obtain the first six Hermite polynomials and compare with Exercise 1.7.4.

9.3.2. Show that

$$H_{2n}(0) = \frac{(-1)^n(2n)!}{n!};$$

$$H_{2n+1}(0) = H'_{2n}(0) = 0;$$

$$H'_{2n+1}(0) = \frac{(-1)^n 2(2n+1)!}{n!}.$$

9.3.3. Show that $H_n(-x) = (-1)^n H_n(x)$.

9.3.4. Note that

$$2xt - t^2 + 2yt - t^2 = 2\left(\frac{x+y}{\sqrt{2}}\right)(\sqrt{2}\,t) - (\sqrt{2}\,t)^2,$$

and work out the addition formula

$$2^{n/2}H_n\left(\frac{x+y}{\sqrt{2}}\right) = \sum_{k=0}^{n} \binom{n}{k} H_k(x)H_{n-k}(y).$$

9.3.5. Solve Hermite's differential equation by the method of Frobenius.

9.3.6. Prove by induction that

$$\frac{d^k}{dx^k} H_n(x) = \frac{2^k n!\, H_{n-k}(x)}{(n-k)!}.$$

9.3.7. Show that

$$H_{n+1}(x) = 2xH_n(x) - H'_n(x).$$

9.3.8. Using the Rodrigues' formula, obtain $H_0(x)$, $H_1(x)$, and $H_2(x)$.

9.3.9. Write the generating function in the form

$$e^{2xt} = e^{t^2} \sum_{n=0}^{\infty} \frac{H_n(x)t^n}{n!},$$

expand the exponentials in a Maclaurin series, and equate coefficients to obtain

$$x^n = \sum_{k=0}^{[n/2]} \frac{n!\, H_{n-2k}(x)}{2^n k!\,(n-2k)!}$$

9.3.10. Obtain the differential equation (9.1.5) from (9.3.3) and (9.3.4).

9.3.11. Obtain (9.3.4) using Rodrigues' formula.

9.3.12. Obtain Rodrigues' formula from the generating function.

9.3.13. Obtain the generating function from the recurrence relation (9.3.4).

9.4 ORTHOGONALITY

To show that the Hermite polynomials are orthogonal, it is convenient to work with Hermite's equation in self-adjoint form,

$$(e^{-x^2}y')' + 2ne^{-x^2}y = 0,$$

which is obtained from (1.9.5), noting that $r(x) = w(x) = e^{-x^2}$, $p(x) = 0$, $\lambda_n = 2n$, and taking the interval as $(-\infty, \infty)$. By (1.9.8) we have

$$2(m-n)(H_n, H_m) = e^{-x^2}(H'_n H_m - H'_m H_n)\Big|_{-\infty}^{\infty}$$

The right-hand side is zero and hence we have the orthogonality relation

$$(H_n, H_m) = 0, \quad m \neq n. \tag{9.4.1}$$

For the case where $m = n$ we consider the generating function (9.2.3) written as

$$\exp(2xt - t^2) = \sum_{n=0}^{\infty} \frac{H_n(x)t^n}{n!}$$

and

$$\exp(2xs - s^2) = \sum_{m=0}^{\infty} \frac{H_m(x)s^m}{m!}.$$

Multiplication of these expressions yields

$$\exp[2x(t + s) - (t^2 + s^2)] = \sum_{m=0}^{\infty} \sum_{n=0}^{\infty} \frac{H_n(x)H_m(x)t^n s^m}{n!\, m!}.$$

Multiplying both sides by e^{-x^2}, integrating over $(-\infty, \infty)$, and using (9.4.1), this relation simplifies to

$$\int_{-\infty}^{\infty} \exp[2x(t + s) - (t^2 + s^2) - x^2]\, dx = \sum_{n=0}^{\infty} \left\{ \int_{-\infty}^{\infty} \left[\frac{H_n(x)}{n!}\right]^2 e^{-x^2}\, dx \right\} (ts)^n.$$

This integral may be evaluated by completing the square on x and noting that [see (5.1.4)]

$$\int_{-\infty}^{\infty} e^{-x^2}\, dx = \sqrt{\pi}.$$

The result is

$$\sqrt{\pi} \exp(2ts) = \sum_{n=0}^{\infty} \left\{ \int_{-\infty}^{\infty} \left[\frac{H_n(x)}{n!}\right]^2 e^{-x^2}\, dx \right\} (ts)^n.$$

Expanding $\exp(2ts)$ in a power series and equating coefficients of $(ts)^n/n!$, we obtain

$$\|H_n\|^2 = \sqrt{\pi}\, n!\, 2^n. \tag{9.4.2}$$

Combining (9.4.1) and (9.4.2), we see that the Hermite polynomials satisfy the orthogonality relation

$$(H_n, H_m) = \int_{-\infty}^{\infty} e^{-x^2} H_n(x) H_m(x)\, dx = \sqrt{\pi}\, 2^n n!\, \delta_{mn}. \tag{9.4.3}$$

9.5 EXPANSION OF FUNCTIONS IN TERMS OF $H_n(x)$

For the set $\{H_n(x)\}$ the generalized Fourier series (1.3.1) of a function $f(x)$ is

$$f(x) = \sum_{n=0}^{\infty} C_n H_n(x), \tag{9.5.1}$$

where the C_n, given by (1.3.2), are

$$C_n = \frac{(f, H_n)}{\|H_n\|^2}. \tag{9.5.2}$$

This expression, written out fully, is given by

$$C_n = \frac{1}{\sqrt{\pi}\, n!\, 2^n} \int_{-\infty}^{\infty} e^{-x^2} f(x) H_n(x)\, dx. \tag{9.5.3}$$

The conditions under which (9.5.1) converges uniformly are quite general and hold for most functions encountered in engineering problems. For these conditions and the proof of their sufficiency, the reader is referred to [Sa], pages 367–371.

EXERCISES

9.5.1. The functions $\phi_n(x) = e^{-x^2/2} H_n(x)$ are called *Hermite's orthogonal functions.* Obtain the differential equation and the orthogonality relation for $\phi_n(x)$. Ans. $\phi''_n + (2n + 1 - x^2)\phi_n = 0$; $\int_{-\infty}^{\infty} \phi_n \phi_m\, dx = \sqrt{\pi}\, 2^n n!\, \delta_{mn}$.

9.5.2. Frequently, in statistics, the Hermite polynomials are defined by the generating function

$$\exp\,(xt - \tfrac{1}{2}t^2) = \sum_{n=0}^{\infty} \frac{h_n(x)t^n}{n!}$$

Develop the various relations which the polynomials $h_n(x)$ satisfy, including the relation between $h_n(x)$ and $H_n(x)$.

9.5.3. Differentiate Hermite's differential equation k times to obtain the differential equation for $D^k H_n(x) = H_n^{(k)}(x)$. From this equation obtain the relation of Exercise 9.3.6.

9.5.4. Differentiate the generating function k times and from this relation show that

$$\frac{H_{n+k}^{(k)}(x)}{2^k (n + k)!} = \frac{H_n(x)}{n!}\,.$$

9.5.5. Define functions $H_n{}^a(x)$ by the generating function

$$\exp\,[-t^2 + 2t(x - a)] = \sum_{n=0}^{\infty} \frac{H_n{}^a(x)t^n}{n!} = e^{-2ta} \sum_{m=0}^{\infty} \frac{H_m(x)t^m}{m!}\,.$$

Show that

$$H_n{}^a(x) = H_n(x - a) = \sum_{m=0}^{n} \frac{(-2a)^{n-m} n!}{m!\,(n - m)!} H_m(x)$$

$$= 2^{-n/2} \sum_{m=0}^{n} \binom{n}{m} H_m(\sqrt{2}\, x) H_{n-m}(-\sqrt{2}\, a).$$

Work out the orthogonality relation for the $\{H_n{}^a(x)\}$.

9.5.6. Evaluate the integral $\int_{-\infty}^{\infty} e^{-x^2} H_n(x)\, dx$.

9.5.7. Evaluate the integral $\int_{-\infty}^{\infty} x e^{-x^2} H_n(x)\, dx$.

9.5.8. Obtain the differential equation for $H_n(\lambda x)$. Change the dependent variable $H_n(\lambda x) = y = uv$ and determine u so that v' is absent from the equation.

9.5.9. Expand $e^{-\alpha x}$ in a series of Hermite polynomials.

9.5.10. Expand $\sin \alpha x$ and $\cos \alpha x$ in a series of Hermite polynomials.

9.5.11. Given $F(z) = \int_{-\infty}^{\infty} e^{-(z-x)^2} f(x)\, dx$. Solve for $f(x)$ by expanding it and the exponential function in a series of Hermite polynomials.

$$\text{Ans.}\quad f(x) = \sum_{n=0}^{\infty} C_n H_n(x);\; C_n = \frac{F^{(n)}(0)}{n!\,\sqrt{\pi}\,2^n}.$$

9.5.12. Use Rodrigues' formula to obtain the orthogonality relation.

9.6 GENERAL SOLUTION OF HERMITE'S EQUATION

To solve Hermite's equation

$$y'' - 2xy' + 2\alpha y = 0 \tag{9.6.1}$$

by the method of Frobenius, as was considered in Exercise 9.3.5, it is sufficient to assume a solution of the type

$$y = \sum_{k=0}^{\infty} a_k x^k,$$

which results in the recurrence relation

$$a_{k+2} = \frac{-2(\alpha - k)a_k}{(k+1)(k+2)} \tag{9.6.2}$$

and the two independent solutions,

$$y_1 = a_0 \sum_{k=0}^{\infty} \frac{(-4)^k(\alpha/2 - k + 1)_k x^{2k}}{(2k)!} \tag{9.6.3}$$

and

$$y_2 = a_1 \sum_{k=0}^{\infty} \frac{(-4)^k[(\alpha - 1)/2 - k + 1]_k x^{2k+1}}{(2k+1)!}. \tag{9.6.4}$$

Applying the ratio test to (9.6.2) we see that these two solutions are valid for all values of x.

We note that if $\alpha = 2n$, n a non-negative integer, then y_1 becomes a polynomial of degree $2n$, and if $\alpha = 2n + 1$, y_2 becomes a polynomial of degree $2n + 1$. This is true because

$$\left(\frac{\alpha}{2} - k + 1\right)_k\Bigg|_{\alpha=2n} = \left(\frac{\alpha - 1}{2} - k + 1\right)_k\Bigg|_{\alpha=2n+1} = (n + 1 - k)_k$$
$$= (n + 1 - k)(n + 2 - k)\cdots(n).$$

Thus if α is a non-negative integer in (9.6.1), one of the two solutions, y_1 or y_2, will be a polynomial of degree α and the other will be an infinite series. The polynomial solution, with suitable constant multiplier, must of course be the Hermite polynomial given in (9.2.4).

For a general α, both solutions will be infinite series, and observing (9.6.2) we see that for both y_1 and y_2, for large values of k,

$$a_{k+2} \sim \frac{2}{k} a_k, \quad k \geq N_1. \tag{9.6.5}$$

The function e^{x^2} has the expansion

$$e^{x^2} = \sum_{m=0}^{\infty} b_m x^m = 1 + x^2 + \frac{x^4}{2!} + \cdots + \frac{x^k}{(k/2)!} + \frac{x^{k+2}}{[(k+2)/2]!} + \cdots$$

where k is an even integer. Forming the ratio b_{k+2}/b_k for this latter series, we find that

$$b_{k+2} \sim \frac{2}{k} b_k, \quad k \geq N_2. \tag{9.6.6}$$

Now let $a_N/b_N = c$, where N is greater than N_1 and N_2, and c is a constant. Then by (9.6.5) and (9.6.6) we have

$$\frac{a_{N+2j}}{b_{N+2j}} \sim c, \quad j = 0, 1, 2, \ldots. \tag{9.6.7}$$

Hence, for x large enough so that the terms for $k < N$ in y_1, y_2, and e^{x^2} may be neglected, we have

$$y_1 \sim c_1 e^{x^2}, \quad y_2 \sim c_2 e^{x^2} \tag{9.6.8}$$

The constants $c = c_1$ and $c = c_2$ are different because the coefficients in y_1 and y_2 are different. Thus we see that the series solutions of Hermite's equation behave like e^{x^2} for large x.

In many applications it is required that the solutions y of Hermite's equation be *square-integrable*; that is,

$$\int_{-\infty}^{\infty} e^{-x^2} y^2 \, dx = k,$$

where k is a finite number. Clearly, if y is one of the two series solutions, this requirement cannot be met, in view of (9.6.8). Hence for this case α must be an integer yielding a polynomial solution, and the coefficient of the second solution must be zero. That is, our solution must be

$$y = cH_n(x). \tag{9.6.9}$$

9.7 HERMITE'S ORTHOGONAL FUNCTIONS

The equation

$$y'' + (2n + 1 - x^2)y = 0 \tag{9.7.1}$$

obtained in Exercise 9.5.1 for the *Hermite functions*,

$$y = \phi_n(x) = e^{-x^2/2} H_n(x), \tag{9.7.2}$$

is a special case of the equation

$$y'' + (\lambda - x^2)y = 0. \tag{9.7.3}$$

Equation (9.7.1) is obtained from Hermite's equation with α a non-negative integer by substituting $H_n(x) = e^{x^2/2}y$. If α were not an integer the substitution would result in the more general equation (9.7.3) with $\lambda = 2\alpha + 1$, and the solution would be

$$y = e^{-x^2/2}(c_1 y_1 + c_2 y_2), \tag{9.7.4}$$

where y_1 and y_2 are given in Sec. 9.6. As we have seen in (9.6.8), our solution would then be of the form

$$y \sim c_3 e^{x^2/2}.$$

The only solution of (9.7.3) which is bounded for large values of x will therefore arise when $\alpha = n$ in Hermite's equation. This is equivalent to $\lambda = 2n + 1$ in (9.7.3), for $n = 0, 1, 2, \ldots$. In a problem involving (9.7.3) in which the solutions must be bounded, the eigenvalues must then be odd positive integers, and the solution must be the Hermite functions,

$$y = ce^{-x^2/2}H_n(x). \tag{9.7.5}$$

EXERCISES

9.7.1. Determine a_0 and a_1 so that (9.6.3), for α an even positive integer, and (9.6.4) for α an odd positive integer, become (9.2.4).

9.7.2. Solve the special case of *Schrödinger's equation*,

$$\nabla^2\psi(x, t) - x^2\psi(x, t) = \frac{\partial\psi(x, t)}{\partial t}, \quad -\infty < x < \infty,$$

where $\psi(x, 0) = f(x)$, which is bounded, and $\psi(x, t)$ is square-integrable with respect to x.

9.7.3. Solve for $u(x, t)$ if

$$\begin{aligned} u_{tt}(x, t) &= u_{xx}(x, t) - 2xu_x(x, t), \quad t > 0, \ -\infty < x < \infty, \\ u(x, 0) &= f(x), \quad u_t(x, 0) = 0, \\ |u(x, t)| &< Me^{kx^2} \text{ as } x \to \infty, \quad k < 1. \end{aligned}$$

(See Exercise 8.3.11.)

9.7.4. Solve for $u(x, t)$ if

$$\begin{aligned} u_t(x, t) &= u_{xx}(x, t) - 2xu_x(x, t) + f(t), \quad t > 0, \ -\infty < x < \infty, \\ u(x, 0) &= g(x), \quad |u(x, t)| < Me^{kx^2} \text{ as } x \to \infty, \quad k < 1. \end{aligned}$$

10

Laguerre Polynomials

10.1 DEFINITION

To illustrate still another approach in defining the special functions we define the *generalized* or *associated Laguerre polynomial* $L_n^\alpha(x)$ by the generating function

$$(1 - t)^{-1-\alpha} \exp\left(-\frac{xt}{1-t}\right) = \sum_{n=0}^{\infty} L_n^\alpha(x) t^n. \tag{10.1.1}$$

In this section we will show that $L_n^\alpha(x)$ is a polynomial of degree n by obtaining an explicit expression for it. We should note first that there are several different sets of polynomials defined as $\{L_n^\alpha\}$ by various authors, and in reading the literature on this subject one should check the definition being used. We give two alternative definitions in Exercise 10.2.13.

To obtain an expression for $L_n^\alpha(x)$ we expand the left-hand side of (10.1.1) in powers of t, obtaining as a first step

$$(1 - t)^{-1-\alpha} \exp\left(-\frac{xt}{1-t}\right) = \sum_{k=0}^{\infty} \frac{(-x)^k t^k}{k!\,(1-t)^{k+\alpha+1}}.$$

Expanding $(1 - t)^{-k-\alpha-1}$, we may write this expression as

$$(1 - t)^{-1-\alpha} \exp\left(-\frac{xt}{1-t}\right) = \sum_{k=0}^{\infty} \frac{(-x)^k t^k}{k!} \sum_{n=0}^{\infty} \frac{(1+\alpha+k)_n t^n}{n!}.$$

Using the Cauchy product, the double series can be written as

$$\sum_{n=0}^{\infty} \left[\sum_{k=0}^{n} \frac{(-x)^k (1+\alpha+k)_{n-k}}{k!\,(n-k)!} \right] t^n,$$

and since

$$\begin{aligned}(1+\alpha+k)_{n-k} &= (1+\alpha+k)(1+\alpha+k+1)\cdots(1+\alpha+k+n-k-1)\\ &= \frac{(1+\alpha+k)\cdots(\alpha+n)(1+\alpha)_k}{(1+\alpha)_k} = \frac{(1+\alpha)_n}{(1+\alpha)_k},\end{aligned}$$

we have, upon comparing coefficients of t^n in the two series expansions of the generating function,

$$L_n^{\alpha}(x) = \sum_{k=0}^{n} \frac{(1+\alpha)_n(-x)^k}{k!\,(n-k)!\,(1+\alpha)_k}. \tag{10.1.2}$$

We observe that $L_n^{\alpha}(x)$ is clearly a polynomial of degree n and the coefficient of x^0 is $(1+\alpha)_n/n!$. Since $(1)_n = n!$, the constant term for the case $\alpha = 0$ is 1. This case, $L_n^0(x)$, is defined as the *Laguerre polynomial* of degree n, and is given by (10.1.2) with $\alpha = 0$ as

$$L_n(x) = \sum_{k=0}^{n} \frac{n!\,(-x)^k}{(k!)^2(n-k)!}. \tag{10.1.3}$$

An alternative expression for $L_n(x)$, using the binomial coefficients, is

$$L_n(x) = \sum_{k=0}^{n} \binom{n}{k} \frac{(-x)^k}{k!}. \tag{10.1.4}$$

10.2 RECURRENCE RELATIONS AND DIFFERENTIAL EQUATION

As in previous chapters we obtain recurrence relations by differentiating the generating function (10.1.1). First, differentiating with respect to t we have

$$(1-t)^{-3-\alpha} \exp\left(-\frac{xt}{1-t}\right)[(1+\alpha)(1-t) - x] = \sum_{n=0}^{\infty} nL_n^{\alpha}(x)t^{n-1}, \tag{10.2.1}$$

which can be written

$$[(1+\alpha)(1-t) - x] \sum_{n=0}^{\infty} L_n^{\alpha}(x)t^n = (1-t)^2 \sum_{n=0}^{\infty} (n+1)L_{n+1}^{\alpha}(x)t^n.$$

Equating coefficients of t^n, we have the pure recurrence relation

$$(n+1)L_{n+1}^{\alpha}(x) + (x - 1 - \alpha - 2n)L_n^{\alpha}(x) + (n+\alpha)L_{n-1}^{\alpha}(x) = 0. \tag{10.2.2}$$

Differentiating the generating function with respect to x, we obtain

$$-t(1-t)^{-2-\alpha} \exp\left(-\frac{xt}{1-t}\right) = -\frac{t}{1-t} \sum_{n=0}^{\infty} L_n^{\alpha}(x)t^n = \sum_{n=0}^{\infty} DL_n^{\alpha}(x)t^n. \tag{10.2.3}$$

Multiplying by $1 - t$ and comparing coefficients of t^n, we have

$$L_{n-1}^{\alpha}(x) = \frac{d}{dx}[L_{n-1}^{\alpha}(x) - L_n^{\alpha}(x)]. \tag{10.2.4}$$

We may obtain other recurrence relations from these two basic relations, (10.2.2) and (10.2.4). For example, differentiating (10.2.2) and replacing

DL_{n+1}^{α} by its value from (10.2.4) yields

$$(x - \alpha - n) \frac{d}{dx} L_n^{\alpha}(x) - nL_n^{\alpha}(x) + (n + \alpha) \frac{d}{dx} L_{n-1}^{\alpha}(x) = 0. \quad (10.2.5)$$

Differentiating (10.2.4) and (10.2.5), combining the results to eliminate L_n^{α}, and replacing n by $n + 1$, we obtain the differential equation for the generalized polynomials,

$$x \frac{d^2}{dx^2} L_n^{\alpha}(x) + (\alpha + 1 - x) \frac{d}{dx} L_n^{\alpha}(x) + nL_n^{\alpha}(x) = 0. \quad (10.2.6)$$

This equation, for $\alpha = 0$, is *Laguerre's differential equation* and, for $y = L_n(x)$, is given by

$$xy'' + (1 - x)y' + ny = 0. \quad (10.2.7)$$

To obtain mixed recurrence relations (that is, with both upper and lower indices varying) we multiply (10.1.1) through by $1 - t$:

$$(1 - t)^{-\alpha} \exp\left(-\frac{xt}{1 - t}\right) = \sum_{n=0}^{\infty} L_n^{\alpha}(x)(t^n - t^{n+1}). \quad (10.2.8)$$

We note that the left-hand member is the generating function of $L_n^{\alpha-1}(x)$, and hence (10.2.8) may be written as

$$\sum_{n=0}^{\infty} L_n^{\alpha-1}(x)t^n = \sum_{n=0}^{\infty} L_n^{\alpha}(x)(t^n - t^{n+1}).$$

Equating coefficients of t^n yields the mixed relation

$$L_n^{\alpha-1}(x) = L_n^{\alpha}(x) - L_{n-1}^{\alpha}(x). \quad (10.2.9)$$

We may also obtain a mixed relation by observing that (10.2.3) may be written as

$$-\sum_{n=0}^{\infty} L_{n-1}^{\alpha+1}(x)t^n = \sum_{n=0}^{\infty} DL_n^{\alpha}(x)t^n,$$

from which we have

$$\frac{d}{dx} L_n^{\alpha}(x) = -L_{n-1}^{\alpha+1}(x) \quad (10.2.10)$$

(We have used the convention, $L_n^{\alpha} = 0$ for $n < 0$.)

Adding $n + \alpha$ times the derivative of (10.2.9) to (10.2.5) yields

$$x \frac{d}{dx} L_n^{\alpha}(x) = (n + \alpha) \frac{d}{dx} L_n^{\alpha-1}(x) + nL_n^{\alpha}(x), \quad (10.2.11)$$

and substituting (10.2.10) into this expression, we have

$$x \frac{d}{dx} L_n^{\alpha}(x) + (n + \alpha)L_{n-1}^{\alpha}(x) = nL_n^{\alpha}(x). \quad (10.2.12)$$

Substituting for L_{n-1}^{α} from (10.2.9) this latter relation becomes

$$x\frac{d}{dx}L_n^{\alpha}(x) = (n+\alpha)L_n^{\alpha-1}(x) - \alpha L_n^{\alpha}(x). \tag{10.2.13}$$

Finally, if we eliminate DL_n^{α} between (10.2.13) and (10.2.10), we have the mixed recurrence relation

$$\alpha L_n^{\alpha}(x) - xL_{n-1}^{\alpha+1}(x) - (\alpha+n)L_n^{\alpha-1}(x) = 0. \tag{10.2.14}$$

EXERCISES

10.2.1. Show that

$$L_n^{\alpha}(x) = \sum_{k=0}^{n} \frac{\Gamma(\alpha+n+1)(-x)^k}{k!\,(n-k)!\,\Gamma(\alpha+k+1)}.$$

10.2.2. Solve the Laguerre differential equation by the method of Frobenius.

10.2.3. By expanding $(1-t)^{-1}$ in (10.2.3) prove that

$$\frac{d}{dx}L_n^{\alpha}(x) = -\sum_{k=0}^{n-1} L_k^{\alpha}(x).$$

10.2.4. Verify that

$$\Gamma(1+\alpha)(xt)^{-\alpha/2}e^t J_{\alpha}(2\sqrt{xt}) = \sum_{n=0}^{\infty} \frac{L_n^{\alpha}(x)t^n}{(1+\alpha)_n}.$$

Differentiate with respect to x and use (6.4.4), to obtain (10.2.10). Then use (6.4.3) and obtain (10.2.11).

10.2.5. Differentiate the generating function in Exercise 10.2.4 with respect to t. Combine this result with (6.4.4) and prove that

$$(1+\alpha+n)L_n^{\alpha}(x) - xL_n^{\alpha+1}(x) = (n+1)L_{n+1}^{\alpha}(x).$$

10.2.6. Use (6.4.3) with the derivative obtained in Exercise 10.2.5 to obtain (10.2.9).

10.2.7 Write the first six $L_n^{\alpha}(x)$ for fixed α, using (a) (10.1.1), (b) (10.1.2).

10.2.8. Show that, for $n = 0, 1, 2, \ldots,$

$$H_{2n}(x) = (-1)^n 2^{2n} n!\, L_n^{-1/2}(x^2),$$

$$H_{2n+1}(x) = (-1)^n 2^{2n+1} n!\, xL_n^{1/2}(x^2).$$

10.2.9. Multiply (10.1.1) by $(1-t)^2$ and equate coefficients of t^n. From this and other recurrence relations, obtain

$$xL_n^{\alpha}(x) = -(n+1)L_{n+1}^{\alpha}(x) + (\alpha+2n+1)L_n^{\alpha}(x) - (n+\alpha)L_{n-1}^{\alpha}(x).$$

10.2.10. Show that

$$L_n^{\alpha+1}(x) = \sum_{k=0}^{n} L_k^{\alpha}(x).$$

10.2.11. Express $D^k L_n^{\alpha}(x)$ as a generalized Laguerre polynomial.

10.2.12. Show that

$$D^k[x^m L_n{}^m(x)] = \frac{(1+m)_n}{(m)_{n-k}} x^{m-k} L_n^{m-k}(x), \quad n \geq k,$$
$$= (1+m)_n (m+n-k+1)_{k-n} x^{m-k} L_n^{m-k}(x), \quad n < k.$$

10.2.13. Two other definitions of associated Laguerre polynomials are

(a) $(-1)^k(1-z)^{-1}\left(\frac{z}{1-z}\right)^k \exp\left(-\frac{xz}{1-z}\right) = \sum_{n=k}^{\infty} \frac{L_n{}^k(x)}{n!} z^n$ (see, for example, [MM] and [Sn]).

(b) $(1-t)^{-a-1} \exp\left(-\frac{xt}{1-t}\right) = \sum_{n=0}^{\infty} \frac{L_n{}^\alpha(x)}{\Gamma(n+a+1)} t^n$ (see, for example, [MF]).

Obtain relations between these $L_n{}^m(x)$ and ours defined in (10.1.1).

10.3 RODRIGUES' FORMULA

We observe that the differential equation (10.2.6) for $L_n{}^\alpha(x)$ satisfies the requirements in Sec. 1.11 that the Rodrigues' formula have the form

$$L_n{}^\alpha(x) = C_n x^{-\alpha} e^x D^n(x^{n+\alpha} e^{-x}), \tag{10.3.1}$$

where, in the notation of Sec. 1.11, $f(x) = x$, $g(x) = \alpha + 1 - x$. To evaluate C_n, we equate coefficients of the highest power x^n. From (10.1.2) this coefficient is seen to be $(-1)^n/n!$. Performing the operator shift in (10.3.1) results in

$$L_n{}^\alpha(x) = C_n x^{-\alpha}(D-1)^n(x^{n+\alpha}),$$

and the leading coefficient is seen to be $(-1)^n C_n$. Thus $C_n = 1/n!$ and the Rodrigues' formula is

$$L_n{}^\alpha(x) = \frac{1}{n!} x^{-\alpha} e^x D^n(x^{n+\alpha} e^{-x}). \tag{10.3.2}$$

10.4 ORTHOGONALITY

To show that $\{L_n{}^\alpha(x)\}$ is an orthogonal set we use the method of Sec. 1.9, which requires that Laguerre's equation be put in the self-adjoint form,

$$(x^{\alpha+1} e^{-x} y')' + n x^\alpha e^{-x} y = 0. \tag{10.4.1}$$

In the notation of (1.9.5), $r(x) = x^{\alpha+1} e^{-x}$, $p(x) = 0$, $\lambda_n = n$, $w(x) = x^\alpha e^{-x}$, and $y_n = L_n{}^\alpha(x)$. For this case, (1.9.8) becomes

$$(m-n)(L_n{}^\alpha, L_m{}^\alpha) = [x^{\alpha+1} e^{-x}(L_m{}^\alpha D L_n{}^\alpha - L_n{}^\alpha D L_m{}^\alpha)]\big|_a^b. \tag{10.4.2}$$

Thus we see that if $\alpha > -1$, the right-hand side of (10.4.2) is zero if the interval (a, b) is chosen to be $(0, \infty)$. Hence, we may state the orthogonality relation

$$(L_n^\alpha, L_m^\alpha) = \int_0^\infty x^\alpha e^{-x} L_n^\alpha(x) L_m^\alpha(x)\, dx = 0, \quad m \neq n, \alpha > -1. \quad (10.4.3)$$

To evaluate the norm $\|L_n^\alpha\|$, we first square the generating function (10.1.1) by means of the Cauchy product, obtaining

$$\frac{1}{(1-t)^{2(\alpha+1)}} e^{-2xt/(1-t)} = \sum_{n=0}^{\infty} \sum_{m=0}^{n} L_m^\alpha(x) L_{n-m}^\alpha(x) t^n.$$

Multiplying both sides by $x^\alpha e^{-x}$ and integrating over $(0, \infty)$ yields

$$\frac{1}{(1-t)^{2(\alpha+1)}} \int_0^\infty x^\alpha \exp\left[-x \frac{1+t}{1-t}\right] dx = \sum_{n=0}^{\infty} \sum_{m=0}^{n} t^n (L_m^\alpha, L_{n-m}^\alpha).$$

On the right-hand side, the only non-zero terms are those for which $m = n - m$ or $2m = n$. That is, n is an even integer. Evaluating the integral on the left-hand side (see Exercise 5.1.2), we have

$$\frac{\Gamma(\alpha+1)}{(1-t^2)^{\alpha+1}} = \sum_{n=0}^{\infty} t^{2n} \|L_n^\alpha\|^2. \quad (10.4.4)$$

The left-hand side may be expanded, to give

$$\frac{\Gamma(\alpha+1)}{(1-t^2)^{\alpha+1}} = \Gamma(\alpha+1) \sum_{n=0}^{\infty} \frac{(1+\alpha)_n t^{2n}}{n!},$$

and comparing coefficients of t^{2n} yields

$$\frac{\Gamma(\alpha+1)(1+\alpha)_n}{n!} = \|L_n^\alpha\|^2. \quad (10.4.5)$$

Noting that

$$\Gamma(\alpha+1)(1+\alpha)_n = \Gamma(n+\alpha+1),$$

we may then write the orthogonality relation

$$(L_n^\alpha, L_m^\alpha) = \frac{\Gamma(n+\alpha+1)}{n!} \delta_{mn}, \quad \alpha > -1. \quad (10.4.6)$$

EXERCISES

10.4.1. Prove the orthogonality relation for $L_n^\alpha(x)$ using the Rodrigues' formula.

10.4.2. Evaluate the integral, $\alpha > -1$,

$$\int_0^\infty x^\alpha e^{-x} L_n^\alpha(x)\, dx.$$

10.4.3. Evaluate the integral, $\alpha > -1$,

$$\int_0^\infty x^{\alpha+1}e^{-x}L_n^{\alpha}(x)\,dx.$$

10.4.4. Evaluate the integral, $\alpha > -1$,

$$\int_0^\infty x^{\alpha}e^{-x}L_n^{\alpha}(x)L_n^{\alpha-1}(x)\,dx.$$

10.4.5. Evaluate the integral, $\alpha > -1$,

$$\int_0^\infty x^{\alpha}e^{-x}L_n^{\alpha}(x)DL_{n+2}^{\alpha}(x)\,dx.$$

10.4.6. Integrate by parts to evaluate the integral

$$\int_x^\infty e^{-t}L_n^{\alpha}(t)\,dt.$$

10.4.7. Evaluate the integral, for $n \neq m$,

$$\int_0^x y^{\alpha}e^{-y}L_n^{\alpha}(y)L_m^{\alpha}(y)\,dy.$$

10.4.8. Obtain the Rodrigues' formula from the generating function (10.1.1).

10.5 SIMPLE LAGUERRE POLYNOMIALS $L_n(x)$

The case $\alpha = 0$ is of sufficient importance to merit special consideration. As mentioned before, these polynomials are denoted by $L_n(x)$. We now proceed to list the important properties of these polynomials, obtained directly from previous sections:

$$\frac{1}{1-t}\exp\left(-\frac{xt}{1-t}\right) = \sum_{n=0}^{\infty} L_n(x)t^n, \tag{10.5.1}$$

$$e^tJ_0(2\sqrt{xt}) = \sum_{n=0}^{\infty}\frac{L_n(x)t^n}{n!}, \tag{10.5.2}$$

$$L_n(x) = \sum_{k=0}^{n}\frac{(-1)^k n!\,x^k}{(k!)^2(n-k)!}, \tag{10.5.3}$$

$$(n+1)L_{n+1} + (x-1-2n)L_n + nL_{n-1} = 0, \tag{10.5.4}$$

$$DL_{n-1} - DL_n = L_{n-1}, \tag{10.5.5}$$

$$xD^2L_n + (1-x)DL_n + nL_n = 0, \tag{10.5.6}$$

$$DL_n = -\sum_{k=0}^{n-1} L_k, \tag{10.5.7}$$

$$L_n = \frac{e^x}{n!}D^n(e^{-x}x^n), \tag{10.5.8}$$

$$(L_n, L_m) = \int_0^\infty e^{-x}L_n(x)L_m(x)\,dx = \delta_{mn}. \tag{10.5.9}$$

The reader will observe that in our treatment of the Laguerre polynomials we have reversed the procedure used for the Legendre polynomials; that is, we have developed the associated polynomials first and given the simple ones as a special case. The development, of course, could have been given the other way.

We now represent $D^k L_n(x)$ by $L_n^{(k)}(x)$ and develop a relation between $L_n^{(k)}(x)$ and $L_n{}^k(x)$. Differentiating the generating function for $L_n(x)$ k times, we obtain the one for $L_n^{(k)}(x)$,

$$\frac{(-1)^k t^k}{(1-t)^{k+1}} \exp\left(-\frac{xt}{1-t}\right) = \sum_{n=0}^{\infty} L_n^{(k)}(x) t^n. \tag{10.5.10}$$

Since $L_n(x)$ is a polynomial of degree n, we have $L_n^{(k)}(x) = 0$ for $k > n$, and hence

$$\frac{1}{(1-t)^{k+1}} \exp\left(-\frac{xt}{1-t}\right) = \sum_{n=0}^{\infty} (-1)^k L_{n+k}^{(k)}(x) t^n.$$

Comparing coefficients of t^n in this expansion of the generating function with (10.1.1) we have

$$L_n{}^k(x) = (-1)^k L_{n+k}^{(k)}(x). \tag{10.5.11}$$

Finally we note that for the expansion of a function $f(x)$ in a series of Laguerre polynomials,

$$f(x) = \sum_{n=0}^{\infty} C_n L_n{}^\alpha(x), \tag{10.5.12}$$

the coefficients are given by (1.3.2) as

$$C_n = \frac{(f, L_n{}^\alpha)}{\|L_n{}^\alpha\|^2},$$

or

$$C_n = \frac{n!}{\Gamma(n+\alpha+1)} \int_0^\infty x^\alpha e^{-x} f(x) L_n{}^\alpha(x)\, dx. \tag{10.5.13}$$

For $\alpha = 0$ we have the case for the simple Laguerre polynomials.

As in the case of the Hermite expansion, the conditions on $f(x)$ sufficient for the uniform convergence of (10.5.12) are satisfied for most functions encountered in engineering. These conditions and the proof that they are sufficient are given in [Sa], pages 361–367.

EXERCISES

10.5.1. Use (10.5.8) to obtain (10.5.3).

10.5.2. Use Sister Celine's method (Sec. 4.14) to obtain (10.5.4) from (10.5.3).

10.5.3. Obtain (10.5.1) from (10.5.4). Use the method of Sec. 1.8.

10.5.4. The *Sonine polynomials* $S_m{}^n(x)$ are closely related to the Laguerre polynomials and are defined by (see [Ba], page 451)

$$(1 + t)^{-m-1} \exp\left(\frac{xt}{1 + t}\right) = \sum_{n=0}^{\infty} \Gamma(m + n + 1) S_m{}^n(x) t^n, \quad |t| < 1.$$

Show that

$$L_n{}^m(x) = (-1)^n \Gamma(m + n + 1) S_m{}^n(x), \quad n = 0, 1, 2, \ldots.$$

10.5.5. Obtain expressions for $S_m{}^n(x)$ corresponding to (10.1.2), (10.2.2), (10.2.4), (10.3.2), and (10.4.6).

10.5.6. Show that, for $m + n$ a non-negative integer,

$$S_m{}^n(x) = \frac{(-1)^{m+n} e^x}{\Gamma(m + n + 1)\Gamma(n + 1)} D^{m+n}(e^{-x} x^n).$$

10.5.7. Prove that

$$\frac{(1 - t)^n}{n!} e^{xt} = \sum_{m=-n}^{\infty} x^m S_m{}^n(x) t^{m+n}.$$

10.5.8. Show that, for $m + n$ a non-negative integer.

$$x^m S_m{}^n(x) = S_{-m}^{m+n}(x).$$

10.5.9. Show that $L_n^{(k)}(x) \equiv (d^k/dx^k)\, L_n(x) = (-1)^k L_{n-k}^k(x)$.

10.5.10. Show that the differential equation satisfied by $L_n^{(k)}(x)$ is

$$xy'' + (k + 1 - x)y' + (n - k)y = 0.$$

10.5.11. Show that the generating function for $L_n^{(k)}(x)$ is

$$(-t)^k (1 - t)^{-1-k} \exp\left(-\frac{xt}{1 - t}\right) = \sum_{n=k}^{\infty} L_n^{(k)}(x) t^n.$$

10.5.12. Expand $f(x) = \sum_{n=0}^{\infty} C_n L_n(x)$, where $f(x)$ is

(a) 1,

(b) $x^m, \quad m = 1, 2, 3, \ldots,$

(c) $e^{-ax}, \quad a > -1$.

10.5.13. (a) Show that the method of Frobenius applied to the differential equation (10.2.6) satisfied by $L_n{}^m(x)$ for $m = 0, 1, 2, \ldots,$

$$xy'' + (m + 1 - x)y' + ny = 0,$$

yields a polynomial solution and an infinite series solution if n is a non-negative integer. Show by (3.4.2) that the series solution has an infinite discontinuity at $x = 0$.

(b) Show that if n is not a non-negative integer both solutions are infinite series and behave like e^x as $x \to \infty$. (See Sec. 9.6.)

10.5.14. Solve for $u(x, t)$ if

$$u_t(x, t) = x u_{xx}(x, t) + (1 - x) u_x(x, t), \quad x,t > 0,$$

$$u(x, 0) = g(x),$$

$$|u(x, t)| < Me^{kx} \text{ as } x \to \infty, \quad k < 1.$$

10.5.15. Solve for $u(x, t)$ if

$$e^{-x}u_{tt}(x, t) = \frac{\partial}{\partial x}[xe^{-x}u_x(x, t)], \quad x,t > 0,$$

$$u(x, 0) = f(x), \quad u_t(x, 0) = 0,$$

$$|u(x, t)| < Me^{kx} \text{ as } x \to \infty, \quad k < 1.$$

(See Exercise 8.3.11.)

10.5.16. Solve for $u(x, t)$ if

$$e^{-x}u_{tt}(x, t) = \frac{\partial}{\partial x}[xe^{-x}u_x(x, t)] + f(t)e^{-x}, \quad x,t > 0,$$

$$u(x, 0) = g(x), \quad u_t(x, 0) = 0,$$

$$|u(x, t)| < Me^{kx} \text{ as } x \to \infty, \quad k < 1.$$

10.5.17. Show that $y = e^{-ax}x^b L_p{}^q(cx)$ satisfies

$$y'' + \left[\frac{q + 1 - 2b}{x} + 2a - c\right]y' + \left[a(a - c) + \frac{a(q + 1 - 2b) + c(b + p)}{x} + \frac{b(b - q)}{x^2}\right]y = 0.$$

10.6 EXAMPLE FROM QUANTUM MECHANICS

The fundamental equation in quantum mechanics (see [MM], pages 213–214) is *Schrödinger's equation*,

$$\nabla^2\psi + \epsilon(E - V)\psi = 0, \tag{10.6.1}$$

which governs the *wave function* $\psi(r, \theta, \phi)$, characterizing the motion of an electron with mass m, total energy E, and potential energy V. The constant ϵ is given by

$$\epsilon = \frac{8\pi^2 m}{h^2},$$

where h is *Planck's constant.* We have already considered a special case of (10.6.1) in Exercise 9.7.2.

The solution of (10.6.1) is an eigenvalue problem in which the energy level E is an eigenvalue corresponding to the eigenfunction ψ. The conditions that ψ is required to satisfy are that it be periodic in ϕ, be bounded, tend to zero as $r \to \infty$, and be normalized to unity; that is,

$$\int_0^{2\pi}\int_0^{\pi}\int_0^{\infty} |\psi|^2 r^2 \sin\theta \, dr \, d\theta \, d\phi = 1.$$

For the case of the hydrogen atom, it is known (see [MM], page 363) that $V = -k/r$ for $k > 0$. To solve (10.6.1) for this case, we try a solution of the form

$$\psi = R(r)\Theta(\theta)e^{im\phi}, \tag{10.6.2}$$

which leads to the separated equations

$$\Theta'' + (\cot\theta)\Theta' + \left(\lambda - \frac{m^2}{\sin^2\theta}\right)\Theta = 0 \tag{10.6.3}$$

and

$$R'' + \frac{2}{r}R' + \left(\epsilon E + \frac{\epsilon k}{r} - \frac{\lambda}{r^2}\right)R = 0. \tag{10.6.4}$$

To satisfy the periodicity condition on ψ, we see from (10.6.2) that m must be an integer. Equation (10.6.3) is then recognized as Legendre's associated equation, and the only bounded solutions are

$$\Theta(\theta) = P_n{}^m(\cos\theta), \tag{10.6.5}$$

corresponding to

$$\lambda = n(n+1), \quad n = 0, 1, 2, \ldots .$$

We observe that (10.6.4) is a special case of the equation in Exercise 10.5.17, where

$$q - 2b = 1, \qquad 2a = c, \qquad a(a - c) = \epsilon E,$$
$$a(q + 1 - 2b) + c(b + p) = \epsilon k, \qquad b(b - q) = -n(n+1).$$

Solving this set of equations, we have

$$a^2 = -\epsilon E, \qquad b = n, \qquad c = 2a,$$
$$q = 2n + 1, \qquad p = -n - 1 + \frac{\epsilon k}{2a}.$$

From Exercise 10.5.13 we know that the only solution which does not behave like $e^{2ar}e^{-ar}r^n = r^n e^{ar}$ as $r \to \infty$ is the one involving the Laguerre polynomial with p a non-negative integer. Since this is the only case in which $\psi \to 0$ as $r \to \infty$, we must have

$$p = -n - 1 + \frac{\epsilon k}{2a} = -n - 1 + j, \quad j = n + 1, n + 2, n + 3, \ldots .$$

Therefore we must have $\epsilon k/2a = j$, or

$$E = E_j = \frac{-ak}{2j}.$$

The solution of (10.6.4) is then

$$R_{nj} = e^{-a_j r} r^n L_{j-n-1}^{2n+1}(2a_j r),$$

where

$$a_j{}^2 = -\epsilon E_j,$$

or

$$a_j = \frac{\epsilon k}{2j}, \quad E_j = -\frac{\epsilon k^2}{4j^2}, \quad n = 0, 1, 2, \ldots, \quad j = n + 1, n + 2, \ldots .$$

EXERCISES

10.6.1. Show that $y = e^{-ax^2}x^bL_p{}^q(cx^2)$ satisfies

$$y'' + \left[(4a - 2c)r + \frac{2q - 2b + 1}{r}\right]y'$$

$$+ \left[\frac{b(b - 2q)}{r^2} + (4a - 4ab + 4aq + 2bc + 4pc) + 4a(a - c)r^2\right]y = 0.$$

10.6.2. Solve Schrödinger's equation with $V = \frac{1}{2}M\omega^2r^2$.

Ans. $\psi_{mnk} = C_{mnk}e^{im\phi}P_n{}^m(\cos\theta)r^ne^{-\beta r^2/2}L^{n+1/2}_{(k-n-1)/2}(\beta r^2)$;
$E_k = (\omega h/2\pi)(k + \frac{1}{2})$; $k = 1, 2, 3, \ldots$; $m = 0, 1, \ldots, n$;
$n = k - 1, k - 3, k - 5, \ldots, 1$ or 0; $\beta = 2\pi\omega M/h$.

10.6.3. Given $\phi(t) = \int_0^\infty (1 - t)^{-1-\alpha}e^{-xt/(1-t)}\psi(x)\,dx$, find $\psi(t)$ if $\alpha > -1$.

$$\text{Ans.}\quad \psi(t) = \sum_{n=0}^{\infty} \frac{\phi^{(n)}(0)}{\Gamma(n + \alpha + 1)}\, t^\alpha e^{-t}L_n{}^\alpha(t).$$

10.6.4. Find $\psi(t)$ if

$$\phi(t) = \int_0^\infty \frac{J_\alpha(2\sqrt{xt})\psi(x)\,dx}{(xt)^{\alpha/2}}, \quad \alpha > -1.$$

$$\text{Ans.}\quad \psi(t) = \sum_{n=0}^{\infty} a_nt^\alpha e^{-t}L_n{}^\alpha(t), \text{ where } a_n = (d^n/dt^n)[e^t\phi(t)]_{t=0}.$$

10.6.5. Solve Schrödinger's equation with $V = M\omega^2x^2 + b^2/x^2$. This is the one-dimensional case in rectangular coordinates.

Ans. $\psi_n = C_ne^{-\pi M\omega x^2/h}x^{\alpha+1/2}L^\alpha_{n+\alpha}(2\pi M\omega x^2/h)$, where $\alpha^2 = 8\pi^2Mb^2/h^2 + \frac{1}{4}$.

10.6.6. Show that a solution of the diffusion equation,

$$\nabla^2\psi(r, t) = \frac{\partial^2\psi}{\partial r^2} + \frac{1}{r}\frac{\partial\psi}{\partial r} = \frac{\partial\psi}{\partial t},$$

is

$$\psi(r, t) = e^{-\lambda^2t}J_0(\lambda r),$$

where λ is a constant. Show that this may be written

$$\psi(r, t) = \sum_{n=0}^{\infty} C_nU_n(r, t),$$

where

$$U_n = \sum_{k=0}^{n}\binom{n}{k}\frac{r^{2k}t^{n-k}}{2^{2k}k!} = t^nL_n\left(-\frac{r^2}{4t}\right),$$

and

$$C_n = \frac{(-1)^n\lambda^{2n}}{n!}.$$

Therefore, $U_n(r, t)$ is a solution of the differential equation. Verify this by direct substitution of the closed-form expression. Compare this result with that in Exercise 10.2.4.

11

Chebyshev Polynomials

11.1 DEFINITIONS

Chebyshev in 1857 obtained the polynomials which bear his name while seeking a monic polynomial (one whose leading coefficient is 1) of degree n deviating least from zero on the interval $-1 \leq x \leq 1$. These monic polynomials are constant multiples of the *Chebyshev polynomials of the first kind*, defined as

$$T_n(x) = \cos(n \arccos x), \qquad n = 0, 1, 2, \ldots. \tag{11.1.1}$$

A set of polynomials related to these are the *Chebyshev polynomials of the second kind*, given by

$$\begin{aligned} U_n(x) &= (n+1)^{-1} T'_{n+1}(x) \\ &= (1 - x^2)^{-1/2} \sin[(n+1) \arccos x]. \end{aligned} \tag{11.1.2}$$

It is not at all obvious that $T_n(x)$ and $U_n(x)$ are actually polynomials of degree n. To see this, we let $\cos\theta = x$ and then from (11.1.1) and (11.1.2) we have

$$\begin{aligned} T_n(x) &= \cos n\theta, \\ U_n(x) &= \frac{\sin(n+1)\theta}{\sin\theta}. \end{aligned} \tag{11.1.3}$$

Using Demoivre's theorem,

$$(\cos\theta + i \sin\theta)^n = \cos n\theta + i \sin n\theta,$$

we see that

$$\cos n\theta = \text{real part of } (\cos\theta + i\sin\theta)^n.$$

Now using the binomial expansion we have

$$\begin{aligned} T_n(x) = \cos n\theta &= \operatorname{Re}\left\{\sum_{k=0}^{n} i^k \binom{n}{k} \cos^{n-k}\theta \sin^k\theta\right\} \\ &= \sum_{k=0}^{[n/2]} (-1)^k \binom{n}{2k} \cos^{n-2k}\theta \sin^{2k}\theta. \end{aligned} \tag{11.1.4}$$

In terms of x, (11.1.4) is

$$T_n(x) = \sum_{k=0}^{[n/2]} (-1)^k \binom{n}{2k} x^{n-2k}(1 - x^2)^k, \tag{11.1.5}$$

which is clearly a polynomial of degree n. The expression for $U_n(x)$ is given in Exercise 11.1.3.

Since arccos x is real only for $-1 \leq x \leq 1$, (11.1.1) seems to be inconvenient when $|x| > 1$. An alternative form

$$T_n(x) = \cosh (n \operatorname{arccosh} x) \tag{11.1.6}$$

may be derived, as outlined in Exercise 11.1.4.

EXERCISES

11.1.1. Show that

$$\begin{aligned} T_0(x) &= 1 & U_0(x) &= 1 \\ T_1(x) &= x & U_1(x) &= 2x \\ T_2(x) &= 2x^2 - 1 & U_2(x) &= 4x^2 - 1 \\ T_3(x) &= 4x^3 - 3x & U_3(x) &= 8x^3 - 4x \\ T_4(x) &= 8x^4 - 8x^2 + 1 & U_4(x) &= 16x^4 - 12x^2 + 1 \end{aligned}$$

11.1.2. Show that

(a) $T_n(\pm 1) = (\pm 1)^n$,
(b) $T_{2n}(0) = (-1)^n$,
(c) $T_{2n+1}(0) = 0$,

by using both (11.1.1) and (11.1.5). Show also that $U_n(1) = n + 1$.

11.1.3. Show that $U_n(x)$ is a polynomial of degree n given by

$$U_n(x) = \sum_{k=1}^{[(n+1)/2]} (-1)^{k-1} \binom{n+1}{2k-1} x^{n-2k+2}(1 - x^2)^{k-1}.$$

11.1.4. Using the identity $x = \cos \theta = \cosh i\theta$ and thus $i\theta = \operatorname{arccosh} x$, derive (11.1.6).

11.2 RECURRENCE RELATIONS AND DIFFERENTIAL EQUATIONS

Recurrence relations may be obtained from the trigonometric identities,

$$\cos (n + 1)\theta + \cos (n - 1)\theta = 2 \cos \theta \cos n\theta, \tag{11.2.1}$$

$$\sin (n + 1)\theta + \sin (n - 1)\theta = 2 \cos \theta \sin n\theta. \tag{11.2.2}$$

Equation (11.2.1) may be written, since $T_n(x) = \cos n\theta$, as

$$T_{n+1}(x) = 2xT_n(x) - T_{n-1}(x), \tag{11.2.3}$$

and (11.2.2), after division by $\sin\theta$ and using (11.1.3), becomes

$$U_n(x) = 2xU_{n-1}(x) - U_{n-2}(x). \tag{11.2.4}$$

We note that $T_n(x)$ and $U_n(x)$ satisfy the same recurrence relation.

The differential equation for $T_n(x)$ can be obtained from the equation for $y = \cos n\theta$,

$$\frac{d^2y}{d\theta^2} + n^2y = 0.$$

For $x = \cos\theta$, this equation is

$$(1 - x^2)T''_n(x) - xT'_n(x) + n^2T_n(x) = 0, \tag{11.2.5}$$

which is *Chebyshev's equation.*

The differential equation for $U_n(x)$, obtained by differentiating (11.2.5) and using (11.1.2), is

$$(1 - x^2)U''_n(x) - 3xU'_n(x) + n(n + 2)U_n(x) = 0. \tag{11.2.6}$$

The details are left as an exercise.

As we shall see in Exercise 11.2.5, the general solution of Chebyshev's equation (11.2.5) is

$$y = C_1T_n(x) + C_2\sqrt{1 - x^2}\,U_{n-1}(x), \quad n = 1, 2, 3, \ldots. \tag{11.2.7}$$

The case $n = 0$ is given by

$$y = C_3T_0(x) + C_4 \arccos x. \tag{11.2.8}$$

EXERCISES

11.2.1. Use the identity

$$\cos(n + 1)\theta = \cos n\theta \cos\theta - \sin n\theta \sin\theta$$

to obtain

$$T_{n+1}(x) = xT_n(x) - (1 - x^2)U_{n-1}(x).$$

11.2.2. Use $\sin(n+1)\theta = \sin n\theta \cos\theta + \cos n\theta \sin\theta$ to obtain

$$U_n(x) = xU_{n-1}(x) + T_n(x).$$

11.2.3. Using $U_{n-1}(x) = (\sin n\theta)/\sin\theta$ and $y = \sin n\theta$, derive (11.2.6) from $y'' + n^2y = 0$.

11.2.4. Solve Chebyshev's equation by the method of Frobenius.

11.2.5. Show that $\sqrt{1 - x^2}\,U_{n-1}(x)$ satisfies Chebyshev's equation for $n = 1, 2, 3, \ldots$, and hence that the general solution is given by (11.2.7).

11.2.6. Find the general solution of Chebyshev's equation for $n = 0$.

11.2.7. Replace θ by $i\theta$ in (11.2.1) to obtain

$$\cosh(n + 1)\theta + \cosh(n - 1)\theta = 2\cosh\theta \cosh n\theta,$$

valid for $|x| > 1$, where $\cosh\theta = x$. Use the definition of $T_n(x)$ for this case, given in Exercise 11.1.4, to establish (11.2.3). Use induction to show that $T_n(x)$ is a polynomial of degree n.

11.2.8. If we define

$$\begin{aligned} T_n(x) &= \cos (n \arccos x), \\ U_n^*(x) &= \sin (n \arccos x), \\ V_n(x) &= \cos (n \arcsin x), \\ W_n(x) &= \sin (n \arcsin x), \end{aligned}$$

show that these four functions all satisfy Chebyshev's equation and the equation

$$(y')^2 + \frac{n^2}{1 - x^2} (y^2 - 1) = 0.$$

(See [Clt].) Show also that

$$\begin{aligned} V_{2n}(x) &= (-1)^n T_{2n}(x), \\ V_{2n+1}(x) &= (-1)^n U_{2n+1}^*(x), \\ W_{2n}(x) &= (-1)^n U_{2n}^*(x), \\ W_{2n+1}(x) &= (-1)^n T_{2n+1}(x). \end{aligned}$$

11.2.9. Show that

$$\begin{aligned} &\text{(a)}\ 2T_n(x)T_m(x) = T_{n+m}(x) + T_{n-m}(x), \quad n \geq m, \\ &\text{(b)}\ U_{n-1}(x)U_{m-1}(x) = \frac{T_{n-m}(x) - T_{n+m}(x)}{1 - T_2(x)}, \quad n \geq m, \\ &\text{(c)}\ 2U_{n-1}(x)T_m(x) = U_{m+n-1}(x) + U_{n-m-1}(x), \quad n \geq m, \\ &\qquad\qquad\qquad\qquad\ = U_{m+n-1}(x) - U_{m-n-1}(x), \quad m \geq n. \end{aligned}$$

11.3 ORTHOGONALITY RELATIONS

We may establish the orthogonality relations for the Chebyshev polynomials by making use of the relations obtained for the trigonometric functions in Chapter 1. It was shown there that

$$\int_0^\pi \cos n\theta \cos m\theta \, d\theta = \frac{\pi}{\epsilon_n} \delta_{mn}, \quad m, n = 0, 1, 2, \ldots, \tag{11.3.1}$$

where

$$\epsilon_n = 2, \quad n \neq 0; \qquad \epsilon_0 = 1. \tag{11.3.2}$$

Making the substitutions

$$\cos \theta = x, \quad T_n(x) = \cos n\theta$$

in (11.3.1) we have the orthogonality relation for the $T_n(x)$,

$$\int_{-1}^1 \frac{T_n(x)T_m(x)\, dx}{\sqrt{1 - x^2}} = \frac{\pi \delta_{mn}}{\epsilon_n}. \tag{11.3.3}$$

Other integral relations involving $T_n(x)$ and $U_n(x)$ will be obtained in the exercises.

EXERCISES

11.3.1. Derive the relation (11.3.3) for $m \neq n$ from Chebyshev's differential equation.

11.3.2. Show that $U_n(x)$ satisfies the differential equation

$$[(1 - x^2)^{3/2}y']' + n(n + 2)\sqrt{1 - x^2}\, y = 0.$$

11.3.3. Using the differential equation in Exercise 11.3.2, obtain

$$[(1 - x^2)^{3/2}(U_m U'_n - U'_m U_n)]' = (m - n)(m + n + 2)\sqrt{1 - x^2}\, U_m U_n.$$

Hence show that

$$\int_{-1}^{1} \sqrt{1 - x^2}\, U_m(x) U_n(x)\, dx = 0, \quad m \neq n.$$

11.3.4. Use the orthogonality relation $\int_0^\pi \sin n\theta \sin m\theta\, d\theta = (\pi/2)\delta_{mn}$ to obtain

$$\int_{-1}^{1} \sqrt{1 - x^2}\, U_m(x) U_n(x)\, dx = \frac{\pi}{2}\delta_{mn}.$$

11.3.5. Show that

$$\int_0^\pi \cos m\theta \sin n\theta\, d\theta = \int_{-1}^{1} T_m(x) U_{n-1}(x)\, dx.$$

Hence show that for $m \geq 0$, $n \geq 1$,

$$\begin{aligned}\int_{-1}^{1} T_m(x) U_{n-1}(x)\, dx &= \tfrac{1}{2}\left\{\frac{1}{m + n}[1 - (-1)^{m+n}] + \frac{1}{n - m}[1 - (-1)^{n-m}]\right\}, \quad n \neq m \\ &= 0, \quad n = m.\end{aligned}$$

11.3.6. Prove that when $m + n - 1$ is odd, the last integral in Exercise 11.3.5 is zero, and when $m + n - 1$ is even, the integral is

$$\frac{1}{m + n} + \frac{1}{n - m} = \frac{2n}{n^2 - m^2}.$$

The formula may then be written

$$\begin{aligned}\int_{-1}^{1} T_m(x) U_{n-1}(x)\, dx &= 0, \quad n = m \text{ or } m + n - 1 \text{ odd}, \\ &= \frac{2n}{n^2 - m^2}, \quad m + n - 1 \text{ even}.\end{aligned}$$

11.3.7. Illustrate the formulas in Exercises 11.3.3, 11.3.4, and 11.3.5 using T_0, T_1, T_2, U_0, U_1.

11.4 GENERATING FUNCTIONS

We may obtain a generating function for the $T_n(x)$ by considering the generating function for $\{\cos n\theta\}$ given in (1.8.4) as

$$\frac{1 - t\cos\theta}{1 - 2t\cos\theta + t^2} = \sum_{n=0}^{\infty} (\cos n\theta)t^n. \tag{11.4.1}$$

Noting that $\cos n\theta = T_n(x)$ when $x = \cos\theta$, the desired generating function is

$$\frac{1 - xt}{1 - 2xt + t^2} = \sum_{n=0}^{\infty} T_n(x)t^n. \tag{11.4.2}$$

Another generating function for $T_n(x)$ which is sometimes used is

$$\frac{1 - t}{1 - 2xt + t^2} = \sum_{n=0}^{\infty} \epsilon_n T_n(x)t^n, \tag{11.4.3}$$

where ϵ_n is given by (11.3.2). This expression can be obtained from (11.4.2) by writing

$$\begin{aligned} \sum_{n=0}^{\infty} T_n(x)t^n &= \tfrac{1}{2}\left[1 + 1 + \sum_{n=1}^{\infty} 2T_n(x)t^n\right] \\ &= \tfrac{1}{2}\left[1 + \sum_{n=0}^{\infty} \epsilon_n T_n(x)t^n\right]. \end{aligned} \tag{11.4.4}$$

When (11.4.4) is substituted into (11.4.2), the result in (11.4.3) is obtained.

To obtain the generating function for $U_n(x)$, we begin by noting from Exercise 1.8.2 that

$$\frac{t\sin\theta}{1 - 2t\cos\theta + t^2} = \sum_{n=1}^{\infty} (\sin n\theta)t^n.$$

Dividing by $\sin\theta$ and making the substitutions $x = \cos\theta$ and $U_{n-1}(x) = (\sin n\theta)/\sin\theta$, this relation becomes

$$\frac{1}{1 - 2xt + t^2} = \sum_{n=0}^{\infty} U_n(x)t^n. \tag{11.4.5}$$

The generating function for $T_n(x)$ may also be obtained directly from the recurrence relation (11.2.3) by using the method given in Chapter 1. For convenience we write (11.2.3) as

$$T_{n+2}(x) = 2xT_{n+1}(x) - T_n(x), \quad n \geq 0. \tag{11.4.6}$$

In seeking a generating function of the form

$$G[T_n(x)] = \sum_{n=0}^{\infty} T_n(x)t^n,$$

we see that we need to multiply (11.4.6) by t^n and sum over n to obtain

$$\sum_{n=0}^{\infty} T_{n+2}(x)t^n = 2x\sum_{n=0}^{\infty} T_{n+1}(x)t^n - \sum_{n=0}^{\infty} T_n(x)t^n. \tag{11.4.7}$$

The next step is to express each term in this equation in terms of $G(T_n)$. The first term can be written as

$$G(T_{n+2}) = \sum_{n=0}^{\infty} T_{n+2}(x)t^n$$

which, after shifting the index, becomes

$$G(T_{n+2}) = t^{-2}[-T_0(x) - T_1(x)t + G(T_n)].$$

Proceeding in a similar manner, we obtain for the second term in (11.4.7)

$$2x\sum_{n=0}^{\infty} T_{n+1}(x)t^n = 2xt^{-1}[-T_0(x) + G(T_n)].$$

Substituting these expressions into (11.4.7) and solving for $G(T_n)$ we obtain (11.4.2).

An explicit expression for $U_n(x)$ may be obtained by expanding the generating function (11.4.5) as

$$\begin{aligned}\frac{1}{1-2xt+t^2} &= \sum_{n=0}^{\infty}(2xt-t^2)^n \\ &= \sum_{n=0}^{\infty}\sum_{k=0}^{n}(-1)^k\binom{n}{k}(2x)^{n-k}t^{n+k} \\ &= \sum_{n=0}^{\infty}\sum_{k=0}^{[n/2]}(-1)^k\binom{n-k}{k}(2x)^{n-2k}t^n.\end{aligned}$$

When coefficients of t^n are equated in these expansions of $G(U_n)$, it follows that

$$U_n(x) = \sum_{k=0}^{[n/2]}(-1)^k\binom{n-k}{k}(2x)^{n-2k}. \tag{11.4.8}$$

Since $U_n(x) = (n+1)^{-1}T'_{n+1}(x)$, an explicit expression for $T_n(x)$ can be obtained by integrating (11.4.8) over the interval $(0, x)$. This yields

$$T_{n+1}(x) = T_{n+1}(0) + \frac{n+1}{2}\sum_{k=0}^{[n/2]}(-1)^k\binom{n-k}{k}\frac{(2x)^{n-2k+1}}{n-2k+1}.$$

Replacing n by $n-1$ and noting from Exercise 11.1.2 that

$$T_{2n}(0) = (-1)^n, \quad T_{2n+1}(0) = 0,$$

we may finally write

$$T_n(x) = \frac{n}{2}\sum_{k=0}^{[n/2]}\frac{(-1)^k(n-k-1)!}{k!\,(n-2k)!}(2x)^{n-2k}, \quad n > 0. \tag{11.4.9}$$

11.5 RODRIGUES' FORMULA

To obtain a Rodrigues' formula for $T_n(x)$ we will use the method of Sec. 1.11, noting from Chebyshev's equation (11.2.5) that $f(x) = 1 - x^2$ and $g(x) = -x$. Following the procedure of Sec. 1.11 we have

$$T_n(x) = C_n\sqrt{1 - x^2}\, D_n(1 - x^2)^{n-1/2}. \tag{11.5.1}$$

To evaluate C_n we expand the right member and use the relation $T_n(1) = 1$. Using Leibnitz's rule we write

$$\begin{aligned} D^n(1 - x^2)^{n-1/2} &= \sum_{k=0}^{n}\binom{n}{k}D^k(1 - x)^{n-1/2}D^{n-k}(1 + x)^{n-1/2} \\ &= \sum_{k=0}^{n}\binom{n}{k}(n - k + \tfrac{1}{2})_k(k + \tfrac{1}{2})_{n-k}(-1)^k(1 - x)^{n-k-1/2}(1 + x)^{k-1/2}. \end{aligned}$$

Substituting in (11.5.1) we have

$$T_n(x) = C_n\sum_{k=0}^{n}\binom{n}{k}(n - k + \tfrac{1}{2})_k(k + \tfrac{1}{2})_{n-k}(-1)^k(1 - x)^{n-k}(1 + x)^k, \tag{11.5.2}$$

which is a polynomial of degree n. Now using $T_n(1) = 1$, we see that

$$1 = C_n(\tfrac{1}{2})_n(-2)^n,$$

since the only non-zero term in (11.5.2) occurs when $k = n$. Simplifying the value of C_n and substituting into (11.5.1) we have the Rodrigues' formula

$$T_n(x) = \frac{(-2)^n n!}{(2n)!}\sqrt{1 - x^2}\, D^n(1 - x^2)^{n-1/2}. \tag{11.5.3}$$

EXERCISES

11.5.1. Using the method in Sec. 11.4, begin with (11.2.4) and obtain (11.4.5).

11.5.2. Obtain the recurrence relation (11.4.6) and the differential equation for $T_n(x)$ from the generating function (11.4.3).

11.5.3. Using (11.4.2), show that

(a) $T_{2n}(0) = (-1)^n$,
(b) $T_{2n+1}(0) = 0$,
(c) $T_n(1) = 1$,
(d) $T_n(-1) = (-1)^n$.

11.5.4. Use partial fractions to write the generating function (11.4.2) as a sum of two functions and show that

$$T_n(x) = \tfrac{1}{2}[(x + \sqrt{x^2 - 1})^n + (x - \sqrt{x^2 - 1})^n].$$

11.5.5. Show that if $x = \cos\theta$ in Exercise 11.5.4, then $T_n(x) = \cos n\theta$; and if $x = \cosh\theta$, then $T_n(x) = \cosh n\theta$.

11.5.6. Find a Rodrigues' formula for $U_n(x)$.

$$\text{Ans.}\quad U_n(x) = \frac{(-2)^n(n+1)!}{(2n+1)!\sqrt{1-x^2}} D^n(1-x^2)^{n+1/2}.$$

11.5.7. Obtain the recurrence relation for the set $\{T_n(x)/n!\}$, and from this relation derive the generating function

$$\sum_{n=0}^{\infty} \frac{T_n(x)t^n}{n!} = e^{xt}\cosh(t\sqrt{x^2-1}).$$

11.5.8. Follow the procedure in Exercise 11.5.7 to obtain the generating function

$$\sum_{n=0}^{\infty} \frac{U_n(x)t^{n+1}}{(n+1)!} = \frac{e^{xt}\sinh(t\sqrt{x^2-1})}{\sqrt{x^2-1}}.$$

11.6 ZEROS OF $T_n(x)$ AND ASSOCIATED PROPERTIES

As mentioned earlier the Chebyshev polynomials $T_n(x)$, when properly normalized, have the property of deviating least from zero on $[-1, 1]$ among all polynomials of degree n. As seen from (11.4.9), the coefficient of x^n in $T_n(x)$ is 2^{n-1}, and the normalized polynomial is $2^{1-n}T_n(x)$. We state this property as follows:

Theorem 11.6.1. The polynomial $2^{1-n}T_n(x)$, $n \geq 1$, $T_0(x) = 1$, is the monic polynomial of degree n whose maximum absolute value on the interval $[-1, 1]$ deviates least from zero.

To show this, we let $\theta = \arccos x$ and consider the points

$$x_k = \cos\frac{k\pi}{n}, \quad k = 0, 1, 2, \ldots, n.$$

Since $T_n(x) = \cos(n \arccos x)$, $T_n(x)$ achieves its greatest deviation from zero at these $n + 1$ points in $[-1, 1]$, and this deviation is given by

$$2^{1-n}T_n(x_k) = (-1)^k 2^{1-n}, \quad k = 0, 1, 2, \ldots, n.$$

Suppose now that there is a polynomial

$$2^{1-n}R_n(x) = x^n + a_{n-1}x^{n-1} + \cdots + a_0$$

whose deviation from zero in $[-1, 1]$ is less than that of $2^{1-n}T_n(x)$. Then it follows that

$$(-1)^k[T_n(x_k) - R_n(x_k)] > 0.$$

Now $T_n(x) - R_n(x)$ is a polynomial of degree at most $n - 1$ which changes sign at least n times. Therefore it must have at least n zeros. This is possible only if $T_n(x) \equiv R_n(x)$, and thus the theorem is proved.

The zeros of $T_n(x)$ occur when x is such that

$$n \arccos x = \frac{(2k-1)\pi}{2}, \quad k \text{ an integer},$$

or when

$$x = \cos \frac{(2k-1)\pi}{2n}.$$

This result can be concisely stated as follows:

Theorem 11.6.2. The zeros of $T_n(x)$ are real and distinct, and are located in the open interval $(-1, 1)$.

11.7 EXPANSIONS IN SERIES OF CHEBYSHEV POLYNOMIALS

The generalized Fourier series in $T_n(x)$ has the form

$$f(x) = \sum_{n=0}^{\infty} C_n T_n(x), \tag{11.7.1}$$

where the Fourier coefficients are given by

$$C_n = \frac{(f, T_n)}{\|T_n\|^2} = \frac{\epsilon_n}{\pi}(f, T_n), \quad n = 0, 1, 2, \ldots. \tag{11.7.2}$$

The inner product (f, T_n) is the integral

$$(f, T_n) = \int_{-1}^{1} \frac{f(x)T_n(x)\,dx}{\sqrt{1-x^2}},$$

and the square of the norm $\|T_n\|^2$ is given in (11.3.3).

We have noted in Exercises 4.6.7 and 4.6.8 that $R_m(x)$, a polynomial of degree m, may be written as

$$R_m(x) = \sum_{n=0}^{m} C_n \phi_n(x), \tag{11.7.3}$$

where $\{\phi_n(x)\}$ is a set of non-zero orthogonal polynomials of degree n, and that $(\phi_n, R_m) = 0$ for $m < n$. These results obviously hold for the set $\{T_n(x)\}$.

As an example of a series expansion of the type (11.7.3), let us consider the function,

$$x^2 = \sum_{n=0}^{2} C_n T_n(x).$$

By (11.7.2) the coefficients are given by

$$C_0 = \frac{1}{\pi}\int_{-1}^{1} \frac{x^2\,dx}{\sqrt{1-x^2}} = \tfrac{1}{2},$$

$$C_1 = \frac{2}{\pi}\int_{-1}^{1} \frac{x^3\,dx}{\sqrt{1-x^2}} = 0,$$

$$C_2 = \frac{2}{\pi}\int_{-1}^{1} \frac{x^2(2x^2-1)\,dx}{\sqrt{1-x^2}} = \tfrac{1}{2},$$

and hence we have the expansion

$$x^2 = \tfrac{1}{2}[T_0(x) + T_2(x)].$$

An alternative method of expanding $f(x) = x^n$ is to observe that $2^{1-n}T_n(x)$ is a monic polynomial, so that

$$g_{n-2}(x) = x^n - 2^{1-n}T_n(x)$$

is a polynomial of degree $n - 2$. In the same way we may find

$$g_{n-4}(x) = g_{n-2} - a_{n-2}T_{n-2}(x),$$

a polynomial of degree $n - 4$ or less. The constant a_{n-2} is selected to make the leading coefficients vanish. We may continue the process to eliminate all the functions except the terms involving the $T_k(x)$. Thus we will have

$$\begin{aligned} x^n &= g_{n-2} + 2^{1-n}T_n(x) \\ &= g_{n-4} + a_{n-2}T_{n-2}(x) + 2^{1-n}T_n(x) \\ &= g_{n-6} + a_{n-4}T_{n-4}(x) + a_{n-2}T_{n-2}(x) + 2^{1-n}T_n(x) \\ &= \cdots. \end{aligned}$$

To illustrate this process, let us consider $f(x) = x^4$. We have

$$\begin{aligned} x^4 - \tfrac{1}{8}T_4(x) &= x^2 - \tfrac{1}{8}, \\ x^2 - \tfrac{1}{8} - \tfrac{1}{2}T_2(x) &= \tfrac{3}{8} \\ \tfrac{3}{8} - \tfrac{3}{8}T_0(x) &= 0, \end{aligned}$$

which leads to

$$x^4 = \tfrac{1}{8}T_4(x) + \tfrac{1}{2}T_2(x) + \tfrac{3}{8}T_0(x).$$

For computational purposes it is usually convenient to make the substitution $x = \cos\theta$ in the expression for the Fourier coefficients. They are then given by

$$C_n = \frac{\epsilon_n}{\pi}\int_0^{\pi} f(\cos\theta)\cos n\theta\,d\theta, \quad n = 0, 1, 2, \ldots. \tag{11.7.4}$$

Often the process of determining the C_n's can be vastly simplified by using this formula. For example, the coefficients for the function

$$f(x) = 0, \quad -1 < x < 0$$
$$= 1, \quad 0 < x < 1,$$

are given by (11.7.2) as

$$C_n = \frac{\epsilon_n}{\pi} \int_0^1 \frac{T_n(x)}{\sqrt{1 - x^2}}\, dx,$$

and by (11.7.4) as

$$C_n = \frac{\epsilon_n}{\pi} \int_0^{\pi/2} \cos n\theta \, d\theta.$$

Incidentally, comparing these two expressions for $n = 1, 2, 3, \ldots$, we have the formula

$$\int_0^1 \frac{T_n(x)\, dx}{\sqrt{1 - x^2}} = 0, \quad n \text{ even}$$

$$= \frac{(-1)^{(n+1)/2}}{n}, \quad n \text{ odd}. \tag{11.7.5}$$

EXERCISES

11.7.1. Express x^5 and x^6 as a sum of Chebyshev polynomials of the first kind.

11.7.2. Expand the function $f(x) = \operatorname{sgn} x$ over $[-1, 1]$, where

$$\operatorname{sgn} x = -1, \quad x < 0$$
$$= 0, \quad x = 0$$
$$= 1, \quad x < 0,$$

in a series $\Sigma_{k=0}^{\infty} C_k T_k(x)$.

11.7.3. Approximate $f(x) = |x|$ by $\Sigma_{k=0}^{3} C_k T_k(x)$. Compute the mean-square error. Obtain a general expression for C_k.

11.7.4. Determine the coefficients a_k in an expansion of the form

$$f(x) = \frac{1}{\sqrt{1 - x^2}} \sum_{k=0}^{\infty} a_k T_k(x).$$

11.7.5. Obtain an expansion for e^x over $[-1, 1]$ of the form given in the preceding exercise.

11.7.6. It was shown in Exercise 11.3.3 that $\{U_n(x)\}$ is an orthogonal set on $(-1, 1)$ with weight function $w(x) = \sqrt{1 - x^2}$. Show that the coefficients in the generalized Fourier series

$$f(x) = \sum_{k=0}^{\infty} C_k U_k(x), \quad |x| < 1$$

are

$$C_k = \frac{2}{\pi}\int_{-1}^{1} \sqrt{1-x^2}\, f(x) U_k(x)\, dx$$

$$= \frac{2}{\pi}\int_{0}^{\pi} f(\cos\theta)\sin\theta \sin(k+1)\theta\, d\theta.$$

11.7.7. Show that the coefficients in the Fourier series

$$f(x) = \sqrt{1-x^2}\sum_{k=0}^{\infty} C_k U_k(x)$$

are

$$C_k = \frac{2}{\pi}\int_{-1}^{1} f(x) U_k(x)\, dx, \quad k = 0, 1, 2, \ldots.$$

11.7.8. Determine the coefficients in the approximation

$$|x| \doteq \sum_{k=0}^{3} C_k U_k(x),$$

and obtain a general expression for C_k.

11.7.9. Express the first three Legendre polynomials in terms of $T_0(x)$, $T_1(x)$, $T_2(x)$.

11.7.10. Another method of expanding x^n in a series of the $T_n(x)$ employs the recurrence relation (11.2.3) written

$$xT_n(x) = \tfrac{1}{2}[T_{n+1}(x) + T_{n-1}(x)], \quad n = 1, 2, 3, \ldots,$$

$$xT_0(x) = T_1(x).$$

To illustrate the method, consider x^4:

$$x^4 = x^2(xT_1) = \frac{x^2}{2}(T_2 + T_0) = \frac{x}{4}(T_1 + T_3 + 2T_1)$$

$$= \tfrac{1}{4}(3xT_1 + xT_3) = \tfrac{1}{8}(3T_0 + 3T_2 + T_4 + T_2)$$

$$= \tfrac{1}{8}T_4 + \tfrac{1}{2}T_2 + \tfrac{3}{8}T_0.$$

Use this method to expand x^5 and x^6.

11.7.11. Show that for $m, n = 0, 1, 2, \ldots,$

$$\int_{-1}^{1} \frac{x^m T_n(x)\, dx}{\sqrt{1-x^2}} = \frac{\pi}{2^m}\binom{m}{(m-n)/2}, \quad m > n,\ m - n \text{ even},$$

$$= \frac{2^{1-n}\pi}{\epsilon_n}, \quad m = n,$$

$$= 0, \text{ otherwise}.$$

Suggestion: Let $x = \cos\theta$ and write

$$\cos^m\theta\cos n\theta = \mathrm{Re}\left\{e^{in\theta}\left(\frac{e^{i\theta} + e^{-i\theta}}{2}\right)^m\right\}.$$

11.7.12. Show that the generalized Fourier series of $f(x)$ in terms of $T_n(x)$ and $U_n(x)$ given by (11.7.1) and Exercise 11.7.7 are the half-range sine and cosine series, respectively, of $f(\cos \theta)$.

11.8 AN APPROXIMATION EXAMPLE

Sometimes when a function $f(x)$ is to be approximated by a polynomial of the form

$$f(x) = \sum_{k=0}^{n} a_k x^k + E_n(x), \quad |x| \le 1,$$

where $|E_n(x)|$ does not exceed an allowed limit, it is possible to reduce the degree of the polynomial by a process called *economization of power series*. The procedure is to convert the polynomial to a linear combination of the form

$$\sum_{k=0}^{n} a_k x^k = \sum_{k=0}^{n} b_k T_k(x),$$

yielding the approximation,

$$f(x) \doteq \sum_{k=0}^{n} b_k T_k(x).$$

It may then be possible to drop some of the last terms without permitting the error to exceed the prescribed limit. Since $|T_k(x)| \le 1$, the number of terms which can be omitted is determined by the magnitude of the coefficients b_k.

Consider, for $|x| \le 1$, the Maclaurin series with remainder,

$$\cos x = 1 - \frac{x^2}{2} + \frac{x^4}{24} - (\sin \xi)\frac{x^6}{6!}. \tag{11.8.1}$$

For this expression, $|E_5(x)| \le 1/6! = .0014$. Suppose that the maximum error allowed is $\pm .008$. It is not permissible to drop the term $x^4/24$ without exceeding this allowable error. Let us now convert to a sum of Chebyshev polynomials,

$$\begin{aligned}\cos x &\doteq T_0 - \tfrac{1}{2}\left(\frac{T_0 + T_2}{2}\right) + \tfrac{1}{24}\left(\frac{3T_0 + 4T_2 + T_4}{8}\right)\\ &\doteq \tfrac{49}{64}T_0 - \tfrac{11}{48}T_2 + \tfrac{1}{192}T_4.\end{aligned}$$

The maximum error introduced by eliminating T_4 is $\frac{1}{192} = .0052$, and the maximum error in the approximation

$$\cos x \doteq \tfrac{49}{64}T_0 - \tfrac{11}{48}T_2$$

is $.0014 + .0052 = .0066$, which is allowable. Converting back to a polynomial in x, we have

$$\cos x \doteq \tfrac{191}{192} - \tfrac{11}{24}x^2$$

with $|E| \leq .0066$. Thus two terms are sufficient rather than the three necessary in (11.8.1).

11.9 BOUNDARY-VALUE PROBLEMS

An example of a boundary-value problem involving Chebyshev's equation, though a highly artificial one, is the problem of the vibrating string fastened at $x = \pm 1$, with variable density $\delta(x)$ and tension $T(x)$. Suppose $\delta(x) = 1/\sqrt{1 - x^2}$ and $T(x) = \sqrt{1 - x^2}$, so that by (8.3.3) we have the system

$$\frac{1}{\sqrt{1 - x^2}}\,\psi_{tt}(x, t) = \frac{\partial}{\partial x}\,[\sqrt{1 - x^2}\;\psi_x(x, t)],$$
$$\psi(-1, t) = \psi(1, t) = 0. \tag{11.9.1}$$

Let us suppose also that the string is initially at rest with a displacement $f(x)$; that is,

$$\psi_t(x, 0) = 0, \qquad \psi(x, 0) = f(x).$$

We see that the differential equation is equivalent to

$$\psi_{tt} = (1 - x^2)\psi_{xx} - x\psi_x,$$

and upon assuming a solution, $\psi = X(x)T(t)$, we have

$$T(t) = C_1 \cos \lambda t + C_2 \sin \lambda t,$$
$$(1 - x^2)X'' - xX' + \lambda^2 X = 0.$$

The latter equation is Chebyshev's equation and if we let $x = \cos \theta$ it becomes

$$\frac{d^2X}{d\theta^2} + \lambda^2 X = 0,$$

so that

$$X = C_3 \cos \lambda\theta + C_4 \sin \lambda\theta.$$

The boundary conditions, in terms of θ, are

$$X(0) = X(\pi) = 0,$$

and hence

$$X = C_4 \sin n\theta, \quad n = 1, 2, 3, \ldots,$$

where we must have λ an integer. The value $\lambda = 0$ is trivial, and negative integers do not contribute anything to the solution. Collecting our results, noting that $C_2 = 0$, and returning to functions of x, we have

$$\psi(x, t) = \sum_{n=1}^{\infty} C_n\sqrt{1 - x^2}\; U_{n-1}(x) \cos nt.$$

The condition $\psi(x, 0) = f(x)$ determines C_n (see Exercise 11.7.7).

It is evident that a problem of this type is equivalent to problems involving θ which we have already considered. The transformation $x = \cos \theta$ used throughout the problem would have led to the same result, and the Fourier trigonometric series would have been involved. As the reader may verify in Exercise 11.7.12, the set

$$f(\theta) = \frac{a_0}{2} + \sum_{n=1}^{\infty} a_n \cos n\theta,$$

$$a_n = \frac{2}{\pi}\int_0^{\pi} f(\theta) \cos n\theta \, d\theta,$$

is equivalent to

$$F(x) = f(\arccos x) = \sum_{n=0}^{\infty} a_n T_n(x),$$

$$a_n = \frac{\epsilon_n}{\pi}\int_{-1}^{1} \frac{F(x)}{\sqrt{1-x^2}} T_n(x)\, dx,$$

and the set

$$g(\theta) = \sum_{n=1}^{\infty} b_n \sin n\theta,$$

$$b_n = \frac{2}{\pi}\int_0^{\pi} g(\theta) \sin n\theta \, d\theta,$$

is equivalent to

$$G(x) = g(\arccos x) = \sum_{n=1}^{\infty} b_n \sqrt{1-x^2}\, U_{n-1}(x),$$

$$b_n = \frac{2}{\pi}\int_{-1}^{1} G(x) U_{n-1}(x)\, dx.$$

EXERCISES

11.9.1. Approximate $\sin x$ for $|x| < 1$ by a polynomial with error not to exceed .007 in magnitude. Use a truncated Maclaurin series and "economization" if possible.

11.9.2. Approximate e^{-x} for $|x| < 1$ by a polynomial with error not to exceed .06 in magnitude, using economization if possible.

Ans. $e^{-x} \doteq \frac{91}{64}T_0 - \frac{9}{8}T_1 + \frac{13}{48}T_2 = \frac{221}{192} - \frac{9}{8}x + \frac{13}{24}x^2.$

11.9.3. Solve the system

$$\psi_{tt}(x, t) = (1 - x^2)\psi_{xx}(x, t) - x\psi_x(x, t), \quad -1 < x < 1, t > 0,$$
$$\psi(-1, t) = \psi(1, t) = \psi(x, 0) = 0,$$
$$\psi_t(x, 0) = f(x).$$

Ans. $\psi = \sum_{n=1}^{\infty} C_n \sqrt{1-x^2}\, U_{n-1}(x) \sin nt,$

$$C_n = \frac{2}{n\pi}\int_0^{\pi} f(\cos\theta) \sin n\theta \, d\theta.$$

11.9.4. Solve the system

$$\psi_{tt}(x, t) = (1 - x^2)\psi_{xx}(x, t) - x\psi_x + ax, \quad -1 < x < 1, t > 0,$$
$$\psi(-1, t) = \psi(1, t) = \psi(x, 0) = \psi_t(x, 0) = 0.$$

11.9.5. Show that if cylindrical coordinates are expressed as

$$x = \xi_1\xi_2,$$
$$y = \xi_1\sqrt{1 - \xi_2^2},$$
$$z = \xi_3,$$

then the variables will separate in

$$\nabla^2\psi(\xi_1, \xi_2, \xi_3) = 0,$$

and the separated equation for ξ_2 is Chebyshev's equation.

12

Mathieu Functions

12.1 MATHIEU'S EQUATION

If we apply the method of separation of variables to the wave equation in two space variables, or in the equivalent problem, to the Helmholtz equation with sinusoidal time dependence (see Sec. 8.1),

$$\nabla^2 W + K^2 W = 0, \tag{12.1.1}$$

we seek a solution of the form

$$W(\xi, \eta) = F(\xi)G(\eta) \tag{12.1.2}$$

where ξ, η are the space coordinates in W. If the coordinates under consideration are the elliptic-cylinder coordinates, related to the cartesian coordinates by the equations,

$$\begin{aligned} x &= a \cosh \xi \cos \eta, \\ y &= a \sinh \xi \sin \eta, \end{aligned} \tag{12.1.3}$$

where a is a constant, Eq. (12.1.1) takes the form (see Exercise 1.1.3)

$$\frac{2}{a^2(\cosh 2\xi - \cos 2\eta)}(W_{\xi\xi} + W_{\eta\eta}) + K^2 W = 0. \tag{12.1.4}$$

Substitution of (12.1.2) into (12.1.4) leads to the separated equations

$$\begin{aligned} G'' + (\lambda - 2h \cos 2\eta)G &= 0, \\ F'' - (\lambda - 2h \cosh 2\xi)F &= 0, \end{aligned} \tag{12.1.5}$$

where λ is a separation constant and

$$2h = \frac{a^2K^2}{2}. \tag{12.1.6}$$

Replacing ξ by $i\xi$ in the second of (12.1.5) reduces it to the form of the first of (12.1.5); hence the problem of finding W is essentially that of solving the first of (12.1.5), which we write in the form

$$y'' + (\lambda - 2h \cos 2x)y = 0. \tag{12.1.7}$$

This equation is known as *Mathieu's equation*; and the *Mathieu functions*, with which we shall be concerned in the remainder of this chapter, are solutions of this equation. The second of (12.1.5) is known as the *modified* equation of Mathieu.

As can be seen from the coordinate relations, the variable η represents an angle, and hence in a physical problem we shall be interested in solutions of the first of (12.1.5) which are periodic of period 2π. The variable ξ appears as the argument of hyperbolic functions which have period $2\pi i$, and the substitution $x = i\xi$ necessary to reduce the second of (12.1.5) to (12.1.7) indicates that in seeking solutions in ξ we must also seek solutions of (12.1.7) of period 2π. Therefore in solving (12.1.7) in terms of η or ξ we shall concern ourselves with solutions of period 2π.

12.2 PROPERTIES OF ELLIPTIC-CYLINDER COORDINATES

Before proceeding to the solution of Mathieu's equation, we shall need to consider in more detail the elliptic-cylinder coordinate system. This can best be done by developing the equations (12.1.3), beginning with the known properties of the ellipse.

An ellipse with foci at $(\pm a, 0)$ and eccentricity e is given by

$$\frac{x^2}{\alpha^2} + \frac{y^2}{\beta^2} = 1 \tag{12.2.1}$$

where $e = a/\alpha$ and $\beta^2 = \alpha^2 - a^2$. Since for the ellipse, $1/e > 1$, we may make the substitution $1/e = \cosh \xi$, $0 < \xi < \infty$, and eliminate α and β from (12.2.1), to obtain

$$\frac{x^2}{a^2 \cosh^2 \xi} + \frac{y^2}{a^2 \sinh^2 \xi} = 1. \tag{12.2.2}$$

The degenerate ellipse corresponding to $\xi = 0$ is the line segment joining the foci. We note that for a fixed, (12.2.2) represents a family of confocal ellipses, each member of which is specified by $\xi =$ constant.

The nature of Eq. (12.2.2), a sum of two squares equated to 1, suggests the substitutions

$$\frac{x}{a \cosh \xi} = \cos \eta,$$

$$\frac{y}{a \sinh \xi} = \sin \eta, \quad 0 \le \eta \le 2\pi,$$

which are the two equations of (12.1.3).

Eliminating ξ from these last two equations, using the identity $\cosh^2 \xi - \sinh^2 \xi = 1$, we have for a fixed a the family of confocal hyperbolas

$$\frac{x^2}{a^2 \cos^2 \eta} - \frac{y^2}{a^2 \sin^2 \eta} = 1. \tag{12.2.3}$$

We note that giving η a constant value specifies a member of the family with eccentricity $1/\cos \eta$.

EXERCISES

12.2.1. Verify Eqs. (12.1.5) and (12.1.6).

12.2.2. Show that the coordinate curves given by (12.2.2) and (12.2.3) are mutually orthogonal.

12.3 SOLUTION OF MATHIEU'S EQUATION

We have seen in Sec. 12.1 that we are interested only in solutions of Mathieu's equation which have period 2π. It is evident that if $y(x, \lambda, h)$ is a solution of (12.1.7), then $y(x + 2\pi, \lambda, h)$ is also a solution. The constraint that these two solutions are equal (i.e., y is periodic) gives us an interrelation between λ and h which we express as

$$\lambda = \sum_{n=0}^{\infty} a_n h^n. \tag{12.3.1}$$

Any periodic solution must be either (a) even, (b) odd, or (c) neither even nor odd. Suppose $y(x)$ is a solution of type (c). Then as can be seen from the differential equation, $y(-x)$ is also a solution and is independent of $y(x)$. The linear combinations

$$y_1(x) = \frac{y(x) + y(-x)}{2},$$

which is an even function, and

$$y_2(x) = \frac{y(x) - y(-x)}{2},$$

which is odd, are also solutions. Thus we may restrict our attentions to either even or odd solutions.

In seeking an even periodic solution we note first that for the case $h = 0$, from (12.1.7) $\lambda = a_0$ which we take as $\lambda = a_0 = m^2$, and the differential equation becomes

$$y'' + m^2 y = 0. \tag{12.3.2}$$

Thus the even periodic solution for this case is $y = \cos mx$ where m is a positive integer. Therefore our even periodic solution $y(x, \lambda, h)$ for a general h should have the properties

$$\begin{aligned} y(x, m^2, 0) &= \cos mx, \\ y(-x, \lambda, h) &= y(x, \lambda, h), \\ y(x + 2\pi, \lambda, h) &= y(x, \lambda, h). \end{aligned} \tag{12.3.3}$$

We observe that the system

$$\begin{aligned} y &= \mathrm{ce}_m\,(x, h) = \cos mx + \sum_{n=0}^{\infty} b_n(h) \cos nx, \\ \lambda &= m^2 + \sum_{k=1}^{\infty} a_k h^k, \\ b_n(0) &= 0, \qquad b_m(h) \equiv 0 \end{aligned} \tag{12.3.4}$$

satisfies the required conditions (12.3.3) and is consistent with (12.3.1). The condition $b_m \equiv 0$ specifies the coefficient of $\cos mx$ as 1 and thus leads us to a particular solution. The notation $\mathrm{ce}_m\,(x, h)$ is standard for the even periodic solution, being an abbreviation for "cosine elliptic." The notation $y = \mathrm{se}_m\,(x, h)$, the "sine elliptic" function, is standard for the odd periodic solution, which we consider later. Our method of solution will consist in determining b_n and a_k so that (12.3.4) satisfies Mathieu's equation.

Substitution of the first of (12.3.4) into Mathieu's equation,

$$y'' + (\lambda - 2h \cos 2x)y = 0,$$

yields, after some simplification,

$$\begin{aligned} (\lambda - m^2) \cos mx + \sum_{n=0}^{\infty} (\lambda - n^2) b_n \cos nx - h \sum_{n=0}^{\infty} b_{n-2} \cos nx \\ - hb_0 \cos 2x - hb_1 \cos x - h \sum_{n=0}^{\infty} b_{n+2} \cos nx \\ - h \cos (m + 2)x - h \cos (m - 2)x = 0. \end{aligned}$$

Equating to zero the coefficient of $\cos nx$ and solving for b_{n+2} we have the recurrence relation

$$b_{n+2} = \frac{(\lambda - n^2)(\delta_{mn} + b_n)}{h} - b_{n-2} - b_0\delta_{n2} - b_1\delta_{n1} - \delta_{m+2,\,n} - \delta_{|m-2|,\,n}. \tag{12.3.5}$$

We also use the convention $b_n = 0$ for n negative.

To illustrate the method we consider $\mathrm{ce}_1\,(x)$, in which case (12.3.4) becomes

$$\begin{aligned} \mathrm{ce}_1\,(x, h) &= \cos x + \sum_{n=0}^{\infty} b_n \cos nx, \quad b_1 = 0, \\ \lambda &= 1 + \sum_{k=1}^{\infty} a_k h^k. \end{aligned} \tag{12.3.6}$$

From (12.3.5) for $n = 0$ we have $b_2 = \lambda b_0/h$, or

$$b_2 = \frac{(1 + a_1 h + a_2 h^2 + \cdots)b_0}{h}.$$

The condition $b_2(0) = 0$ requires either that $b_0 = 0$ or that $b_0 = h^2 f_1(h)$, where $f_1(0)$ is finite. If b_0 is not chosen to be zero, in the computation of b_4 it will be necessary that $b_0 = h^3 f_2(h)$; in the computation of b_6 we will need $b_0 = h^4 f_3(h)$; and so on. To avoid this difficulty, and since we want only a particular solution we take $b_0 = 0$. This choice, as can be seen from (12.3.5) makes $b_{2k} = 0$, $k = 0, 1, 2, \ldots$.

Continuing the process we have

$$b_3 = \frac{\lambda - 1}{h} - 1 = (a_1 - 1) + a_2 h + a_3 h^2 + \cdots,$$

and $b_3(0) = 0$ requires that $a_1 = 1$. Also, we have

$$b_5 = \frac{(\lambda - 9)b_3}{h} - 1 = \frac{(-8 + h + a_2 h^2 + \cdots)(a_2 h + a_3 h^2 + \cdots)}{h} - 1$$
$$= (-8a_2 - 1) + (a_2 - 8a_3)h + ({a_2}^2 + a_3 - 8a_4)h^3 + \cdots.$$

The condition $b_5(0) = 0$ requires that $a_2 = -\frac{1}{8}$. For $n \geq 5$, the recurrence relation (12.3.5) becomes

$$b_{n+2} = \frac{(\lambda - n^2)b_n}{h} - b_{n-2}. \tag{12.3.7}$$

The work from this point can be simplified by using l'Hospital's rule in imposing the condition $b_{n+2}(0) = 0$, which gives from (12.3.7)

$$b'_n(0) = 0, \quad n \geq 5. \tag{12.3.8}$$

Applying the condition (12.3.8) to the expression for b_5 yields $a_3 = -\frac{1}{64}$.

Continuing in this manner we may obtain as many terms as we wish in (12.3.6). The result for the first few terms is

$$\begin{aligned} \text{ce}_1(x, h) &= \cos x - \tfrac{1}{8}h(1 + \tfrac{1}{8}h + \tfrac{1}{192}h^2 + \cdots)\cos 3x \\ &\quad + \tfrac{1}{192}h^2(1 + \tfrac{1}{6}h + \cdots)\cos 5x + \cdots, \\ \lambda &= 1 + h - \tfrac{1}{8}h^2 - \tfrac{1}{64}h^3 - \tfrac{1}{1536}h^4 + \cdots. \end{aligned} \tag{12.3.9}$$

Our method has the advantage of giving directly the so-called *preferred* form (see [Mc-2], page 20), as contrasted with the form

$$\text{ce}_1(x, h) = \cos x - \tfrac{1}{8}h\cos 3x + \tfrac{1}{64}h^2(-\cos 3x + \tfrac{1}{3}\cos 5x) - \cdots,$$

obtained by one of the classical methods, which we outline in Exercises 12.3.4 and 12.3.5.

To obtain the odd periodic solutions $\mathrm{se}_m(x, h)$, we determine α_k and β_n so that Mathieu's equation is satisfied by the system

$$y = \mathrm{se}_m(x, h) = \sin mx + \sum_{n=1}^{\infty} \beta_n(h) \sin nx,$$

$$\lambda = m^2 + \sum_{k=1}^{\infty} \alpha_k h^k, \tag{12.3.10}$$

$$\beta_n(0) = 0, \quad \beta_m(h) \equiv 0.$$

For $h = 0$ this solution reduces to the required value, $y = \sin mx$.

It will be seen, upon working out $\mathrm{ce}_m(x, h)$ and $\mathrm{se}_m(x, h)$ for various values of m, that we will have only even or only odd harmonics according as m is even or odd. In the case of even harmonics, our solutions are periodic of period π.

EXERCISES

12.3.1. Show that

$$\mathrm{ce}_0(x, h) = 1 + \frac{\lambda}{h}\cos 2x + \left[\frac{\lambda(\lambda - 4)}{h^2} - 2\right]\cos 4x + \left[\frac{\lambda(\lambda - 4)(\lambda - 16)}{h^3} - \frac{2(\lambda - 16) + \lambda}{h}\right]\cos 6x + \cdots,$$

where $\lambda = -\frac{1}{2}h^2 + \frac{7}{128}h^4 - \frac{29}{2304}h^6 + \cdots$.

12.3.2. Show that the result in Exercise 12.3.1 can be written

$$\mathrm{ce}_0(x, h) = 1 + (-\tfrac{1}{2}h + \tfrac{7}{128}h^3 + \cdots)\cos 2x + (\tfrac{1}{32}h^2 - \tfrac{5}{1152}h^4 + \cdots)\cos 4x + \cdots.$$

12.3.3. Obtain the recurrence relation for $\mathrm{se}_m(x, h)$ corresponding to (12.3.5) and show that

$$\mathrm{se}_1(x, h) = \sin x - \tfrac{1}{8}h(1 - \tfrac{1}{8}h + \tfrac{1}{192}h^2 - \cdots)\sin 3x + \tfrac{1}{64}h^2(\tfrac{1}{3} - \tfrac{1}{18}h + \cdots)\sin 5x + \cdots$$

where $\lambda = 1 - h - \frac{1}{8}h^2 + \frac{1}{64}h^3 + \cdots$.

12.3.4. Show that if $y = \cos mx + \sum_{k=1}^{\infty} C_k(x)h^k$, $\lambda = m^2 + \sum_{k=1}^{\infty}\alpha_k h^k$ is to satisfy (12.1.7), then

$$(D^2 + m^2)C_1 = \cos(m + 2)x + \cos(m - 2)x - \alpha_1 \cos mx,$$

$$(D^2 + m^2)C_k = (2\cos 2x - \alpha_1)C_{k-1} - \sum_{n=2}^{k-1}\alpha_n C_{k-n} - \alpha_k \cos mx, \quad k > 1.$$

12.3.5. Show that

$$\frac{1}{D^2 + m^2}\cos mx = \frac{x}{2m}\sin mx$$

and

$$\frac{1}{D^2 + m^2} \cos nx = \frac{1}{-n^2 + m^2} \cos nx.$$

Use these formulas to solve the system of equations in Exercise 12.3.4 for the case $m = 2$, noting that, for a periodic solution, the α_k must be determined to suppress the terms involving $x \sin 2x$.

$$\text{Ans.}\quad \text{ce}_2(x, h) = \cos 2x - \tfrac{1}{8}h(\tfrac{2}{3}\cos 4x - 2) + \tfrac{1}{384}h^2 \cos 6x$$
$$- \tfrac{1}{512}h^3(\tfrac{1}{45}\cos 8x + \tfrac{43}{27}\cos 4x + \tfrac{40}{3}) + \cdots,$$
$$\lambda = 4 + \tfrac{5}{12}h^2 - \tfrac{763}{13824}h^4 + \cdots.$$

12.3.6. Obtain $\text{se}_2(x, h)$ by a method similar to that in Exercises 12.3.4 and 12.3.5.

$$\text{Ans.}\quad \text{se}_2(x, h) = \sin 2x - \tfrac{1}{12}h \sin 4x + \tfrac{1}{384}h^2 \sin 6x$$
$$- \tfrac{1}{512}h^3(\tfrac{1}{45}\sin 8x - \tfrac{5}{27}\sin 4x) + \cdots,$$
$$\lambda = 4 - \tfrac{1}{12}h^2 + \tfrac{5}{13824}h^4 + \cdots.$$

12.4 NATURE OF THE GENERAL SOLUTIONS

If $h \neq 0$, the functions $\text{ce}_m(x, h)$ and $\text{se}_m(x, h)$ cannot form the general solution of Mathieu's equation because, as we have seen, their computations require different values of λ. The standard notation for the general solution of Mathieu's equation is either

$$y = C_1 \,\text{ce}_m(x, h) + C_2 \,\text{fe}_m(x, h) \tag{12.4.1}$$

or

$$y = C_3 \,\text{se}_m(x, h) + C_4 \,\text{ge}_m(x, h) \tag{12.4.2}$$

corresponding respectively to the cases where one independent solution is even and of period 2π, or odd and of period 2π. In this section we will discuss the nature of the second solutions, $\text{fe}_m(x, h)$ and $\text{ge}_m(x, h)$.

We begin by noting from Exercise 3.4.11 that the Wronskian of the two independent solutions is given by

$$W(x) = y_1 y'_2 - y'_1 y_2 = W_0 \tag{12.4.3}$$

where W_0 is a constant different from zero. We see that fe_m cannot be even, for then the Wronskian of ce_m and fe_m would be the difference of two odd functions and could not satisfy (12.4.3). By similar reasoning ge_m cannot be odd. Since in Sec. 12.3 we have ruled out any solutions except those which are even or odd, fe_m must therefore be odd and ge_m must therefore be even.

The functions fe_m and ge_m cannot be periodic of period 2π for if, say, fe_m has period 2π, then since it is odd it must be either se_{2k} or se_{2k+1}. These two possibilities, paired with ce_m for m even and odd, give us four cases to consider. It may be shown that each case leads to a contradiction. One such case is considered in Exercises 12.4.1 and 12.4.2.

Since fe_m and ge_m are not periodic, they will not appear in the solution of boundary-value problems which we shall consider. For this reason we shall not be concerned with their computation. The interested reader is referred to [Mc-2] for a discussion of these functions.

In the notation of Sec. 12.1 the functions G and F in (12.1.5) will be given by

$$\begin{aligned} G_1(\eta) &= C_1 \,\mathrm{ce}_m(\eta, h) + C_2 \,\mathrm{fe}_m(\eta, h), \\ F_1(\xi) &= C_3 \,\mathrm{Ce}_m(\xi, h) + C_4 \,\mathrm{Fe}_m(\xi, h), \end{aligned} \tag{12.4.4}$$

or for a different λ,

$$\begin{aligned} G_2(\eta) &= C_5 \,\mathrm{se}_m(\eta, h) + C_6 \,\mathrm{ge}_m(\eta, h), \\ F_2(\xi) &= C_7 \,\mathrm{Se}_m(\xi, h) + C_8 \,\mathrm{Ge}_m(\xi, h). \end{aligned} \tag{12.4.5}$$

The capital letters used for the modified functions are standard, and the periodic modified functions, as we have seen, are obtained from ce_m and se_m. The standard relationships are

$$\begin{aligned} \mathrm{Ce}_m(\xi, h) &= \mathrm{ce}_m(i\xi, h), \\ \mathrm{Se}_m(\xi, h) &= -i \,\mathrm{se}_m(i\xi, h). \end{aligned} \tag{12.4.6}$$

Finally, to obtain $W = FG$ in (12.1.2) we may use F_1G_1 in (12.4.4) or F_2G_2 in (12.4.5), each of which represents an infinite set of solutions corresponding to the infinite set of values that λ may have. We may not, however, use F_1G_2 or F_2G_1 since the factors in each of these products are based on different eigenvalues. In a physical problem in which periodicity is required, the coefficients with even subscripts in (12.4.4) and (12.4.5) must be taken as zero so that the possible eigenfunctions are the two sets,

$$\begin{aligned} W_c &= A \,\mathrm{Ce}_m(\xi, h) \,\mathrm{ce}_m(\eta, h), \\ W_s &= B \,\mathrm{Se}_m(\xi, h) \,\mathrm{se}_m(\eta, h). \end{aligned} \tag{12.4.7}$$

To be more specific about the modified functions, we write Eqs. (12.3.4) and (12.3.10) respectively as

$$\begin{aligned} \mathrm{ce}_m(x, h) &= \sum_{k=0}^{\infty} b_{km} \cos kx, \\ \mathrm{se}_m(x, h) &= \sum_{k=1}^{\infty} \beta_{km} \sin kx, \end{aligned} \tag{12.4.8}$$

so that by (12.4.6) we have

$$\begin{aligned} \mathrm{Ce}_m(x, h) &= \sum_{k=0}^{\infty} b_{km} \cosh kx, \\ \mathrm{Se}_m(x, h) &= \sum_{k=1}^{\infty} \beta_{km} \sinh kx. \end{aligned} \tag{12.4.9}$$

EXERCISES

12.4.1. If

$$\mathrm{ce}_{2n+1}(x, h) = \sum_{r=0}^{\infty} A_{2r+1} \cos (2r + 1)x$$

and

$$\mathrm{se}_{2k+1}(x, h) = \sum_{r=0}^{\infty} B_{2r+1} \sin (2r + 1)x,$$

show that

$$\begin{aligned} (\lambda - 1 - h)A_1 - hA_3 &= 0, \\ [\lambda - (2r + 1)^2]A_{2r+1} - h(A_{2r+3} + A_{2r-1}) &= 0, \quad r \geq 1, \\ (\lambda - 1 + h)B_1 - hB_3 &= 0, \\ [\lambda - (2r + 1)^2]B_{2r+1} - h(B_{2r+3} + B_{2r-1}) &= 0, \quad r \geq 1. \end{aligned}$$

12.4.2. Using the results of Exercise 12.4.1, show that if

$$\Delta_r = \begin{vmatrix} A_{2r+1} & A_{2r+3} \\ B_{2r+1} & B_{2r+3} \end{vmatrix},$$

then $\Delta_r = \Delta_{r-1} = \cdots = \Delta_0 = 2A_1B_1$. Use this to deduce that the two series in Exercise 12.4.1 cannot converge and therefore cannot together represent solutions of Mathieu's equation.

12.5 ORTHOGONALITY OF THE PERIODIC SOLUTIONS

If y_1 and y_2 are distinct solutions to Mathieu's equation corresponding to λ_1 and λ_2, respectively, but for the same value of h, then we have

$$y''_i + (\lambda_i - 2h \cos 2x)y_i = 0, \quad i = 1, 2.$$

Applying Eq. (1.9.8) to these two relations yields

$$(\lambda_2 - \lambda_1)\int_a^b y_1 y_2 \, dx = y'_1 y_2 - y'_2 y_1 \Big|_a^b.$$

If y_1 and y_2 are periodic solutions, then the right member is zero if we choose $a = 0$, $b = 2\pi$, and thus the functions y_1 and y_2 are orthogonal over $(0, 2\pi)$:

$$\int_0^{2\pi} y_1 y_2 \, dx = 0. \tag{12.5.1}$$

This relation holds for all the combinations of $\mathrm{ce}_m(x, h)$ and $\mathrm{se}_m(x, h)$ such as, for example,

$$\int_0^{2\pi} \mathrm{ce}_m(x, h)\, \mathrm{se}_n(x, h)\, dx = 0. \tag{12.5.2}$$

To obtain the orthogonality relations for ce_m and se_m we make use of

Eqs. (12.4.8), arriving at

$$\int_0^{2\pi} \mathrm{ce}_m{}^2 (x, h)\, dx = \int_0^{2\pi} \left[\sum_{k=0}^{\infty} b_{km} \cos kx \right]^2 dx$$
$$= \sum_{k=0}^{\infty} b_{km}^2 \int_0^{2\pi} \cos^2 kx\, dx = \sum_{k=0}^{\infty} \pi b_{km}^2,$$

and

$$\int_0^{2\pi} \mathrm{se}_m{}^2 (x, h)\, dx = \sum_{k=1}^{\infty} \pi \beta_{km}^2.$$

These orthogonality relations are then

$$\int_0^{2\pi} \mathrm{ce}_m (x, h)\, \mathrm{ce}_n (x, h)\, dx = \left[\sum_{k=0}^{\infty} \pi b_{km}^2 \right] \delta_{mn} \tag{12.5.3}$$

and

$$\int_0^{2\pi} \mathrm{se}_m (x, h)\, \mathrm{se}_n (x, h)\, dx = \left[\sum_{k=1}^{\infty} \pi \beta_{km}^2 \right] \delta_{mn}. \tag{12.5.4}$$

It is known (see [Mc-2], page 37) that the series (12.4.8) converge uniformly for all x, so that the above operations are valid.

A more useful orthogonality relation is the one involving the function $W(\xi, \eta)$ defined by one or the other of Eqs. (12.4.7) and also by the differential equation (12.1.4), which we write, using (12.1.6), as

$$LW = W_{\xi\xi} + W_{\eta\eta} + 2h(\cosh 2\xi - \cos 2\eta) W = 0. \tag{12.5.5}$$

Let us suppose that $W_{nm}(\xi, \eta)$ is the eigenfunction corresponding to the eigenvalue h_{nm} and let us suppose further that the h_{nm} are such that for a given ξ_0,

$$W_{nm}(\xi_0, \eta) = 0. \tag{12.5.6}$$

This is equivalent, by (12.4.7), to supposing that either

$$\mathrm{Ce}_m (\xi_0, h_{nm}) = 0,$$

or

$$\mathrm{Se}_m (\xi_0, h_{nm}) = 0, \quad n,m = 0, 1, 2, \ldots .$$

(It is known that both the Mathieu and the modified Mathieu functions have an infinite number of zeros and that they are real and simple. See [Mc-2], Chapter 12.)

If from (12.5.5) we form the combination

$$W_{rs} L W_{nm} - W_{nm} L W_{rs},$$

rearrange, and integrate with respect to ξ and η between the limits 0, ξ_0 and 0, 2π respectively, we have

$$\int_0^{2\pi} \left[W_{rs} \frac{\partial W_{nm}}{\partial \xi} - W_{nm} \frac{\partial W_{rs}}{\partial \xi} \right]_0^{\xi_0} d\eta + \int_0^{\xi_0} \left[W_{rs} \frac{\partial W_{nm}}{\partial \eta} - W_{nm} \frac{\partial W_{rs}}{\partial \eta} \right]_0^{2\pi} d\xi$$
$$= 2(h_{rs} - h_{nm}) \int_0^{2\pi} \int_0^{\xi_0} W_{nm} W_{rs} (\cosh 2\xi - \cos 2\eta)\, d\xi\, d\eta.$$

The left member of this equation is zero, the first integral vanishing because of (12.5.6) and (12.4.9), and the second vanishing because of the periodicity of W. Hence if $h_{rs} \neq h_{nm}$, the integral on the right is zero.

The norm $\| W_{nm} \|$ (see Sec. 1.10) is given by

$$\| W_{nm} \|^2 = \int_0^{2\pi} \int_0^{\xi_0} (\cosh 2\xi - \cos 2\eta) W_{nm}^2 \, d\xi \, d\eta, \tag{12.5.7}$$

which is a positive quantity since $\cosh 2\xi - \cos 2\eta \geq 0$ and $W_{nm}^2 \geq 0$. The orthogonality relation is then

$$\int_0^{2\pi} \int_0^{\xi_0} (\cosh 2\xi - \cos 2\eta) W_{nm} W_{rs} \, d\xi \, d\eta = \| W_{nm} \|^2 \, \delta_{nm}^{rs}. \tag{12.5.8}$$

12.6 AN EXAMPLE

Consider the problem of determining the temperature distribution in an infinite right elliptic cylinder whose lateral surfaces are kept at zero temperature and whose initial temperature distribution is independent of z. Suppose that the equation of the cross-section in the x-y plane is

$$\frac{x^2}{9} + \frac{y^2}{4} = 1,$$

which is represented by

$$\xi = \xi_0 = \cosh^{-1} \frac{\sqrt{5}}{3}$$

in elliptic-cylinder coordinates. This ellipse is the member of the family of concentric ellipses with foci at $(\pm\sqrt{5}, 0)$ and eccentricity $\sqrt{5}/3$.

Letting $U(\xi, \eta, t)$ be the temperature, we may separate the time factor by assuming that

$$U(\xi, \eta, t) = W(\xi, \eta) T(t).$$

The two separated equations obtained when this expression is substituted into the heat equation are

$$\nabla^2 W(\xi, \eta) + \lambda^2 W = 0,$$

$$T'(t) = -\lambda^2 k T(t).$$

The first of these is the Helmholtz equation which is treated in Sec. 12.1. Since we require that the solution be periodic in ξ and η, $W(\xi, \eta)$ must be of the form [see (12.4.7)]

$$W_c(\xi, \eta) = A \operatorname{Ce}_m (\xi, h) \operatorname{ce}_m (\eta, h),$$

$$W_s(\xi, \eta) = B \operatorname{Se}_m (\xi, h) \operatorname{se}_m (\eta, h),$$

where by (12.1.6) we have

$$2h = \frac{5\lambda^2}{2}.$$

To satisfy the condition that the lateral surface be kept at zero temperature we must have

$$W_c(\xi_0, \eta) = A \operatorname{Ce}_m(\xi_0, h) \operatorname{ce}_m(\eta, h) = 0,$$

$$W_s(\xi_0, \eta) = B \operatorname{Se}_m(\xi_0, h) \operatorname{se}_m(\eta, h) = 0,$$

for all η, $0 \leq \eta \leq 2\pi$, and hence $h = h_{nm}$ or $h = h^*_{nm}$ must be a root of

$$\operatorname{Ce}_m(\xi_0, h_{nm}) = 0$$

$$\operatorname{Se}_m(\xi_0, h^*_{nm}) = 0, \quad n,m = 0, 1, 2, \ldots.$$

The temperature function then has the form

$$U(\xi, \eta, t) = \sum_{n,m=0}^{\infty} [A_{nm} \operatorname{Ce}_m(\xi, h_{nm}) \operatorname{ce}_m(\eta, h_{nm}) e^{-(4/5)h_{nm}kt} + B_{nm} \operatorname{Se}_m(\xi, h^*_{nm}) \operatorname{se}_m(\eta, h^*_{nm}) e^{-(4/5)h^*_{nm}kt}].$$

If the initial temperature is given by

$$U(\xi, \eta, 0) = f(\xi, \eta),$$

the coefficients are found to be, using (12.5.8),

$$A_{nm} = \frac{\int_0^{2\pi}\int_0^{\xi_0} (\cosh 2\xi - \cos 2\eta) \operatorname{Ce}_m(\xi, h_{nm}) \operatorname{ce}_m(\eta, h_{nm}) f(\xi, \eta)\, d\xi\, d\eta}{\|\operatorname{Ce}_m(\xi, h_{nm}) \operatorname{ce}_m(\eta, h_{nm})\|^2},$$

$$B_{nm} = \frac{\int_0^{2\pi}\int_0^{\xi_0} (\cosh 2\xi - \cos 2\eta) \operatorname{Se}_m(\eta, h^*_{nm}) \operatorname{se}_m(\eta, h^*_{nm}) f(\xi, \eta)\, d\xi\, d\eta}{\|\operatorname{Se}_m(\eta, h^*_{nm}) \operatorname{se}_m(\eta, h^*_{nm})\|^2}.$$

EXERCISES

12.6.1. Supply the details in the derivation of (12.5.8).

12.6.2. Obtain C_{nm} if $f(\xi, \eta) = \Sigma_{m=0}^{\infty} \Sigma_{n=0}^{\infty} C_{nm} W_{nm}(\xi, \eta)$, where W_{nm} is defined by (12.5.6) and one of (12.4.7).

12.6.3. Show that

$$\frac{1}{\pi}\int_0^{2\pi} \operatorname{ce}_m{}^2(x, h) \cos 2x\, dx = \sum_{n=0}^{\infty} b_{nm} b_{n+2,m} + b_{2m} b_{0m} + \tfrac{1}{2} b_{1m}^2.$$

12.6.4. Evaluate

$$\frac{1}{\pi}\int_0^{2\pi} \operatorname{se}_m{}^2(x, h) \cos 2x\, dx.$$

12.6.5. Find the transverse displacements of an elliptic membrane, fixed on the boundary and subjected to an initial displacement $f(\xi, \eta)$ and zero initial velocity.

12.6.6. Find the transverse displacements of an elliptic membrane, fixed on the boundary with zero initial displacement and initial velocity $g(\xi, \eta)$.

12.6.7. An infinite right elliptical cylinder has its curved surface, $x^2 + 4y^2 = 4$, kept at zero temperature. If the initial rate of change of the temperature is $f(\xi, \eta)$, find the temperature in the cylinder.

13

Other Special Functions

13.1 HYPERGEOMETRIC FUNCTION

In this chapter we will conclude our study of the properties of the special functions by considering the *hypergeometric functions*, the *Jacobi polynomials*, and the *Bessel polynomials*. We define the first of these as the solutions of *Gauss's* or the *hypergeometric equation*,

$$x(1 - x)y'' + [\gamma - (\alpha + \beta + 1)x]y' - \alpha\beta y = 0, \qquad (13.1.1)$$

where α, β, γ are constants. Since $x = 0$ and $x = 1$ are regular singular points, we may use the method of Frobenius to obtain a solution of the type

$$y = \sum_{n=0}^{\infty} a_n x^{n+m}.$$

The method yields the indicial equation

$$a_0 m(m + \gamma - 1) = 0 \qquad (13.1.2)$$

and the recurrence relation

$$a_{n+1} = \frac{(n + m + \alpha)(n + m + \beta)}{(n + m + 1)(n + m + \gamma)} a_n. \qquad (13.1.3)$$

Carrying out the solution for the root $m = 0$ of (13.1.2), and taking $a_0 = 1$, we obtain the series commonly designated

$$y_1 = F(\alpha, \beta, \gamma, x) = 1 + \frac{\alpha\beta}{1!\,\gamma} x + \frac{\alpha(\alpha + 1)\beta(\beta + 1)}{2!\,\gamma(\gamma + 1)} x^2 + \cdots$$
$$+ \frac{\alpha(\alpha + 1)\cdots(\alpha + n - 1)\beta(\beta + 1)\cdots(\beta + n - 1)}{n!\,\gamma(\gamma + 1)\cdots(\gamma + n - 1)} x^n + \cdots, \qquad (13.1.4)$$

where γ is not zero or a negative integer. The series (13.1.4) is called the hypergeometric function or hypergeometric series because it is a generalization of the elementary geometric series, which may be obtained from (13.1.4); for if $\alpha = 1$, $\beta = \gamma$, then

$$F(1, \beta, \beta, x) = \sum_{n=0}^{\infty} x^n.$$

Recalling our definition of $(\alpha)_k$,

$$\begin{aligned}(\alpha)_k &= \alpha(\alpha+1)\cdots(\alpha+k-1), \quad k = 1, 2, 3, \ldots, \\ (\alpha)_0 &= 1,\end{aligned} \tag{13.1.5}$$

we may write (13.1.4) in the more compact form

$$F(\alpha, \beta, \gamma, x) = \sum_{k=0}^{\infty} \frac{(\alpha)_k(\beta)_k x^k}{k!\,(\gamma)_k}. \tag{13.1.6}$$

To check (13.1.4) for convergence, we note first that the series becomes a polynomial if α or β is a negative integer. Otherwise, since by (13.1.3)

$$\left|\frac{a_{n+1}}{a_n}\right| \sim 1,$$

the ratio test gives the interval of convergence as $|x| < 1$.

To obtain the other solution of (13.1.1), we use the other root of the indicial equation, $m = 1 - \gamma$. For this case it may be shown directly from the differential equation that the second solution, valid for γ not an integer greater than 1, is given by

$$y_2 = x^{1-\gamma}F(\alpha - \gamma + 1, \beta - \gamma + 1, 2 - \gamma, x). \tag{13.1.7}$$

Another method of obtaining the second solution will be indicated in Exercise 13.1.1.

The ratio test yields the same interval of convergence $|x| < 1$ for (13.1.7) since it was applied to the general coefficients (13.1.3). Therefore the general solution of the hypergeometric equation is

$$y = c_1F(\alpha, \beta, \gamma, x) + c_2x^{1-\gamma}F(\alpha - \gamma + 1, \beta - \gamma + 1, 2 - \gamma, x), \tag{13.1.8}$$

valid for $|x| < 1$, when γ is not an integer.

Equations of the form

$$(x - a)(x - b)y'' + (cx + d)y' + ey = 0, \tag{13.1.9}$$

where a, b, c, d, e are constants, and $a \neq b$, may be put in the form (13.1.1), and hence the solutions of (13.1.9) may be expressed in terms of $F(\alpha, \beta, \gamma, x)$. To see this, we note that the transformation

$$x - a = (b - a)z \tag{13.1.10}$$

changes (13.1.9) to

$$z(1 - z)y'' + \left[\frac{ac + d}{a - b} - cz\right]y' - ey = 0, \tag{13.1.11}$$

so that by comparison with (13.1.1) we have

$$\gamma = \frac{ac + d}{a - b}, \qquad \alpha + \beta + 1 = c, \qquad \alpha\beta = e \tag{13.1.12}$$

For example, the Legendre polynomials may be written (see Exercise 13.1.4), for n a positive integer, in the form

$$P_n(x) = F(-n, n + 1, 1, \tfrac{1}{2}[1 - x]). \tag{13.1.13}$$

EXERCISES

13.1.1. Obtain (13.1.7) by letting $y_2 = x^{1-\gamma}v$ and determining v from (13.1.1).

13.1.2. If we write (13.1.1) in operator form,

$$L(\alpha, \beta, \gamma, x, D)y = 0,$$

show that, for $n = 0, 1, 2, \ldots,$

$$L(\alpha + n, \beta + n, \gamma + n, x, D)\frac{d^n y}{dx^n} = 0.$$

13.1.3. Show that
(a) $F(\alpha, \beta, \beta, x) = (1 - x)^{-\alpha}$,
(b) $\log(1 + x) = xF(1, 1, 2, -x)$,
(c) $e^x = \lim_{\alpha\to\infty} F(\alpha, \beta, \beta, x/\alpha)$.

13.1.4. Derive (13.1.13).

13.1.5. Show that $T_n(x) = F(-n, n, \frac{1}{2}, \frac{1}{2}[1 - x])$. (See Chapter 11.)

13.1.6. Obtain a general solution in terms of hypergeometric functions of

$$(x - 3)(x - 1)y'' + (3x - 2)y' - 15y = 0.$$

13.1.7. Show that

$$(1 - x^2)y'' + \left[\frac{2\gamma - 1}{x} - (2\alpha + 2\beta + 1)x\right]y' - 4\alpha\beta y = 0$$

has as a solution

$$y = c_1F(\alpha, \beta, \gamma, x^2) + c_2x^{2-2\gamma}F(\alpha - \gamma + 1, \beta - \gamma + 1, 2 - \gamma, x^2),$$

γ not an integer, $|x| < 1$.

13.1.8. Show that $(1 - x^2)y'' + 2y = 0$ has solutions

$$y_1 = F(\tfrac{1}{2}, -1, \tfrac{1}{2}, x^2) = 1 - x^2,$$

$$y_2 = xF(1, -\tfrac{1}{2}, \tfrac{3}{2}, x^2) = -\sum_{n=0}^{\infty} \frac{x^{2n+1}}{4n^2 - 1}.$$

13.1.9. Show that Legendre's equation has solutions

$$y_1 = F(\tfrac{1}{4} + \tfrac{1}{2}\sqrt{\tfrac{1}{4} + n(n+1)},\, \tfrac{1}{4} - \tfrac{1}{2}\sqrt{\tfrac{1}{4} + n(n+1)},\, \tfrac{1}{2},\, x^2),$$

$$y_2 = xF(\tfrac{3}{4} + \tfrac{1}{2}\sqrt{\tfrac{1}{4} + n(n+1)},\, \tfrac{3}{4} - \tfrac{1}{2}\sqrt{\tfrac{1}{4} + n(n+1)},\, \tfrac{3}{2},\, x^2),$$

and obtain the first few Legendre polynomials from these expressions.

13.1.10. Show that Chebyshev's equation has solutions

$$y_1 = F\left(\frac{n}{2}, -\frac{n}{2}, \frac{1}{2}, x^2\right),$$

$$y_2 = xF\left(\frac{n+1}{2}, \frac{-n+1}{2}, \frac{3}{2}, x^2\right).$$

13.1.11. Show that

$$(1 - x^3)y'' + \left[\frac{3\gamma - 2}{x} - (3\alpha + 3\beta + 1)x^2\right]y' - 9\alpha\beta xy = 0$$

has solutions

$$y_1 = F(\alpha, \beta, \gamma, x^3) \text{ and } y_2 = x^{3-3\gamma}F(\alpha - \gamma + 1, \beta - \gamma + 1, 2 - \gamma, x^3),$$

γ not an integer, $|x| < 1$.

13.1.12. Show that $y_1 = F(\alpha, \tfrac{2}{3} - \alpha, \tfrac{2}{3}, x^3)$ and $y_2 = xF(\alpha + \tfrac{1}{3}, -\alpha + 1, \tfrac{4}{3}, x^3)$ are solutions of

$$[(1 - x^3)y']' + 3\alpha(3\alpha - 2)xy = 0.$$

13.1.13. Let $y_1 = G_n(x) = F(-n, \tfrac{1}{3}[3n + 2], \tfrac{2}{3}, x^3)$, n a positive integer, in Exercise 13.1.12. Show that $G_n(x)$ is a polynomial of degree $3n$ satisfying

$$[(1 - x^3)y']' + 3n(3n + 2)xy = 0.$$

Write out G_0, G_1, G_2, G_3.

13.1.14. Show that $\int_0^1 xG_nG_m\,dx = 0$, $n \neq m$, where $G_n(x)$ is given in Exercise 13.1.13.

13.1.15. Show that

$$(1 - x^m)y'' + \left[\frac{m\gamma + 1 - m}{x} - (m\alpha + m\beta + 1)x^{m-1}\right]y' - m^2\alpha\beta x^{m-2}y = 0$$

has solutions

$$y_1 = F(\alpha, \beta, \gamma, x^m), \qquad y_2 = x^{m-m\gamma}F(\alpha - \gamma + 1, \beta - \gamma + 1, 2 - \gamma, x)$$

if γ is not an integer and $|x| < 1$. Note that Exercises 13.1.7 and 13.1.11 are special cases.

13.1.16. In Exercise 13.1.15, let $\alpha + \beta = \gamma = (m - 1)/m$ to obtain

$$[(1 - x^m)y']' + m\alpha(m\alpha - m + 1)x^{m-2}y = 0,$$

with solutions

$$y_1 = F\left(\alpha, \frac{m-1}{m} - \alpha, \frac{m-1}{m}, x^m\right),$$

$$y_2 = xF\left(\alpha + \frac{1}{m}, -\alpha + 1, \frac{m+1}{m}, x^m\right).$$

13.1.17. Let $\alpha = -n$, n a positive integer, in Exercise 13.1.16, and show that, for

$$[(1 - x^k)y']' + kn(kn + k - 1)x^{k-2}y = 0,$$

y_1 is a polynomial solution which we designate as

$$G_{kn}(x) = F\left(-n, \frac{kn + k - 1}{k}, \frac{k - 1}{k}, x^k\right).$$

Show that for $k \geq 2$,

$$\int_0^1 x^{k-2}G_{kn}(x)G_{km}(x)\,dx = 0, n \neq m,$$

and if k is an even integer ≥ 2,

$$\int_{-1}^1 x^{k-2}G_{kn}(x)G_{km}(x)\,dx = 0, n \neq m.$$

13.1.18. Show that

$$F'(\alpha, \beta, \gamma, x) = \frac{\alpha\beta}{\gamma} F(\alpha + 1, \beta + 1, \gamma + 1, x);$$

extend this result to

$$\frac{d^n}{dx^n} F(\alpha, \beta, \gamma, x) = \frac{(\alpha)_n(\beta)_n}{(\gamma)_n} F(\alpha + n, \beta + n, \gamma + n, x).$$

13.1.19. Assuming that n is a non-negative integer, obtain a Rodrigues' formula for $F(-n, \beta, \gamma, x)$, using the method of Sec. 1.11.

Ans. $F(-n, \beta, \gamma, x) = C_n x^{1-\gamma}(1 - x)^{\gamma+n-\beta} D^n[x^{n+\gamma-1}(1 - x)^{\beta-\gamma}]$.

13.2 JACOBI POLYNOMIALS

In the previous section we have seen that the hypergeometric equation, which we repeat in the form

$$z(1 - z)y'' + [c - (a + b + 1)z]y' - aby = 0, \tag{13.2.1}$$

has a polynomial solution

$$y = F(-n, b, c, z) \tag{13.2.2}$$

if $a = -n$, n a non-negative integer.

We may put (13.2.1) in the form

$$(1 - x^2)y'' + [a + b + 1 - 2c - (a + b + 1)x]y' - aby = 0 \tag{13.2.3}$$

if we let $z = \frac{1}{2}(1 - x)$, a substitution which is suggested by the forms of Eqs. (13.1.9) and (13.1.10). The primes in this case indicate differentiation with respect to x. The solution (13.2.2) now becomes, for $a = -n$,

$$y = F\left(-n, b, c, \frac{1 - x}{2}\right). \tag{13.2.4}$$

Later in this section we shall be interested in reducing (13.2.3) to self-adjoint form (see Sec. 1.9), and to do so we shall need to integrate the term

$$\frac{a+b+1-2c-(a+b+1)x}{(1-x)(1+x)} = \frac{-c}{1-x} + \frac{a+b+1-c}{1+x}. \quad (13.2.5)$$

To make the integration more convenient to perform we shall make the substitutions

$$c = 1 + \alpha, \qquad a + b + 1 - c = 1 + \beta,$$

and upon solving for a, b, and c, we have

$$a = -n, \qquad b = \alpha + \beta + n + 1, \qquad c = 1 + \alpha.$$

When these values are substituted into (13.2.1), we have

$$(1-x^2)y'' + [\beta - \alpha - (\alpha + \beta + 2)x]y' + n(n + \alpha + \beta + 1)y = 0, \quad (13.2.6)$$

with a polynomial solution of degree n given by

$$y = kF\left(-n, \alpha + \beta + n + 1, 1 + \alpha, \frac{1-x}{2}\right). \quad (13.2.7)$$

When the constant k is given the particular value $k = (1+\alpha)_n/n!$, the function y is known as the *Jacobi polynomial* of degree n, for which a more or less standard notation is

$$P_n^{(\alpha,\beta)}(x) = \frac{(1+\alpha)_n}{n!} F\left(-n, \alpha + \beta + n + 1, 1 + \alpha, \frac{1-x}{2}\right). \quad (13.2.8)$$

The symbol $P_n^{(\alpha,\beta)}$ is appropriate because, as we may see from (13.2.6) and (13.2.8),

$$P_n^{(0,0)}(x) = P_n(x), \quad (13.2.9)$$

where $P_n(x)$ is the Legendre polynomial of degree n.

We leave as an exercise (see Exercise 13.4.2) to show that the Chebyshev polynomials may be written in terms of Jacobi polynomials

$$P_n^{(-1/2,-1/2)}(x) = \frac{(\frac{1}{2})_n}{n!} T_n(x). \quad (13.2.10)$$

13.3 RODRIGUES' FORMULA FOR JACOBI POLYNOMIALS

We may obtain a Rodrigues' formula by using the general results in Exercise 13.1.19 for hypergeometric polynomials. Replacing x by $\frac{1}{2}(1-x)$ in that result, and making the substitutions for β and γ as indicated in (13.2.8), we have

$$P_n^{(\alpha,\beta)}(x) = K_n(1-x)^{-\alpha}(1+x)^{-\beta}D^n[(1-x)^{\alpha+n}(1+x)^{\beta+n}]. \quad (13.3.1)$$

We have lumped all the constants into a new constant K_n.

To determine K_n we use Leibnitz's formula, to obtain

$$(1-x)^{-\alpha}(1+x)^{-\beta}D^n[(1-x)^{n+\alpha}(1+x)^{n+\beta}]$$
$$=\sum_{k=0}^{n}\binom{n}{k}(-1)^k(\alpha+n-k+1)_k(1-x)^{n-k}(\beta+k+1)_{n-k}(1+x)^k, \quad (13.3.2)$$

from which we see that, for $x = 1$, the only non-zero term in the series occurs when $k = n$. This term is therefore the coefficient of K_n in (13.3.1) when $x = 1$. From (13.2.8) we see that

$$P_n^{(\alpha,\beta)}(1) = \frac{(1+\alpha)_n}{n!}, \quad (13.3.3)$$

and therefore

$$\frac{(1+\alpha)_n}{n!} = K_n(1+\alpha)_n(-1)^n 2^n.$$

Solving for K_n and substituting in (13.3.1), we have the Rodrigues' formula

$$P_n^{(\alpha,\beta)}(x) = \frac{(-1)^n}{2^n n!}(1-x)^{-\alpha}(1+x)^{-\beta}D^n[(1-x)^{\alpha+n}(1+x)^{\beta+n}]. \quad (13.3.4)$$

13.4 ORTHOGONALITY OF THE JACOBI POLYNOMIALS

Equation (13.2.6) may be put in the self-adjoint form (1.9.2) by evaluating the function $r(x)$ given by (1.9.4), which in this case, with the aid of (13.2.5) and its subsequent substitutions, may be shown to be

$$r(x) = (1-x)^{\alpha+1}(1+x)^{\beta+1}.$$

We note by comparison of (1.9.2) with (13.2.6) that $p(x) = 0$, $w(x) = (1-x)^\alpha(1+x)^\beta$, and $\lambda_n = n(n+\alpha+\beta+1)$. Therefore, by (1.9.8), the integral relation of the Jacobi polynomials is

$$(m-n)(m+n+\alpha+\beta+1)\int_a^b(1-x)^\alpha(1+x)^\beta P_n^{(\alpha,\beta)}(x)P_m^{(\alpha,\beta)}(x)\,dx$$
$$= \{(1-x)^{\alpha+1}(1+x)^{\beta+1}[P_m^{(\alpha,\beta)}(x)DP_n^{(\alpha,\beta)}(x) - P_n^{(\alpha,\beta)}(x)DP_m^{(\alpha,\beta)}(x)]\}_a^b.$$

If $a = -1$, $b = 1$, $n \neq m$, $\alpha > -1$, $\beta > -1$, then the right member is zero and $\{P_n^{(\alpha,\beta)}(x)\}$ is an orthogonal set over $(-1, 1)$ with weight function $w(x) = (1-x)^\alpha(1+x)^\beta$.

To evaluate the norm of $P_n^{(\alpha,\beta)}(x)$, we use the Rodrigues' formula and write

$$\|P_n^{(\alpha,\beta)}\|^2 = \frac{(-1)^n}{2^n n!}\int_{-1}^{1}P_n^{(\alpha,\beta)}(x)D^n[(1-x)^{\alpha+n}(1+x)^{\beta+n}]\,dx. \quad (13.4.1)$$

Upon performing an integration by parts we obtain

$$I = P_n^{(\alpha,\beta)}(x)D^{n-1}[(1-x)^{\alpha+n}(1+x)^{\beta+n}]\Big|_{-1}^{1}$$
$$-\int_{-1}^{1}DP_n^{(\alpha,\beta)}D^{n-1}[(1-x)^{\alpha+n}(1+x)^{\beta+n}]\,dx,$$

where I is the integral in the right member of (13.4.1). Again using Leibnitz's formula we see that $D^k[(1-x)^{\alpha+n}(1+x)^{\beta+n}]$ vanishes at $x = 1$ and $x = -1$ for $0 < k \le n - 1$, so that the integrated part is zero. Hence, after integrating by parts n times, we obtain

$$I = (-1)^n \int_{-1}^{1} D^n P_n^{(\alpha,\beta)}(x)[(1-x)^{\alpha+n}(1+x)^{\beta+n}]\,dx. \tag{13.4.2}$$

Since $P_n^{(\alpha,\beta)}(x)$ is a hypergeometric function, we may use (13.2.8) and Exercise 13.1.18 to obtain

$$D^n P_n^{(\alpha,\beta)}(x) = \frac{(1+\alpha+\beta+n)_n}{2^n} F\left(0, \alpha+\beta+2n+1, 1+\alpha+n, \frac{1-x}{2}\right), \tag{13.4.3}$$

which is equivalent to

$$D^n P_n^{(\alpha,\beta)}(x) = \frac{(1+\alpha+\beta+n)_n}{2^n}. \tag{13.4.4}$$

Substituting this value into (13.4.2), we have

$$I = (-1)^n \frac{(1+\alpha+\beta+n)_n}{2^n} \int_{-1}^{1} (1-x)^{\alpha+n}(1+x)^{\beta+n}\,dx.$$

The integration can now be carried out in terms of the beta function by the use of Exercise 5.1.7, resulting in

$$I = (-1)^n(1+\alpha+\beta+n)_n 2^{\alpha+\beta+n+1} B(\alpha+n+1, \beta+n+1).$$

Substituting this value of I into (13.4.1) and replacing the beta function by its equivalent in Exercise 5.1.4, we have the relation for the norm,

$$\|P_n^{(\alpha,\beta)}\|^2 = \frac{2^{\alpha+\beta+1}(1+\alpha+\beta+n)_n\Gamma(\alpha+n+1)\Gamma(\beta+n+1)}{n!\,\Gamma(\alpha+\beta+2n+2)} \tag{13.4.5}$$

We note finally that (13.4.5) may be simplified and combined with the orthogonality property already shown to obtain the orthogonality relation

$$(P_n^{(\alpha,\beta)}(x), P_m^{(\alpha,\beta)}(x)) = \frac{2^{\alpha+\beta+1}\Gamma(\alpha+n+1)\Gamma(\beta+n+1)\,\delta_{mn}}{n!\,(\alpha+\beta+2n+1)\Gamma(\alpha+\beta+n+1)}, \tag{13.4.6}$$

where the interval is $(-1, 1)$ and $w(x) = (1-x)^\alpha(1+x)^\beta$. The expression on the left of (13.4.6) is explicitly given by

$$(P_n^{(\alpha,\beta)}(x), P_m^{(\alpha,\beta)}(x)) = \int_{-1}^{1} (1-x)^\alpha(1+x)^\beta P_n^{(\alpha,\beta)}(x)P_m^{(\alpha,\beta)}(x)\,dx. \tag{13.4.7}$$

EXERCISES

13.4.1. Show that $y = (1-x)^{-\alpha}(1+x)^{-\beta}D^n[(1-x)^{n+\alpha}(1+x)^{n+\beta}]$ satisfies Eq. (13.2.6).

13.4.2. Establish (13.2.10).

13.4.3. Using (13.1.6) and (13.2.8), obtain

$$P_n^{(\alpha,\beta)}(x) = \frac{(1+\alpha)_n}{n!} \sum_{m=0}^{n} \sum_{k=m}^{n} \frac{(-n)_k(\alpha+\beta+n+1)_k(-1)^m \binom{k}{m} x^m}{k!\, 2^k (1+\alpha)_k}.$$

13.4.4. The differential equation in Exercise 7.7.7 arises in quantum mechanics in the study of the angular momentum of a symmetric top. The function U is required to be continuous and square-integrable (Sec.9.6). For $\lambda = -p(p+1)$ where $p = \max(|m|, |n|)$, m and n integers, the substitutions $t = \frac{1}{2}(1-\cos\theta)$ and $F(\theta) = t^\alpha(1-t)^\beta y$, $\alpha = \frac{1}{2}|m-n|$, $\beta = \frac{1}{2}|m+n|$ yield Jacobi's equation for y. Take the special case $m - n \geq 0$, $m + n \geq 0$ and show that

$$y = CP_N^{(a,b)}(t),$$

where

$$N = p - m, \quad a = m - n, \quad b = m + n.$$

(See [MM], pages 368–371.)

13.4.5. Show that $P_n^{(\alpha,\beta)}(-x) = (-1)^n P_n^{(\beta,\alpha)}(x)$.

13.4.6. Obtain $P_n^{(\alpha,\beta)}(0)$ and $P_n^{(\alpha,\beta)}(-1)$.

13.4.7. If we define the *Gegenbauer polynomials* as

$$C_n^{\nu}(x) = \frac{(2\nu)_n P_n^{(\nu-1/2,\nu-1/2)}(x)}{(\nu+\frac{1}{2})_n},$$

show that $C_n^{\nu}(x)$ satisfies

$$(1-x^2)y'' - (2\nu+1)xy' + n(n+2\nu)y = 0.$$

Note that $P_n(x) = C_n^{1/2}(x)$.

13.4.8. Show that

$$C_n^{\nu}(x) = \frac{(-1)^n(2\nu)_n}{2^n n!\,(\nu+\frac{1}{2})_n}(1-x^2)^{-\nu+1/2} D^n (1-x^2)^{\nu+n-1/2}$$

and that

$$\int_{-1}^{1}(1-x^2)^{\nu-1/2}\, C_n^{\nu}(x)\, C_m^{\nu}(x)\, dx = \|C_n^{\nu}\|^2 \delta_{mn},$$

where

$$\|C_n^{\nu}\|^2 = \frac{2^{2\nu-1}(2\nu)_n \Gamma^2(\nu+\frac{1}{2})}{n!\,(\nu+n)\Gamma(2\nu)}.$$

13.4.9. Show that

$$2(n+1)P_{n+1}^{(\alpha,\beta)}(x) = (n+\beta+1)(x-1)P_n^{(\alpha+1,\beta)}(x) + (n+\alpha+1)(x+1)P_n^{(\alpha,\beta+1)}(x).$$

13.4.10. Show that

$$(x^2-1)DP_n^{(\alpha,\beta)}(x) = 2(n+1)P_{n+1}^{(\alpha-1,\beta-1)}(x) - [(\alpha+\beta)x + (\alpha-\beta)]P_n^{(\alpha,\beta)}(x).$$

13.4.11. Show that

$$D^k P_n^{(\alpha,\beta)}(x) = \frac{1}{2^k}(1+\alpha+\beta+n)_k P_{n-k}^{(\alpha+k,\beta+k)}(x), \quad k = 1, 2, 3, \ldots, n.$$

13.4.12. Show that

$$2n(\alpha+\beta+n)(\alpha+\beta+2n-2)P_n^{(\alpha,\beta)}(x)$$
$$=(\alpha+\beta+2n-1)[\alpha^2-\beta^2+x(\alpha+\beta+2n)(\alpha+\beta+2n-2)]P_{n-1}^{(\alpha,\beta)}(x)$$
$$-2(\alpha+n-1)(\beta+n-1)(\alpha+\beta+2n)P_{n-2}^{(\alpha,\beta)}(x).$$

13.4.13. Show that if $f(x) \sim \sum_{n=0}^{\infty} C_n P_n^{(\alpha,\beta)}(x)$, then

$$\|P_n^{(\alpha,\beta)}\|^2 C_n = \frac{(-1)^n}{2^n n!}\int_{-1}^{1} f(x) D^n[(1-x)^{\alpha+n}(1+x)^{\beta+n}]\,dx.$$

13.4.14. The polynomials $P_n^{(\alpha,\alpha)}(x)$ are called *ultraspherical polynomials*. Special cases, as we have seen, are the Legendre, Chebyshev, and Gegenbauer polynomials. Show that their orthogonality relations and Rodrigues' formulas are special cases of (13.4.6) and (13.3.4).

13.4.15. Obtain the differential equation, the orthogonality relation, and the Rodrigues' formula for $P_n^{(\alpha,\alpha)}(x)$.

13.5 BESSEL POLYNOMIALS

As a last example of a set of special functions, we consider a comparatively new set of orthogonal polynomials, first defined by Krall and Frink in 1949 (see [KF]) as *Bessel polynomials*. According to their definition, these are the polynomial solutions $\{y_n(x)\}$ of degree n of the system

$$x^2y'' + (2x+2)y' = n(n+1)y,$$
$$y(0) = 1, \quad n = 0, 1, 2, \ldots. \tag{13.5.1}$$

Even though $x = 0$ is an irregular singular point we may obtain the polynomial solution by the method of Frobenius using the form

$$y = \sum_{m=0}^{\infty} a_m x^{m+k}.$$

We shall find that for a_0 arbitrary, we have $k = 0$ and

$$a_{m+1} = \frac{(n-m)(n+m+1)}{2(m+1)}\,a_m,$$

which results in

$$y_n(x) = \sum_{m=0}^{n} \frac{(n+m)!\,x^m}{2^m m!\,(n-m)!}, \tag{13.5.2}$$

the Bessel polynomial of degree n. We have taken $a_0 = 1$ to satisfy the requirement $y_n(0) = 1$.

The name, Bessel polynomial, doubtless was chosen because of the relation of the set $\{y_n(x)\}$ to the spherical Bessel functions $J_{n+1/2}(x)$, $n = 0, \pm 1, \pm 2, \ldots$ (see Sec. 6.5). Krall and Frink, in considering the wave equation in spherical coordinates, showed by means of several changes of variables that the Bessel polynomials were solutions of the radial equation (the equation involving r), and that they were related to the spherical Bessel functions by the equation

$$y_n\left(\frac{1}{ir}\right) = \sqrt{\frac{\pi r}{2}}\, e^{ir}[i^{-n-1}J_{n+1/2}(r) + i^n J_{-n-1/2}(r)]. \qquad (13.5.3)$$

We undertake to establish this equation by another method in the exercises (see Exercises 13.6.3 and 13.6.4).

The reader may note that the differential equation in (13.5.1) is not new to us, having arisen in Exercise 1.11.4(f) in connection with Rodrigues' formulas in general. In that exercise we saw that the Bessel polynomials are given by the Rodrigues' formula

$$y_n(x) = C_n e^{2/x} D^n(x^{2n}e^{-2/x}). \qquad (13.5.4)$$

To evaluate C_n we observe that

$$D(x^{2n}e^{-2/x}) = (2x^{2n-2} + 2nx^{2n-1})e^{-2/x},$$

the lower-degree term in the polynomial coming from $x^{2n}De^{-2/x}$. The lowest-degree term in the polynomial for $D^2(x^{2n}e^{-2/x})$ will be $4x^{2n-4}$, coming from $2x^{2n-2}De^{-2/x}$. In general this lowest-degree term for the operator D^k will be $2^k x^{2n-2k}$, so that for $k = n$, the constant term is 2^n. Since $y_n(0) = 1$, (13.5.4) must then be

$$y_n(x) = 2^n e^{2/x} D^n(x^{2n}e^{-2/x}), \quad n = 0, 1, 2, \ldots, \qquad (13.5.5)$$

The Bessel polynomials constitute an orthogonal set, in the sense that the inner product is defined as a contour integral with x considered as a complex variable. This type of orthogonality is outside our range of interests here, but for the interested reader we give the result obtained by Krall and Frink,

$$(y_n, y_m) = \int_C e^{-2/x} y_n(x) y_m(x)\, dx = (-1)^{n+1} \frac{4\pi i\, \delta_{mn}}{2n+1}. \qquad (13.5.6)$$

The integration is around the unit circle with center at the origin ($x = e^{i\theta}$, $0 \le \theta < 2\pi$). We give some other results in the exercises at the end of the chapter.

13.6 SOME RELATED POLYNOMIALS

A set of polynomials which are related to the Bessel polynomials is the set

$$\theta_n(x) = x^n y_n\left(\frac{1}{x}\right), \qquad (13.6.1)$$

considered by Burchnall in 1951 (see [Bur]). These are also polynomials of degree n, as is seen from the series representation

$$\theta_n(x) = \sum_{m=0}^{n} \frac{(n + m)!\, x^{n-m}}{2^m m!\, (n - m)!}, \tag{13.6.2}$$

which is obtained from (13.5.2).

The differential equation satisfied by the Burchnall polynomials may be obtained from (13.5.1) by first replacing x by $1/t$, resulting in

$$t^2 y''\left(\frac{1}{t}\right) - 2t^2 y'\left(\frac{1}{t}\right) = n(n + 1)y\left(\frac{1}{t}\right), \tag{13.6.3}$$

and then letting $y(1/t) = t^{-n}\theta_n(t)$. The result is

$$x\theta''_n(x) - 2(n + x)\theta'_n(x) + 2n\theta_n(x) = 0, \tag{13.6.4}$$

where we have replaced t by x.

The Rodrigues' formula may be obtained by the method of Sec. 1.11 as was done for the Bessel polynomials. We leave as an exercise (see Exercise 13.6.7) to show that the result is

$$\theta_n(x) = \frac{1}{(-2)^n}\, x^{2n+1} e^{2x} D^n(x^{-n-1} e^{-2x}). \tag{13.6.5}$$

Employing the operator shift, we may write this equation in the form

$$\theta_n(x) = \frac{1}{(-2)^n}\, x^{2n+1}(D - 2)^n(x^{-n-1}). \tag{13.6.6}$$

EXERCISES

13.6.1. Show that

$$\begin{aligned} y_0(x) &= 1 \\ y_1(x) &= 1 + x \\ y_2(x) &= 1 + 3x + 3x^2 \\ y_3(x) &= 1 + 6x + 15x^2 + 15x^3 \\ y_4(x) &= 1 + 10x + 45x^2 + 105x^3 + 105x^4. \end{aligned}$$

13.6.2. Use induction to show that

$$\sum_{k=0}^{n} \frac{(-1)^k}{2k + 1}\binom{n}{k} = \frac{(n!)^2 2^{2n}}{(2n + 1)!}.$$

13.6.3. Show that $r^{-1/2}e^{-ir}y_n(1/ir)$ satisfies Bessel's equation of order $n + \frac{1}{2}$, and hence that

$$y_n\left(\frac{1}{ir}\right) = r^{1/2}e^{ir}[A_nJ_{n+1/2}(r) + B_nJ_{-n-1/2}(r)].$$

Suggestion: Start with (13.6.3).

13.6.4. Equate appropriate coefficients in Exercise 13.6.3, making use of Exercise 13.6.2, to obtain (13.5.3).

13.6.5. Show that

$$J_{n+1/2}(r) = \frac{1}{\sqrt{2\pi r}}\left[i^{-n-1}e^{ir}y_n\left(-\frac{1}{ir}\right) + i^{n+1}e^{-ir}y_n\left(\frac{1}{ir}\right)\right],$$

$$J_{-n-1/2}(r) = \frac{1}{\sqrt{2\pi r}}\left[i^{n}e^{ir}y_n\left(-\frac{1}{ir}\right) + i^{-n}e^{-ir}y_n\left(\frac{1}{ir}\right)\right].$$

13.6.6. Show that

$$y_n\left(\frac{1}{x}\right) = (-1)^n x^{n+1}e^x\left(\frac{1}{x}\frac{d}{dx}\right)^n\left(\frac{e^{-x}}{x}\right)$$

and that

$$\theta_n(x) = (-1)^n x^{2n+1}e^x\left(\frac{1}{x}\frac{d}{dx}\right)^n\left(\frac{e^{-x}}{x}\right)$$

(see Sec. 6.5).

13.6.7. Obtain (13.6.5), except for the constant factor, by the method of Sec. 1.11. Use the form corresponding to (13.6.6) to equate appropriate coefficients to obtain the constant factor.

13.6.8. Assume a recurrence relation of type

$$y_n(x) + (Ax + B)y_{n-1}(x) + (Cx + D)y_{n-2}(x) = 0$$

and show that

$$y_n(x) - (2n - 1)xy_{n-1}(x) - y_{n-2}(x) = 0.$$

From this, show that

$$\theta_n(x) - (2n - 1)\theta_{n-1}(x) - x^2\theta_{n-2}(x) = 0.$$

13.6.9. Show that

$$\begin{aligned}
\theta_0(x) &= 1\\
\theta_1(x) &= x + 1\\
\theta_2(x) &= x^2 + 3x + 3\\
\theta_3(x) &= x^3 + 6x^2 + 15x + 15\\
\theta_4(x) &= x^4 + 10x^3 + 45x^2 + 105x + 105.
\end{aligned}$$

13.6.10. Using the generating functions for the spherical Bessel functions, which are known (see [W], page 140) to be

$$\sqrt{\frac{2}{\pi x}}\cos\sqrt{x^2 - 2xt} = \sum_{n=0}^{\infty}\frac{J_{n-1/2}(x)}{n!}t^n,$$

$$\sqrt{\frac{2}{\pi x}}\sin\sqrt{x^2 + 2xt} = \sum_{n=0}^{\infty}\frac{J_{1/2-n}(x)}{n!}t^n,$$

show that

$$\exp\,(ix - i\sqrt{x^2 + 2ixt}) = \sum_{n=0}^{\infty} \frac{y_{n-1}(1/ix)}{n!}\, t^n.$$

13.6.11. It is known from complex-variable theory that the integral in (13.5.6) has the value $(y_n, y_m) = 2\pi i a_{-1}$, where

$$e^{-2/x} y_n y_m = \sum_{k=-\infty}^{\infty} a_k x^k.$$

Use this to show that (13.5.6) holds for $m,n = 0, 1, 2$.

13.6.12. Show that $(y_0, y_n) = 0$, $n > 0$, leads to the formula

$$\sum_{k=0}^{n} \frac{(-1)^k (n + k)!}{k!\,(k + 1)!\,(n - k)!} = 0,$$

or

$$\sum_{k=0}^{n} \frac{(-1)^k}{k + 1} \binom{n}{k} \binom{n + k}{k} = 0.$$

14

Laplace and Fourier Transforms

14.1 INTRODUCTION

In the next two chapters we will give an introduction to operational mathematics with a view toward exhibiting some other properties of the special functions and their applications. We will also consider some boundary-value problems which are more readily solved by operational methods.

The operational mathematics which we shall consider will be confined to a study of linear integral transformations of the form

$$T\{f(t)\} = \int_a^b K(s, t) f(t)\, dt = F(s) \qquad (14.1.1)$$

and their applications to engineering problems. The function $F(s)$ is known as the *image* or *transform* of $f(t)$, and $K(s, t)$ is called the *kernel* of the transformation. The limits a and b, finite or infinite, represent the end points of the interval over which $f(t)$ must be defined.

One important use of transforms is their application to differential equations. When (14.1.1) is applied to an ordinary differential equation, an algebraic equation in s is obtained. The effect of applying the transform to an nth-order partial differential equation is to exclude a particular independent variable temporarily, leaving a differential equation of order $n - 1$ with parameter s.

In order for a transform to be useful in solving differential equations, it must have an inverse. That is, if $T\{f(t)\} = F(s)$ is known, it is desired to find the *inverse transform* $f(t) = T^{-1}\{F(s)\}$. This is usually accomplished by the use of tables, complex inversion integrals, or summing an infinite series. We must also be able to obtain transforms of certain derivatives of $f(t)$ in terms of s, $F(s)$, and known constants. For example, assuming that $T\{f(t)\}$ and $T\{f'(t)\}$ exist, by definition (14.1.1) we have

$$T\{f'(t)\} = \int_a^b K(s, t) f'(t)\, dt.$$

Integration by parts yields

$$T\{f'(t)\} = f(t)K(s,t)\Big|_a^b - \int_a^b K_t(s,t)f(t)\,dt. \tag{14.1.2}$$

To obtain from (14.1.2) the type of transform for $f'(t)$ as outlined above, it is necessary that the integrated part exist and that

$$K_t(s,t) = g(s)K(s,t), \tag{14.1.3}$$

or

$$K(s,t) = C(s)e^{tg(s)}. \tag{14.1.4}$$

When $K(s, t) = e^{-st}$ and the limits are $a = 0$, $b = \infty$, the transformation $T\{f\}$ is known as the *Laplace* transformation. It is the best known and most widely used of the integral transforms, particularly in boundary-value problems. There are a number of problems, however, in which it is advantageous to use other kernels and hence other transforms, many of which arise from Sturm-Liouville systems. In this chapter and in Chapter 15 we will consider the Laplace transform and some of the integral transforms whose kernels are the special functions we have considered throughout the book.

14.2 LAPLACE TRANSFORM

As we noted in the previous section, the Laplace transformation is given by

$$\mathscr{L}\{f(t)\} = \int_0^\infty e^{-st}f(t)\,dt = F(s). \tag{14.2.1}$$

The parameter s is generally complex, but for the present time it is more convenient to consider it as real. The existence of the Laplace transform will depend on the function $f(t)$ and on the parameter s. For example, let $f(t) = u(t)$, the *unit step function*, defined as

$$\begin{aligned} u(t) &= 0, \quad t < 0 \\ &= 1, \quad t > 0. \end{aligned}$$

Then

$$\mathscr{L}\{u(t)\} = \int_0^\infty e^{-st}\,dt = \frac{1}{s}, \quad s > 0.$$

Also, as another example,

$$\mathscr{L}\{e^{at}\} = \frac{1}{s-a}, \quad s > a.$$

Since (14.2.1) is a linear transformation, we have

$$\mathscr{L}\{c_1f_1(t) + c_2f_2(t)\} = c_2\mathscr{L}\{f_1(t)\} + c_2\mathscr{L}\{f_2(t)\},$$

and hence

$$\mathscr{L}\{Au(t) + Be^{at}\} = \frac{A}{s} + \frac{B}{s - a}, \quad s > 0 \quad \text{and} \quad s > a.$$

We shall now develop sufficient conditions for the existence of the Laplace transform of $f(t)$. We begin by defining a function $f(t)$ to be of *exponential order* $e^{\alpha t}$ as t tends to infinity if constants M, α, and T can be found such that

$$|f(t)| < Me^{\alpha t} \text{ when } t \geq T.$$

Any bounded function is necessarily of exponential order; for if $|f(t)| < M$, it follows that $|f(t)| < Me^{\alpha t}$ for any $\alpha \geq 0$. The function e^{2t} is also of exponential order since $|e^{2t}| < 2e^{2t}$.

If $f(t)$, in addition to being of exponential order, is sectionally continuous for all finite intervals in the domain $t \geq 0$, then we can find M such that

$$|e^{-st}f(t)| < Me^{-(s-\alpha)t}.$$

Since

$$\int_0^\infty Me^{-(s-\alpha)t}\, dt = \frac{M}{s - \alpha}, \quad s > \alpha, \tag{14.2.2}$$

we see by the Weierstrass M-test for integrals (see Exercise 14.2.9) that the Laplace transform of $f(t)$ exists. Also, using (14.2.2), we see that

$$\lim_{s \to \infty} F(s) = 0.$$

Thus sufficient conditions for the existence of the Laplace transform are that the function $f(t)$ be of exponential order and be sectionally continuous on every interval $0 \leq t \leq T$ for every positive number T. These conditions are simple to apply and are satisfied by most functions arising in applications. However, they are sufficient and not necessary as is seen by the classic example

$$\mathscr{L}\{t^{-1/2}\} = \sqrt{\frac{\pi}{s}}. \tag{14.2.3}$$

This is a special case of the general formula,

$$\mathscr{L}\{t^n\} = \frac{\Gamma(n + 1)}{s^{n+1}}, \quad n > -1, \tag{14.2.4}$$

which has been obtained previously, in Exercise 5.1.2.

The inverse transform $\mathscr{L}^{-1}\{F(s)\}$ may be expressed explicitly as a contour integral by considering s as a complex variable. However, for our purposes it is sufficient to obtain the inverse transforms by matching entries in a table. A short table of Laplace transforms (Table 1) is given in Appendix C for this purpose.

An example of an inverse transform which we have already obtained is

$$u(t) = \mathscr{L}^{-1}\left\{\frac{1}{s}\right\}.$$

We note that the function

$$f(t) = u(t), \quad t \neq 2$$

$$f(2) = 6,$$

has the same transform as $u(t)$. Thus $\mathscr{L}^{-1}\{1/s\}$ is not unique. In general, the inverse transform is not unique, as we shall see.

A theorem due to Lerch states that if two functions have the same transform they differ by a *null* function $N(t)$, which has the property that, for every $T > 0$,

$$\int_0^T N(t)\,dt = 0.$$

As a consequence of Lerch's theorem, we see that if a given $F(s)$ has an inverse transform which is a continuous function of t, then it is the only continuous function whose transform is $F(s)$. For our purposes this ambiguity of the inverse transform will present no difficulties.

EXERCISES

14.2.1. Find $\mathscr{L}\{f(t)\}$, where $f(t)$ is as follows:

(a) $\sin kt$.

(b) $\cosh kt$.

(c) $e^{at} \cos bt$.

(d) $\cos^2 kt$.

(e) $f(t) = \sin t$ when $0 < t < 1$, $f(t) = 0$ when $t > 1$.

14.2.2. Show that if $f(t)$ is sectionally continuous on every interval $r \leq t \leq T$, r and T positive, is of exponential order, and has an infinite discontinuity at $t = 0$ such that $|t^n f(t)|$ is bounded there for $0 < n < 1$, then $F(s)$ exists.

14.2.3. Show that $\mathscr{L}^{-1}$ is a linear operator.

14.2.4. Find the inverse transforms of the following:

(a) $\dfrac{s}{(s-1)(s-2)}$.

(b) $\dfrac{1}{s(s^2+1)}$.

(c) $\dfrac{2s^2+s+5}{(s-1)(s^2+2s+5)}$.

(d) $\dfrac{2}{s^2(s^2+1)}$.

Suggestion: Use partial fractions.

14.2.5. Show that $e^{at}\cos bt$ and t^n, $n \geq 0$, are both of exponential order. Determine α in each case.

14.2.6. Show that e^{t^2} is not of exponential order.

14.2.7. Show that $F(s - a) = \mathscr{L}\{e^{at}f(t)\}$ and $(1/a)F(s/a) = \mathscr{L}\{f(at)\}$. Find $\mathscr{L}\{e^{3t}t^n\}$ and $\mathscr{L}^{-1}\{1/s(s - 2)^2\}$.

14.2.8. Show that $e^{-as}F(s) = \mathscr{L}\{f(t - a)u(t - a)\}$. Find $\mathscr{L}^{-1}\{e^{-2s}/s^2\} = f(t)$. Sketch the graph of $f(t)$.

14.2.9. The improper integral $\int_c^\infty f(x, t)\,dt$ converges uniformly to a function $g(x)$ in $a \leq x \leq b$ if for any $\epsilon > 0$, there exists $T_\epsilon > 0$ such that $|\int_T^\infty f(x, t)\,dt| < \epsilon$ whenever $T > T_\epsilon$, for all x in $a \leq x \leq b$. The Weierstrass M-test for integrals says that if there exists $M(t)$ such that $|f(x, t)| \leq M(t)$ for all x in $a \leq x \leq b$, and $\int_c^\infty M(t)\,dt$ exists, then $\int_c^\infty f(x, t)\,dt$ converges uniformly. Show that if $f(t)$ is sectionally continuous on every finite interval and of exponential order $e^{\alpha t}$, (14.2.1) converges uniformly for $s \geq \alpha_1 > \alpha$.

14.2.10. If on $a \leq x \leq b$, $c \leq t$, f, f_x are continuous, $\int_c^\infty f(x, t)\,dt$ converges, and $\int_c^\infty f_x(x, t)\,dt$ converges uniformly, then

$$\frac{d}{dx}\int_c^\infty f(x, t)\,dt = \int_c^\infty f_x(x, t)\,dt.$$

(See [Bu], pages 149–156.) Use this to show that, for $f(t)$ defined in Exercise 14.2.9,

$$F^{(n)}(s) = \mathscr{L}\{(-t)^n f(t)\}, \quad n = 0, 1, 2, \ldots.$$

14.2.11. Show that $\mathscr{L}\{L_n(t)\} = \dfrac{1}{s}\left(\dfrac{s - 1}{s}\right)^n$, where $L_n(t)$ is the Laguerre polynomial.

14.2.12. Show that if $f(x) = \int_0^\infty [(\sin xt)/t]\,dt$, then $f(x) = \pi/2$, $x \neq 0$; $f(0) = 0$. Suggestion: Take $\mathscr{L}\{f(x)\}$ and assume that the transform and integral operators are interchangeable.

14.3 SOLUTIONS OF DIFFERENTIAL EQUATIONS

If $f(t)$ is of exponential order, is continuous, with $f'(t)$ sectionally continuous on every finite interval $0 \leq t \leq T$, then we may obtain $\mathscr{L}\{f'(t)\}$ in terms of $F(s)$. By definition,

$$\mathscr{L}\{f'(t)\} = \lim_{T\to\infty}\int_0^T e^{-st}f'(t)\,dt = \lim_{T\to\infty}\sum_{k=0}^{n}\int_{t_k}^{t_{k+1}} e^{-st}f'(t)\,dt,$$

where t_k, $k = 1, 2, \ldots, n - 1$, are points of discontinuity of f', and $t_0 = 0$, $t_{n+1} = T$. Integrating by parts yields

$$\mathscr{L}\{f'(t)\} = \lim_{T\to\infty}\sum_{k=0}^{n}\left\{e^{-st}f(t)\,\Big|_{t_k+0}^{t_{k+1}-0} + s\int_{t_k}^{t_{k+1}} e^{-st}f(t)\,dt\right\}.$$

Since $f(t)$ is continuous, $f(t_k + 0) = f(t_k - 0)$ and hence

$$\mathscr{L}\{f'(t)\} = sF(s) - f(0). \tag{14.3.1}$$

This is a special case of the more general theorem given as follows:

Theorem 14.3.1. Let $f^{(k)}(t)$, $k = 0, 1, 2, \ldots, n-1$, be of exponential order $e^{\alpha t}$ as $t \to \infty$ and continuous, with $f^{(n)}(t)$ sectionally continuous, on every finite interval $0 \le t \le T$. Then for $s > \alpha$ and $n = 1, 2, 3, \ldots,$

$$\mathscr{L}\{f^{(n)}(t)\} = s^n F(s) - s^{n-1}f(0) - s^{n-2}f'(0) - \cdots - f^{(n-1)}(0). \tag{14.3.2}$$

This may be proved by induction on n starting with (14.3.1).

As an example of the use of Theorem 14.3.1, we consider $f(t) = t^n$, noting that $f^{(n)}(t) = n!$, $f(0) = f'(0) = \cdots = f^{(n-1)}(0) = 0$. Substituting into (14.3.2), we have

$$\mathscr{L}\{t^n\} = \frac{n!}{s^{n+1}}.$$

As another example we will solve the system

$$y''(t) + 3y'(t) + 2y(t) = e^t, \qquad y(0) = 0, \qquad y'(0) = 1.$$

Transforming both sides of the equation we have

$$s^2 Y(s) - 1 + 3s Y(s) + 2Y(s) = \frac{1}{s-1},$$

from which we obtain

$$Y(s) = \frac{s}{(s+2)(s+1)(s-1)}.$$

Taking the inverse transform, we have

$$y(t) = -\tfrac{2}{3}e^{-2t} + \tfrac{1}{2}e^{-t} + \tfrac{1}{6}e^t.$$

To illustrate the Laplace transform method for a partial differential equation, we consider the system for a vibrating string,

$$\begin{aligned} y_{xx}(x, t) &= y_{tt}(x, t), \\ y(0, t) &= y(1, t) = y_t(x, 0) = 0, \\ y(x, 0) &= \sin \pi x. \end{aligned}$$

We may transform with respect to either variable, but because of the initial conditions we will choose t.

We note that formally, if

$$\mathscr{L}\{y(x, t)\} = \int_0^\infty y(x, t)e^{-st}\, dt = Y(x, s),$$

then we have

$$\mathscr{L}\{y_{tt}\} = s^2 Y(x, s) - sy(x, 0) - y_t(x, 0),$$

and

$$\mathscr{L}\{y_{xx}\} = \int_0^\infty y_{xx}e^{-st}\, dt = \frac{\partial^2}{\partial x^2}\int_0^\infty ye^{-st}\, dt = Y_{xx}(x, s).$$

The transformed boundary-value problem is then

$$Y_{xx}(x, s) - s^2 Y(x, s) = -s \sin \pi x,$$
$$Y(0, s) = Y(1, s) = 0.$$

Solution of this system yields

$$Y(x, s) = \frac{s}{s^2 + \pi^2} \sin \pi x,$$

and taking the inverse transform, we have

$$y(x, t) = \sin \pi x \cos \pi t.$$

This function clearly satisfies the given system.

EXERCISES

14.3.1. If $f(t)$ in Theorem 14.3.1 has a finite jump, $f(t_0 + 0) - f(t_0 - 0) = K$, $t_0 > 0$, find $\mathscr{L}\{f'(t)\}$.

14.3.2. Using (14.3.2) find

(a) $\mathscr{L}\{\cos kt\}$,

(b) $\mathscr{L}\{e^{at}\}$,

(c) $\mathscr{L}\{f(t)\}$, where $f(t) = t$, $0 < t < 1$; $f(t) = 0$, $t > 1$.

14.3.3. Solve the following systems:

(a) $y'' + 4y = 0$, $y(0) = 1$, $y'(0) = 0$.

(b) $y'' + 4y' + 3y = 1$, $y(0) = y'(0) = 0$.

(c) $y'' + 2y' + y = e^{-t}$, $y(0) = 0$, $y'(0) = 1$.

(d) $y''' + 2y'' + 2y' + y = e^{2t}$, $y''(0) = 1$, $y'(0) = y(0) = 0$.

(e) $y'' + 3y' + 2y = 1$, $y(0) = y(1) = 0$.

(f) $x' + y'' - y = 0$, $x'' + y' = e^t$, $x'(0) = 2$, $x(0) = y(0) = y'(0) = 0$.

(g) $y_x(x, t) + xy_t(x, t) = 0$, $y(x, 0) = 0$, $y(0, t) = e^{-t}$.

14.4 CONVOLUTION

The method that we have given, up to this point, for finding inverse transforms is limited. For example, it cannot be used to find

$$f(t) = \mathscr{L}^{-1}\left\{\frac{1}{(s^2 + 1)^2}\right\} \tag{14.4.1}$$

because partial-fraction expansion does not simplify the expression $F(s)$. This transform has the property, however, of being the product of two known transforms, which leads us to seek, in general, a formula for $\mathscr{L}^{-1}\{F(s)G(s)\}$.

By Exercise 14.2.8 we have that

$$e^{-s\tau}G(s) = \int_0^\infty e^{-st} g(t - \tau) u(t - \tau)\, dt. \tag{14.4.2}$$

Next we write, by definition,

$$F(s)G(s) = \int_0^\infty e^{-s\tau} f(\tau)G(s)\,d\tau,$$

which becomes, upon substitution for $e^{-s\tau}G(s)$ from (14.4.2),

$$F(s)G(s) = \int_0^\infty f(\tau)\left[\int_0^\infty e^{-st} g(t-\tau)u(t-\tau)\,dt\right]d\tau.$$

Interchanging the order of integration, and noting that $u(t-\tau) = 0$ for $\tau > t$, we have

$$F(s)G(s) = \int_0^\infty e^{-st}\left[\int_0^t f(\tau)g(t-\tau)\,d\tau\right]dt. \tag{14.4.3}$$

By the definition of the Laplace transform, this is

$$F(s)G(s) = \mathscr{L}\left\{\int_0^t f(\tau)g(t-\tau)\,d\tau\right\}. \tag{14.4.4}$$

This integral, known as the *convolution* integral, appears frequently and is symbolized by

$$f(t) * g(t) = \int_0^t f(\tau)g(t-\tau)\,d\tau. \tag{14.4.5}$$

This result can be obtained rigorously if the functions f and g are sectionally continuous and of exponential order.

EXERCISES

14.4.1. Show that $f * g = g * f$.

14.4.2. Show by use of (14.4.4) that

$$\mathscr{L}^{-1}\left\{\frac{1}{(s^2 + k^2)^2}\right\} = \frac{1}{2k^3}(\sin kt - kt\cos kt).$$

14.4.3. Show that if $y'' + 2y' + y = f(t)$, $y(0) = y'(0) = 0$, then

$$y = te^{-t} * f(t) = e^{-t}\int_0^t f(\tau)(t-\tau)e^\tau\,d\tau.$$

Use *Leibnitz's rule*,

$$\frac{d}{dx}\int_{\alpha(x)}^{\beta(x)} f(x, t)\,dt = \int_\alpha^\beta f_x(x, t)\,dt + f[x, \beta(x)]\beta'(x) - f[x, \alpha(x)]\alpha'(x),$$

to verify the solution.

14.4.4. Show that $\mathscr{L}\{\int_0^t f(\tau)\,d\tau\} = F(s)/s$.

14.4.5. Show that $\mathscr{L}^{-1}\{1/s\sqrt{s+1}\} = \text{erf}(\sqrt{t})$, the *error function* defined in (7.7.6).

14.4.6. Given that $\mathscr{L}\{(k/2\sqrt{\pi t^3})e^{-k^2/4t}\} = e^{-k\sqrt{s}}$, $k > 0$, $s > 0$ (see [C-2], pages 65, 66), show that

$$\mathscr{L}\left\{\frac{1}{\sqrt{\pi t}}\, e^{-k^2/4t}\right\} = \frac{1}{\sqrt{s}}\, e^{-k\sqrt{s}},$$

$$\mathscr{L}\left\{\text{erfc}\left(\frac{k}{2\sqrt{t}}\right)\right\} = \frac{1}{s}\, e^{-k\sqrt{s}}.$$

[See (7.7.7).]

14.4.7. Solve the following systems:

(a) $y'(t) = 1 - e^t \int_0^t e^{-x} y(x)\, dx$, $y(0) = 0$.

(b) $y'' + k^2 y = f(t)$, $y(0) = y'(0) = 0$.

(c) $y_t(x, t) = y_{xx}(x, t)$, $y(x, 0) = \lim_{x\to\infty} y(x, t) = 0$, $y(0, t) = 1$, $x > 0$, $t > 0$ [see (7.7.3)]. Ans. $y = \text{erfc}(x/2\sqrt{t})$.

(d) $y_t(x, t) = y_{xx}(x, t)$, $y(x, 0) = ax$, $y(0, t) = b$, $y(x, t) \sim ax$ for x large, $x > 0$, $t > 0$.

14.4.8. Show that if $f(t)$ is periodic of period T and sectionally continuous over $0 \le t \le T$, then

$$\mathscr{L}\{f(t)\} = \frac{\int_0^T e^{-st} f(t)\, dt}{1 - e^{-sT}}.$$

14.4.9. Find $\mathscr{L}\{f(t)\}$ if

(a) $f(t) = t, \quad 0 < t < a,$
$\quad = 2a - t, \quad a < t < 2a,$
$f(t + 2a) = f(t).$ Ans. $F(s) = (1/s^2)\tanh(as/2)$.

(b) $f(t) = k, \quad 0 < t < a,$
$\quad = -k, \quad a < t < 2a,$
$f(t + 2a) = f(t).$

14.4.10. Find $y(1, t)$ if

$$y_{tt}(x, t) = y_{xx}(x, t),\ y(x, 0) = y_t(x, 0) = y(0, t) = 0,$$
$$y_x(1, t) = 1.$$

14.4.11. Show from the series expression for $J_0(t)$ that

$$\mathscr{L}\{J_0(t)\} = \frac{1}{\sqrt{s^2 + 1}}.$$

14.4.12. Solve the system $ty'' + y' + ty = 0$, $y(0) = 1$. Use the idea in Exercise 14.2.10.

14.4.13. Show by finding $\mathscr{L}\{J'_0(t)\}$ that $L\{J_1(t)\} = (\sqrt{s^2+1} - s)/\sqrt{s^2+1}$. Extend by induction to $\mathscr{L}\{J_n(t)\} = (\sqrt{s^2+1} - s)^n/\sqrt{s^2+1}$, $n = 0, 1, 2, \ldots$.

14.4.14. Show that $\mathscr{L}\{I_n(t)\} = (s - \sqrt{s^2-1})^n/\sqrt{s^2-1}$, $n = 0, 1, 2, \ldots$, where $I_n(t)$ is the modified Bessel function.

14.4.15. Solve for $u(x, t)$:

$$k\frac{\partial^2 u}{\partial x^2} = \frac{\partial u}{\partial t}, \quad -\infty < x < a, t > 0,$$

$$u(x, 0) = 0, \qquad u_x(a, t) = \frac{b}{k}[c - u(a, t)],$$

$$\lim_{x\to-\infty} u(x, t) = 0.$$

Ans. $u = c\left\{-\exp\left[\frac{b}{k}(a - x) + \frac{b^2t}{k}\right]\operatorname{erfc}\left(b\sqrt{\frac{t}{k}} + \frac{a - x}{2\sqrt{kt}}\right) + \operatorname{erfc}\left(\frac{a - x}{2\sqrt{kt}}\right)\right\}.$

14.4.16. Solve for $u(x, t)$:

$$u_t(x, t) = ku_{xx}(x, t) - hu(x, t), \quad x > 0, t > 0,$$

$$u(x, 0) = u_0,\ u(0, t) = 0,$$

$$\lim_{x\to\infty} u_x(x, t) = 0,$$

where k, h, u_0 are constants. Ans. $u(x, t) = u_0e^{-ht}\operatorname{erf}(x/2\sqrt{kt})$.

14.4.17. Solve for $u(x, t)$:

$$u_t(x, t) = ku_{xx}(x, t) + h, \quad x > 0, t > 0.$$

$$u(0, t) = u(x, 0) = \lim_{x\to\infty} u_x(x, t) = 0.$$

Ans. $u(x, t) = ht - h\int_0^t \operatorname{erfc}\left(\frac{x}{2\sqrt{k\tau}}\right)d\tau = h\int_0^t \operatorname{erf}\left(\frac{x}{2\sqrt{k\tau}}\right)d\tau.$

14.4.18. A semi-infinite distortionless transmission line ($RC = LG$) has the initial conditions,

$$v(x, 0) = v_t(x, 0) = i(x, 0) = i_t(x, 0) = 0, \quad x > 0.$$

In addition,

$$v(0, t) = f(t) \quad \text{and} \quad \lim_{x\to\infty} v(x, t) = 0, \quad t > 0.$$

Find $v(x, t)$ (see Exercise 8.3.12).

Ans. $v(x, t) = \exp[-xR\sqrt{C/L}]f(t - x\sqrt{LC})u(t - x\sqrt{LC})$.

14.5 FOURIER TRANSFORM

In Sec. 2.7 we have seen that a function $f(t)$ can be represented by its Fourier integral if it satisfies the conditions sufficient for the validity of its Fourier series on the fundamental interval and if, in addition, $f(t)$ is *absolutely integrable*; that is,

$$\int_{-\infty}^{\infty} |f(t)|\, dt < M.$$

For $f(t)$ even and for $f(t)$ odd, the Fourier integral becomes the cosine and sine integrals, respectively, which are given in Exercise 2.7.1. We may use the result of this exercise to obtain the *Fourier sine* and *cosine transformations* (unbounded interval), which are defined respectively as

$$S_\nu\{f(t)\} = F_s(\nu) = \int_0^\infty f(t) \sin \nu t \, dt, \tag{14.5.1}$$

and

$$C_\nu\{f(t)\} = F_c(\nu) = \int_0^\infty f(t) \cos \nu t \, dt. \tag{14.5.2}$$

The inverse transforms, given by Exercise 2.7.1, are

$$f(t) = S_\nu^{-1}\{F_s(\nu)\} = \frac{2}{\pi} \int_0^\infty F_s(\nu) \sin \nu t \, d\nu \tag{14.5.3}$$

and

$$f(t) = C_\nu^{-1}\{F_c(\nu)\} = \frac{2}{\pi} \int_0^\infty F_c(\nu) \cos \nu t \, d\nu. \tag{14.5.4}$$

We note here that the inversion integrals are real, whereas the inversion integral of the Laplace transform is a contour integral. In some cases this makes the Fourier transforms more advantageous than the Laplace transform.

A transform which is more widely used than either $F_s(\nu)$ or $F_c(\nu)$ is the *Fourier transform*, which we define as

$$\mathscr{F}\{f(t)\} = F(i\omega) = \int_{-\infty}^\infty f(t) e^{-i\omega t} \, dt. \tag{14.5.5}$$

The conditions sufficient for its existence are the same as those for the cosine and sine transforms, since all three are derived from the Fourier series. From our definition of $F(i\omega)$, we may obtain the inverse transform by using Exercise 2.7.4. The result is

$$f(t) = \mathscr{F}^{-1}\{F(i\omega)\} = \frac{1}{2\pi} \int_{-\infty}^\infty F(i\omega) e^{i\omega t} \, d\omega. \tag{14.5.6}$$

We note that if $f(t) = 0$ for $t < 0$, and $i\omega$ is replaced by s, the Fourier transform becomes the Laplace transform. Thus the Fourier transform for this case corresponds to letting $\mathrm{Re}\{s\} = 0$ in (14.2.1) and results in restricting the class of functions whose transforms exist. For example, the unit step function $u(t)$ has a Laplace transform $1/s$, valid for $\mathrm{Re}\{s\} > 0$, but it has no Fourier transform since

$$\mathscr{F}\{u(t)\} = \int_0^\infty e^{-i\omega t} \, dt$$

does not exist. We note that the sufficient condition $\int_0^\infty |u(t)| \, dt < M$ is not satisfied. Useful results may, however, be obtained by *defining* the transform of $u(t)$ as

$$\mathscr{F}\{u(t)\} = \lim_{\substack{a \to 0 \\ a > 0}} \mathscr{F}\{u(t) e^{-at}\},$$

which results in

$$\mathscr{F}\{u(t)\} = \frac{1}{i\omega}. \tag{14.5.7}$$

Many authors use this device to define Fourier transforms which do not exist in the usual sense as

$$\mathscr{F}\{f(t)\} = \lim_{\substack{a\to 0\\ a>0}} \mathscr{F}\{f(t)e^{-at}\}. \tag{14.5.8}$$

This definition results in such formulas as

$$\mathscr{F}\{(\sin kt)u(t)\} = \frac{k}{k^2 - \omega^2} = \frac{k}{(i\omega)^2 + k^2} \tag{14.5.9}$$

and

$$\mathscr{F}\{(\cos kt)u(t)\} = \frac{i\omega}{k^2 - \omega^2} = \frac{i\omega}{(i\omega)^2 + k^2}. \tag{14.5.10}$$

With this modified definition the analogy between Laplace and Fourier transforms is still more complete.

14.6 PROPERTIES OF THE FOURIER TRANSFORM

Because of the similarities noted in the previous section between the definitions of $\mathscr{F}\{f(t)\}$ and $\mathscr{L}\{f(t)\}$, the Fourier transform has many properties similar to those of the Laplace transform. For example, if $f(t)$, $f'(t), \ldots, f^{(n-1)}(t)$ are continuous and $f^{(n)}(t)$ is sectionally continuous on $-\infty < t < \infty$, with the properties

$$\int_{-\infty}^{\infty} |f^{(k)}(t)|\, dt < M, \quad k = 0, 1, 2, \ldots, n,$$

then

$$\mathscr{F}\{f^{(n)}(t)\} = (i\omega)^n \mathscr{F}\{f(t)\}, \quad n = 0, 1, 2, \ldots. \tag{14.6.1}$$

This can be established for $n = 1$ by integration by parts and extended to n by induction. In the process we need the property

$$\lim_{t\to\pm\infty} f(t) = 0,$$

which is true of functions of this class. This property is considered in Exercise 14.6.2.

In applications where the Fourier transform is used, it is important to be able to find the inverse transform of a product of two transforms just as in the applications of the Laplace transform. In the case of the Fourier transform, the convolution integral which is to be used is

$$f(t) * g(t) = \int_{-\infty}^{\infty} f(\tau)g(t - \tau)\, d\tau. \tag{14.6.2}$$

We see that if we assume that the functions $f(t)$ and $g(t)$ are zero for $t < 0$, then the convolution integral reduces to the one considered for the Laplace transform.

The Fourier transform of $f * g$ is

$$\mathscr{F}\{f * g\} = \int_{-\infty}^{\infty} e^{-i\omega t}\left[\int_{-\infty}^{\infty} f(\tau)g(t - \tau)\, d\tau\right] dt,$$

which becomes, upon interchange of the order of integration,

$$\mathscr{F}\{f * g\} = \int_{-\infty}^{\infty} f(\tau)\left[\int_{-\infty}^{\infty} e^{-i\omega t}g(t - \tau)\, dt\right] d\tau.$$

The inner integral is $\mathscr{F}\{g(t - \tau)\}$ and, by Exercise 14.6.3, is $e^{-i\omega t}G(i\omega)$. Hence the convolution theorem for the Fourier transform is

$$\mathscr{F}\{f * g\} = G(i\omega)\int_{-\infty}^{\infty} f(\tau)e^{-i\omega t}\, d\tau$$

or

$$\mathscr{F}\{f * g\} = \mathscr{F}\left\{\int_{-\infty}^{\infty} f(\tau)g(t - \tau)\, d\tau\right\} = F(i\omega)G(i\omega). \qquad (14.6.3)$$

Other properties, such as translation and change of scale, are considered in the exercises.

EXERCISES

14.6.1. Find the Fourier transforms (if they exist) of

(a) $e^{-k|t|}$, $k > 0$,

(b) e^{-t},

(c) $f(t) = 1$, $-1 < t < 1$; $f(t) = 0$ elsewhere,

(d) $e^{-at}u(t)$, $a > 0$. Ans. $1/(a + i\omega)$.

14.6.2. Show that for $f(t)$ considered in Exercise 14.6.1, $\lim_{t\to\pm\infty} f(t) = 0$.

14.6.3. Derive the translation property, $\mathscr{F}\{f(t - \tau)\} = e^{-i\omega t}F(i\omega)$.

14.6.4. Show that $\mathscr{F}\{f(at)\} = (1/a)F(i\omega/a)$, $a > 0$.

14.6.5. Show that if $g(t) = \int_a^t f(\tau)\, d\tau$ satisfies the sufficient conditions that its Fourier transform exist, then

$$\mathscr{F}\{g(t)\} = \mathscr{F}\left\{\int_a^t f(\tau)\, d\tau\right\} = \frac{1}{i\omega} F(i\omega).$$

14.6.6. Verify (14.5.7), (14.5.9), and (14.5.10).

14.6.7. Given that

$$\begin{aligned} u_{tt} &= c^2 u_{xx}, \\ u(x, 0) &= f(x), \qquad -\infty < x < \infty, \\ u_t(x, 0) &= 0, \end{aligned}$$

show by Fourier transform methods that

$$u(x, t) = \tfrac{1}{2}[f(x + ct) + f(x - ct)].$$

14.6.8. Show that

$$\mathscr{F}^{-1}\left\{\frac{1}{2\pi}\int_{-\infty}^{\infty} F(i\xi)G(i\omega - i\xi)\, d\xi\right\} = f(t)g(t).$$

14.6.9. Given that $F(\omega) = \mathscr{F}\{f(t)\}$. Show that $2\pi f(-\omega) = \mathscr{F}\{F(t)\}$.

14.6.10. Note that if $g(t)$ is real, then $\overline{G(i\omega)} = G(-i\omega)$. Show from Exercise 14.6.8 that

$$\int_{-\infty}^{\infty} f(t)g(t)\, dt = \frac{1}{2\pi}\int_{-\infty}^{\infty} F(i\omega)\, \overline{G(i\omega)}\, d\omega,$$

or, for $g(t) = f(t)$, that

$$\int_{-\infty}^{\infty} [f(t)]^2\, dt = \frac{1}{2\pi}\int_{-\infty}^{\infty} |F(i\omega)|^2\, d\omega.$$

These are forms of *Parseval's formula*, analogous to Parseval's equality.

14.6.11. Show that $\mathscr{F}\{(-it)^n f(t)\} = d^n F(i\omega)/d\omega^n$, $n = 0, 1, 2, \ldots$.

14.6.12. Show that if $F(i\omega) = \mathscr{F}\{e^{-at^2}\}$, $a > 0$, then $F' + (\omega/2a)F = 0$. (*Hint*: Integrate $F(i\omega)$ by parts and use Exercise 14.6.11. Then show that $F(i\omega) = \sqrt{\pi/a}\, e^{-\omega^2/4a}$.)

14.6.13. Find the heat distribution $U(x, t)$ in an infinite medium, $-\infty < x < \infty$, if $U(x, 0) = f(x)$. Ans. $U(x, t) = 1/2\sqrt{\pi k t} \int_{-\infty}^{\infty} f(z) e^{-(x-z)^2/4kt}\, dz$.

14.6.14. Show by integrating $F_s(\nu)$ and $F_c(\nu)$ by parts twice that

$$S_\nu\{f''(t)\} = -\nu^2 S_\nu\{f(t)\} + \nu f(0),$$
$$C_\nu\{f''(t)\} = -\nu^2 C_\nu\{f(t)\} - f'(0).$$

14.6.15. Find the heat distribution $U(x, t)$ in a medium $x > 0$ if $U(x, 0) = 0$ and $U(0, t) = U_0$, a constant. Use an appropriate Fourier transform and compare with Exercise 14.4.7(c).

Ans. $U = \dfrac{2U_0}{\pi}\displaystyle\int_0^{\infty} \frac{(1 - e^{-\nu^2 kt})(\sin \nu x)\, d\nu}{\nu}$.

14.7 SYSTEM FUNCTIONS

One of the more important uses of the Fourier transform is its application to analysis of *linear systems*, such as electrical circuits, servomechanisms, and harmonic oscillators, in which an *input*, or *excitation*, $e(t)$ is related to an *output*, or *response*, $r(t)$ by a linear integrodifferential equation,

$$\phi(D)r(t) = (a_n D^n + a_{n-1}D^{n-1} + \cdots + a_{-m}D^{-m})r(t) = e(t). \quad (14.7.1)$$

We may transform (14.7.1) from the time domain to the s domain by taking its Laplace transform, which, considering all initial conditions to be zero, results in

$$R(s) = H(s)E(s), \quad (14.7.2)$$

where

$$H(s) = \frac{1}{a_n s^n + a_{n-1}s^{n-1} + \cdots + a_{-m}s^{-m}} = \frac{1}{\phi(s)}.$$

The function $H(s)$ is defined as the *system function* and may be thought of as describing the system under consideration in (14.7.2).

In electrical engineering many problems arise in which it is desirable to consider the system function as a function of frequency, $H = H(i\omega)$, where $i = \sqrt{-1}$ and ω might be the frequency associated with an excitation of the form $e(t) = k \cos \omega t$. For this case the system function is complex, in rectangular form

$$H(i\omega) = R(\omega) + iX(\omega), \tag{14.7.3}$$

and in polar form

$$H(i\omega) = A(\omega)e^{i\phi(\omega)}. \tag{14.7.4}$$

The *amplitude* or *magnitude response* is

$$A(\omega) = \sqrt{R^2(\omega) + X^2(\omega)}, \tag{14.7.5}$$

and the *phase response* is

$$\phi(\omega) = \arctan \frac{X(\omega)}{R(\omega)} \tag{14.7.6}$$

These forms of the system function may be obtained directly from the time-domain equation (14.7.1) by applying the Fourier transform rather than the Laplace transform.

14.8 FILTER THEORY

As an example of a linear system, let us consider the system described by the differential equation

$$v_1(t) = (1 + TD)v_2(t),$$

where T is a constant. Assuming zero initial conditions, the transformed equation is

$$V_1(i\omega) = (1 + i\omega T)V_2(i\omega),$$

and if the system function is the ratio of V_2 to V_1, we have

$$H(i\omega) = \frac{V_2(i\omega)}{V_1(i\omega)} = \frac{1}{1 + i\omega T}. \tag{14.8.1}$$

This system might represent an electrical network in which input voltage is $v_1(t)$, and $v_2(t)$ is an output voltage measured across some element in the network. In polar form, the system function is given in terms of its amplitude and phase responses,

$$A(\omega) = \frac{1}{\sqrt{1 + (\omega T)^2}}, \qquad \phi(\omega) = -\arctan \omega T, \tag{14.8.2}$$

graphs of which are shown in Fig. 14–1.

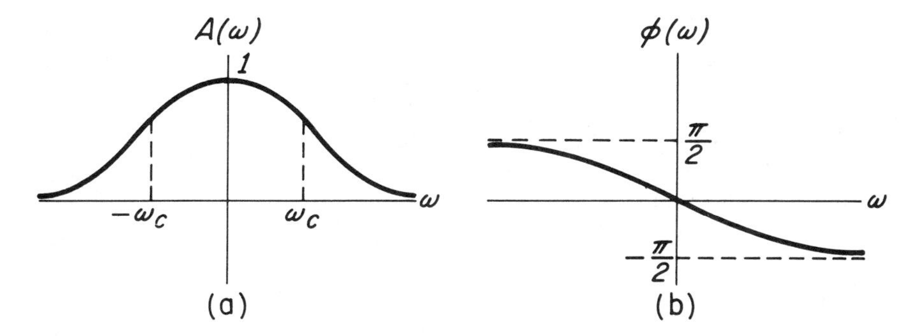

Fig. 14–1. Amplitude and phase responses

The network we have considered might be thought of as a *low-pass filter*, since from Fig. 14–1(a) we see that the low frequencies correspond to high amplitudes, and the amplitudes for high frequencies tend toward zero. The point ω_c is known as the *cut-off* point or *half-power* point and is the value of ω at which $A(\omega)$ takes on $1/\sqrt{2}$ times its maximum value. The interval $0 \leq \omega \leq \omega_c$ is known as the *pass band* and $\omega > \omega_c$ is the *stop band*. An *ideal* low-pass filter has the amplitude response as shown in Fig. 14–2(a), where the low frequencies are passed with constant magnitudes and the higher frequencies are suppressed. An ideal phase response, shown in Fig. 14–2(b), is a straight line, which yields an undistorted output voltage. In the figure we have normalized $A(\omega)$ so that $\omega_c = 1$ and $A(0) = 1$.

The present methods of designing filters require that the system function $H(s)$ be expressed as a ratio of two polynomials in s, so that it is evident that an ideal filter cannot be realized exactly. In this section we consider two methods of approximation based on two of our special functions, the Chebyshev polynomials and the Bessel polynomials.

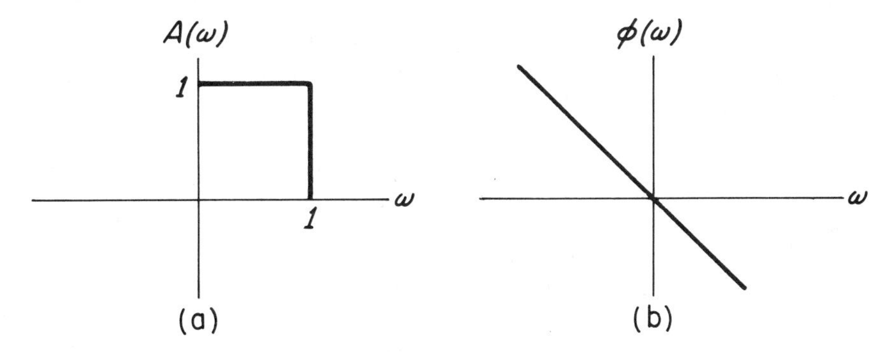

Fig. 14–2. Ideal responses

We consider first the function

$$A(\omega) = \frac{1}{\sqrt{1 + \epsilon^2 T_n^{\,2}(\omega)}}, \tag{14.8.3}$$

where $T_n(\omega)$ is the Chebyshev polynomial of the first kind and $\epsilon < 1$ is a positive real constant. Since we know $|T_n(\omega)| \le 1$, for $0 \le \omega \le 1$, and (see Exercise 11.1.2) that

$$T_n(\pm 1) = (\pm 1)^n, \qquad T_{2n}(0) = (-1)^n, \qquad T_{2n+1}(0) = 0, \tag{14.8.4}$$

we see from (14.8.3) that the maximum response for $0 \le \omega \le 1$ is $A(\omega) = 1$ corresponding to $T_n(\omega) = 0$ and the minimum response is $A(\omega) = 1/\sqrt{1 + \epsilon^2}$ corresponding to $T_n(\omega) = \pm 1$. If n is odd, $A(0)$ is the maximum value, and if n is even, $A(0)$ is the minimum value. For any n, $A(1)$ is the minimum value. Thus, in the pass band $A(\omega)$ will have *ripples* corresponding to the zeros of $T_n(\omega)$, which all occur on $-1 \le \omega \le 1$, but the ripple width, $1 - 1/\sqrt{1 + \epsilon^2}$, may be made small by selecting ϵ small. For $\omega > 1$ the Chebyshev polynomials increase rapidly with ω so that we have a good approximation to the ideal case in the stop band. Methods of obtaining $H(s)$ from its magnitude $A(\omega)$ are well known (see, for example, [VV], Chapter 8).

If it is desired that the output voltage be identical with the input voltage, except delayed T seconds in time, then

$$v_2(t) = v_1(t - T),$$

and the transformed equation may be written

$$H(s) = \frac{V_2(s)}{V_1(s)} = e^{-sT}. \tag{14.8.5}$$

Then we have

$$H(i\omega) = A(\omega)e^{i\phi(\omega)},$$

where

$$A(\omega) = 1, \qquad \phi(\omega) = -\omega T.$$

The phase response is thus the ideal phase response.

We may make a change of variable so as to consider the system function (14.8.5) with $T = 1$ so that

$$H(s) = e^{-s} = \frac{1}{\cosh s + \sinh s}. \tag{14.8.6}$$

This function may be approximated by a method developed by Storch (see [St]) which is based on replacing the denominator in (14.8.6) by $p(s) + q(s)$, a sum of polynomials identified by

$$\frac{p(s)}{q(s)} = \frac{\cosh s}{\sinh s} = \coth s.$$

We may perform the long division by using the Maclaurin series,

$$\cosh s = \sum_{n=0}^{\infty} \frac{s^{2n}}{(2n)!},$$

$$\sinh s = \sum_{n=0}^{\infty} \frac{s^{2n+1}}{(2n+1)!},$$

resulting in a *continued fraction* expansion,

$$\frac{p(s)}{q(s)} = \frac{1}{s} + \cfrac{1}{\cfrac{3}{s} + \cfrac{1}{\ddots \cfrac{2n+1}{s} + \cdots}}. \tag{14.8.7}$$

The method of Storch is to terminate this continued fraction expansion after n terms, simplify it, and equate its numerator to $p(s)$ and its denominator to $q(s)$. Storch has shown that the result is

$$p(s) + q(s) = \theta_n(s),$$

where $\theta_n(s)$ is Burchnall's polynomial, related to the Bessel polynomial $y_n(s)$ of degree n (see Sec. 13.6) by the relation

$$\theta_n(s) = s^n y_n\left(\frac{1}{s}\right).$$

The polynomials are tabulated in Exercise 13.6.9.

An example, as the reader may verify, for a termination of (14.8.7) after 3 terms, is

$$H(s) = \frac{K_0}{\theta_3(s)}.$$

The constant K_0 is inserted so that we may make $H(0) = 1$ as required by (14.8.5). For this case we have

$$H(s) = \frac{15}{s^3 + 6s^2 + 15s + 15}. \tag{14.8.8}$$

The reader is asked in Exercise 14.8.8 to compare the amplitude and phase responses of this function with that of the original system function.

EXERCISES

14.8.1. For the system function $H(s) = 32/(s^2 + 2s + 5)$, find ω_c and sketch the amplitude and phase responses. Note that $A(\omega) = f(\omega^2)$, and that we may find the maximum value by considering $dA^2/d\omega^2 = 0$.

14.8.2. Find ω_c and sketch $A(\omega)$ and $\phi(\omega)$ for $H(s) = 200/(s^2 + 2s + 101)$. Note that there are two positive values of ω_c. The interval $\omega_{c_1} \leq \omega \leq \omega_{c_2}$ is the pass band in this case. Show that $\omega_{c_1}\omega_{c_2} \doteq \omega_m{}^2$, where ω_m is the point at which the maximum value of $A(\omega)$ occurs.

14.8.3. Find the system function if $r(t)$ and $e(t)$ are related by the equation

$$r'' + ar' + br = e(t).$$

Sketch $A(\omega)$ for (a) $a^2 - 2b > 0$, (b) $a^2 - 2b = 0$, (c) $a^2 - 2b < 0$. Show that for (a) and (b) there is only one ω_c and for (c) there is one ω_c if $(2 - \sqrt{2})b < a^2$ and two ω_c's if $(2 - \sqrt{2})b > a^2$.

14.8.4. Find $A(\omega)$ and $\phi(\omega)$ for

$$H(s) = \frac{s^2 - 2s + 2}{s^2 + 2s + 2}.$$

This is an *all-pass* function.

14.8.5. Consider (14.8.3) for $n = 4$ and $\epsilon = .2$. Sketch $A(\omega)$ and find its maximum and minimum values in the pass band. Find the ripple width and ω_c.

14.8.6. Work Exercise 14.8.5 for $n = 10$. Suggestion: Use the form $T_n(x) = \cosh(n \operatorname{arccosh} x)$ for $|x| > 1$.

14.8.7. If the ripple width is to be .02 or less and $\omega_c = 1.1$, find n in (14.8.3).

14.8.8. Sketch the amplitude and phase responses of $H(s)$ in (14.8.8) and compare with those of (14.8.5) for $0 \leq \omega \leq 2$.

14.8.9. Show that the process used to obtain (14.8.8) yields $\theta_4(x)$ and $\theta_5(x)$ as denominators if the continued fraction expansion is terminated after 4 and 5 terms respectively.

14.8.10. Define the delay $T(\omega) = -d\phi(\omega)/d\omega$ and show that the Burchnall polynomial filter gives a good approximation to $T(\omega) \equiv 1$ by finding $T(1)$ for the Burchnall polynomial filter for the cases $n = 2, 3, 4$.

Ans. $T_2(1) = \frac{12}{13}$, $T_3(1) = \frac{276}{277}$, $T_4(1) = \frac{12745}{12746}$.

15

Sturm-Liouville Transforms

15.1 DEFINITION

An integral transform can be determined by any complete set of orthogonal functions $\{\phi_n(x)\}$, $n = 1, 2, 3, \ldots, a \le x \le b$. For if

$$f(x) = \sum_{n=1}^{\infty} C_n \phi_n(x), \tag{15.1.1}$$

then the sequence of coefficients given by

$$C_n = \frac{\int_a^b f(x) w(x) \phi_n(x)\, dx}{\|\phi_n\|^2} \tag{15.1.2}$$

can be defined as an integral transform of $f(x)$ with kernel $w(x)\phi_n(x)$. The series (15.1.1) is then an inversion formula or inverse transform.

We recall from Theorem 1.9.1 that a Sturm-Liouville system,

$$\begin{aligned} [r(x)y']' + [p(x) + \lambda w(x)]y &= 0, \\ Ay'(a) + By(a) &= 0, \\ Cy'(b) + Dy(b) &= 0, \end{aligned} \tag{15.1.3}$$

gives rise to such a set of orthogonal functions $\{\phi_n(x)\}$. When the orthogonal functions arise in this manner, (15.1.2) is known as a *Sturm-Liouville transform* (for a more detailed treatment see [E]).

For our purposes we define

$$F(n) = C_n \|\phi_n\|^2,$$

and the Sturm-Liouville transform of $f(x)$ to be

$$T\{f(x)\} = F(n) = \int_a^b f(x) w(x) \phi_n(x)\, dx, \quad n = 1, 2, 3, \ldots. \tag{15.1.4}$$

The inverse transform will then be

$$T^{-1}\{F(n)\} = f(x) = \sum_{n=1}^{\infty} \frac{F(n)\phi_n(x)}{\|\phi_n\|^2} \tag{15.1.5}$$

We shall now develop an operational property of the Sturm-Liouville transform which will be useful in solving certain types of differential equations. Consider the differential form

$$Lf = \frac{1}{w(x)} \{[r(x)f'(x)]' + p(x)f(x)\} \tag{15.1.6}$$

and its transform

$$T\{Lf\} = \int_a^b [r(x)f'(x)]'\phi_n(x)\,dx + \int_a^b p(x)f(x)\phi_n(x)\,dx.$$

Integrating by parts twice in the first integral we obtain

$$T\{Lf\} = r(x)[f'(x)\phi_n(x) - f(x)\phi'_n(x)]\Big|_a^b + \int_a^b \{[r(x)\phi'_n(x)]' + p(x)\phi_n(x)\}f(x)\,dx. \tag{15.1.7}$$

Since $\phi_n(x)$ satisfies (15.1.3) for $\lambda = \lambda_n$, this equation can be written

$$T\{Lf\} = r(x)[f'(x)\phi_n(x) - f(x)\phi'_n(x)]\,\Big|_a^b - \lambda_n T\{f\}. \tag{15.1.8}$$

The expression on the right of (15.1.8) could be expressed in terms of the boundary conditions of (15.1.3). Since there are many variations of these, we will consider them as they arise. However, we note that, for the special case,

$$r(a) = r(b) = 0,$$

in which the two boundary conditions are trivially satisfied, Eq. (15.1.8) is

$$T\{Lf\} = -\lambda_n T\{f\}. \tag{15.1.9}$$

The Sturm-Liouville transform is useful in solving boundary-value problems of the type

$$\frac{1}{w(x)}\left\{\frac{\partial}{\partial x}\left[r(x)\frac{\partial \psi}{\partial x}\right] + p(x)\psi\right\} + L_t\psi = 0, \tag{15.1.10}$$

where L_t is a differential operator independent of x, with two of the boundary conditions of the form

$$\begin{aligned} A\psi_x(a, t) + B\psi(a, t) &= 0, \\ C\psi_x(b, t) + D\psi(b, t) &= 0. \end{aligned} \tag{15.1.11}$$

The transformed equation is then a differential equation or a differential-difference equation with one continuous independent variable fewer and with the discrete variable n.

15.2 FINITE FOURIER SINE AND COSINE TRANSFORMS

As was shown in Exercise 1.9.1, the Sturm-Liouville system

$$\begin{aligned} y'' + \lambda y &= 0, \\ y(0) = y(L) &= 0 \end{aligned} \tag{15.2.1}$$

has the eigenfunctions $\phi_n = \sin(n\pi x/L)$ and eigenvalues $\lambda_n = n^2\pi^2/L^2$, $n = 1, 2, 3, \ldots$. By comparison with (15.1.3) and (15.1.4) we define the *finite Fourier sine transform* of $f(x)$

$$S_n\{f(x)\} = F_s(n) = \int_0^L f(x) \sin \frac{n\pi x}{L}\, dx, \quad n = 1, 2, 3, \ldots. \tag{15.2.2}$$

The inverse transform is, by (15.1.5),

$$S_n^{-1}\{F_s(n)\} = f(x) = \frac{2}{L} \sum_{n=1}^{\infty} F_s(n) \sin \frac{n\pi x}{L}, \tag{15.2.3}$$

which is the Fourier sine series.

The *finite Fourier cosine transform*

$$C_n\{f(x)\} = F_c(n) = \int_0^L f(x) \cos \frac{n\pi x}{L}\, dx, \quad n = 0, 1, 2, \ldots, \tag{15.2.4}$$

may be obtained by considering the Sturm-Liouville system

$$\begin{gathered} y'' + \lambda y = 0 \\ y'(0) = y'(L) = 0 \end{gathered} \tag{15.2.5}$$

and Eqs. (15.1.3) and (15.1.4). This system has eigenfunctions $\phi_n = \cos(n\pi x/L)$ and eigenvalues $\lambda_n = n^2\pi^2/L^2$. The inverse transform is given by

$$C_n^{-1}\{F_c(n)\} = f(x) = \frac{F_c(0)}{L} + \frac{2}{L} \sum_{n=1}^{\infty} F_c(n) \cos \frac{n\pi x}{L}, \tag{15.2.6}$$

which is the Fourier cosine series. Short tables of sine and cosine transforms are given in Tables 2 and 3 of Appendix C, respectively.

Equation (15.1.8) may be used to obtain operational properties of the sine and cosine transforms. We note that for both systems (15.2.1) and (15.2.5) we have $r(x) = w(x) = 1$, $p(x) = 0$. Substituting these values into (15.1.8) for the sine and cosine transforms, we obtain

$$S_n\{f''(x)\} = -\frac{n^2\pi^2}{L^2} F_s(n) + \frac{n\pi}{L}[f(0) - (-1)^n f(L)] \tag{15.2.7}$$

and

$$C_n\{f''(x)\} = -\frac{n^2\pi^2}{L^2} F_c(n) - f'(0) + (-1)^n f'(L). \tag{15.2.8}$$

Since two integrations by parts were used to obtain (15.2.8), sufficient conditions that these operational properties be valid are that f and f' be continuous and f'' sectionally continuous on $(0, L)$.

These formulas may be extended by induction to obtain $S_n\{f^{(2k)}(x)\}$ and $C_n\{f^{(2k)}(x)\}$. Thus, finite Fourier transforms may be used to solve differential equations in which only even-ordered derivatives appear.

As an example of the use of the finite sine transform let us consider the boundary-value problem

$$w_{xx}(x, y) + w_{yy}(x, y) = 0, \quad 0 < x < \pi, y > 0,$$

$$w(0, y) = w(x, 0) = 0, \quad w(\pi, y) = 1, \quad |w(x, y)| < M.$$

Transforming with respect to x, making use of (15.2.7), we have the transformed system

$$-n^2 W_s(n, y) - n(-1)^n + W''_s(n, y) = 0,$$

$$W_s(n, 0) = 0, \quad |W_s(n, y)| < M\pi.$$

The general solution of the differential equation is

$$W_s(n, y) = c_1 e^{ny} + c_2 e^{-ny} - \frac{(-1)^n}{n},$$

and satisfying the boundary conditions we have

$$W_s(n, y) = \frac{1}{n}(-1)^n e^{-ny} - \frac{1}{n}(-1)^n.$$

At this point we could express $w(x, y)$ as an infinite series by means of (15.2.3), but we can also express it in closed form by means of entries 2 and 6 of Table 2. The result is

$$w(x, y) = -\frac{2}{\pi} \arctan \frac{e^{-y} \sin x}{1 + e^{-y} \cos x} + \frac{x}{\pi}.$$

This is essentially the problem worked by separation of variables in Exercise 7.5.4, and illustrates a method of summing certain Fourier series.

In this example we have chosen the sine transform rather than the cosine transform because in the boundary conditions the value of w was specified at $x = 0$ and $x = L$. Thus, (15.2.7) is better suited than (15.2.8). If the first derivatives had been specified at 0 and L, (15.2.8) would indicate that the cosine transform should be used. Of course, if the initial values of w and its derivatives had been specified, the Laplace transform would have been preferable.

Just as in the case of the Laplace transform, convolution theorems exist for the finite sine and cosine transforms. The *convolution* of $f(x)$ and $g(x)$ is defined to be

$$f(x) * g(x) = \int_{-L}^{L} f(x - \xi) g(\xi) \, d\xi, \tag{15.2.9}$$

and if f_1, g_1 and f_2, g_2 are the odd and even extensions of f and g, then the following formulas hold:

$$
\begin{aligned}
2F_s(n)G_s(n) &= -C_n\{f_1(x) * g_1(x)\}, \\
2F_c(n)G_c(n) &= C_n\{f_2(x) * g_2(x)\}, \qquad (15.2.10) \\
2F_s(n)G_c(n) &= S_n\{f_1(x) * g_2(x)\}.
\end{aligned}
$$

We will consider the derivation of one of these in Exercise 15.2.13.

EXERCISES

15.2.1. Work out transforms 1, 2, 4, and 5 given in Table 2 (Appendix C). Use (15.2.7) when it is appropriate.

15.2.2. Work out transforms 1, 3, 4, and 5 in Table 3.

15.2.3. Expand $\log(1 + z)$ in a Maclaurin series where $z = re^{i\theta}$, $|r| < 1$, and equate imaginary parts to obtain

$$\arctan \frac{r \sin \theta}{1 + r \cos \theta} = -\sum_{n=1}^{\infty} \frac{(-r)^n}{n} \sin n\theta.$$

Take the transform of both sides, replacing θ by $\pi x/L$ and r by k to obtain transform 6 in Table 2.

15.2.4. Replace k by $-k$ in the transform of Exercise 15.2.3 and use the two transforms thus obtained to derive transform 7 in Table 2.

15.2.5. Show that $C_n\{f(x) + A\} = F_c(n) + \frac{L}{\pi}A\delta_{n0}$, where A is constant.

15.2.6. Show that

$$S_n\{f'(x)\} = -\frac{n\pi}{L} F_c(n), \quad n = 1, 2, 3, \ldots,$$

$$C_n\{f'(x)\} = \frac{n\pi}{L} F_s(n) - f(0) + (-1)^n f(L), \quad n = 0, 1, 2, \ldots.$$

15.2.7. Use Exercise 15.2.6 and transform 6, Table 2, to obtain transform 6, Table 3.

15.2.8. Solve the system, $u_{xx}(x, y) + u_{yy}(x, y) = 0$, $u(0, y) = u(\pi, y) = 0$, $u(x, 0) = 1$, $|u(x, y)| < M$, $0 \le x \le \pi$, $y \ge 0$, in two ways, one by separation of variables and the other by operational methods.

15.2.9. Solve Exercise 7.5.8 by a transform method.

15.2.10. Use transforms to solve the system

$$
\begin{aligned}
&u_t(x, t) = ku_{xx}(x, t), \\
&u(x, 0) = 0,\ u_x(0, t) = a,\ u_x(L, t) = b,\ a > b, \\
&0 \le x \le L,\ t \ge 0.
\end{aligned}
$$

15.2.11. Use finite transforms twice (that is, take the transform of a transform) to solve Exercise 8.5.13.

15.2.12. Use transform methods to solve Exercise 8.6.11.

15.2.13. Show that
(a) $f_1 * g_1 = \int_0^L g(\xi)\,[f_1(x-\xi) - f_1(x+\xi)]\,d\xi$,
(b) $2F_s(n)\sin nk = C_n\{f_1(x+k) - f_1(x-k)\}$,
(c) $2F_s(n)G_s(n) = -C_n\{f_1(x) * g_1(x)\}$.

15.3 HANKEL TRANSFORM

Bessel's equation (6.1.1) in self-adjoint form with x replaced by αx is

$$(xy')' + \left(\alpha^2 x - \frac{\nu^2}{x}\right)y = 0. \tag{15.3.1}$$

Comparing this equation with (15.1.3) and (15.1.4) we see that the kernel is $w(x)\phi_n(x) = xJ_\nu(\alpha_n x)$, where α_n is a value of α. We will take the interval as $(0, b)$ and $\alpha_n b$ as a positive zero of $J_\nu(\alpha)$, $n = 1, 2, 3, \ldots$. The *finite Hankel transform* is then defined, by use of (15.1.4), as

$$H_{n\nu}\{f(x)\} = F_h(n, \nu) = \int_0^b xf(x)J_\nu(\alpha_n x)\,dx, \tag{15.3.2}$$

with the inverse transform given by (15.1.5) as

$$f(x) = H_{n\nu}^{-1}\{F_h(n, \nu)\} = \frac{2}{b^2}\sum_{n=1}^{\infty}\frac{F_h(n, \nu)J_\nu(\alpha_n x)}{[J'_\nu(\alpha_n b)]^2}. \tag{15.3.3}$$

[See (6.7.9) and (6.7.13), and note that (15.3.3) is simply the generalized Fourier series of Bessel functions.] When $\nu = 0$, we write for the transform

$$H_n\{f(x)\} = F_h(n).$$

To obtain an operational property, we use (15.1.8), which for this case results in

$$H_{n\nu}\left\{f'' + \frac{1}{x}f' - \frac{\nu^2}{x^2}f\right\} = -b\alpha_n f(b)J'_\nu(b\alpha_n) - \alpha_n{}^2F_h(n, \nu). \tag{15.3.4}$$

It is well to note here that ν is not specified at this point and can be chosen to fit best the particular problem under consideration. Of special interest is the case $\nu = 0$, for which (15.3.4) becomes

$$H_n\left\{f''(x) + \frac{1}{x}f'(x)\right\} = b\alpha_n f(b)J_1(b\alpha_n) - \alpha_n{}^2F_h(n). \tag{15.3.5}$$

As an example of the use of the Hankel transform we will consider the problem of heat conduction in a finite circular cylinder, formulated as

$$\begin{aligned} &u_{rr}(r, t) + \frac{1}{r}u_r(r, t) = u_t(r, t) \\ &u(r, 0) = f(r), \quad u(b, t) = 0, \quad 0 \le r \le b,\ t \ge 0. \end{aligned} \tag{15.3.6}$$

We note by comparing the given differential equation with (15.3.5) that the Hankel transform is well suited for this problem.

The transformed equations, using (15.3.2) and (15.3.5) are

$$b\alpha_n u(b, t)J_1(b\alpha_n) - \alpha_n{}^2 U_h(n, t) = \frac{d}{dt} U_h(n, t),$$

which is equivalent to

$$-\alpha_n{}^2 U_h(n, t) = \frac{d}{dt} U_h(n, t),$$

and

$$U_h(n, 0) = F_h(n).$$

The transformed system has as a solution

$$U_h(n, t) = F_h(n)e^{-\alpha_n{}^2 t},$$

and the inverse transform is

$$u(r, t) = \frac{2}{b^2} \sum_{n=1}^{\infty} \frac{F_h(n)e^{-\alpha_n{}^2 t} J_0(\alpha_n r)}{J_1{}^2(\alpha_n b)} .$$

In terms of $f(r)$, this is

$$u(r, t) = \frac{2}{b^2} \sum_{n=1}^{\infty} \frac{J_0(\alpha_n r)}{J_1{}^2(\alpha_n b)} e^{-\alpha_n{}^2 t} \int_0^b x f(x) J_0(\alpha_n x)\, dx,$$

where

$$J_0(\alpha_n b) = 0, \quad n = 1, 2, 3, \ldots .$$

The reader may wish to compare this example with Exercise 8.5.9, which was solved by another method.

In our definition of the Hankel transform we have considered $J_\nu(\alpha_n b) = 0$. In other words, we have taken $C = 0$ in the second boundary condition of (15.1.3). The transform could also be defined if $C \neq 0$ and $D \neq 0$, in which case the boundary condition will be

$$\alpha_n J'_\nu(\alpha_n b) + hJ_\nu(\alpha_n b) = 0, \tag{15.3.7}$$

where $h = D/C$ is taken as non-negative. The transform for this case is considered in Exercise 15.3.5.

EXERCISES

15.3.1. Obtain the following transforms:

(a) $H_n\{1\}$,
(b) $H_n\{x^2\}$,
(c) $H_{n1}\{x^2\}$,
(d) $H_{n4}\{x^4\}$,
(e) $H_n\{\log x\}$,
(f) $H_n\{J_0(\beta x)\}$.

15.3.2. Prove that $H_{n\nu}\{x^\nu\} = (b^{\nu+1}/\alpha_n)J_{\nu+1}(\alpha_n b)$.

15.3.3. Solve for $u(r, t)$, $0 \le r \le b$, $t \ge 0$:

$$u_{rr} + \frac{1}{r}\, u_r = u_t,$$

$$u(r, 0) = u_0, \quad u(b, t) = u_1, \quad u_0, u_1 \text{ constants.}$$

15.3.4. Solve for $u(r, t)$, $0 \le r \le b$, $t \ge 0$:

$$u_{rr} + \frac{1}{r}\, u_r = \frac{1}{c^2}\, u_{tt},$$

$$u(b, t) = 0, \quad u(r, 0) = f(r), \quad u_t(r, 0) = g(r).$$

Ans. $$u(r, t) = \frac{2}{b^2} \sum_{n=1}^{\infty} \left\{ \cos c\alpha_n t \int_0^b r f(r) J_0(\alpha_n r)\, dr \right.$$

$$\left. + \frac{\sin c\alpha_n t}{c\alpha_n} \int_0^b r g(r) J_0(\alpha_n r)\, dr \right\} \frac{J_0(\alpha_n r)}{J_1^2(\alpha_n b)}, \quad \text{where } J_0(\alpha_n b) = 0,\ \alpha_n > 0$$

15.3.5. Develop the Hankel transform and the inversion formula, with the condition $J_\nu(\alpha_n b) = 0$ replaced by (15.3.7).

15.3.6. Use the results of the previous exercise to solve Exercise 8.5.10.

15.3.7. If the system (15.1.3) is

$$y'' + \frac{1}{x}\, y' + \left(\alpha^2 - \frac{k^2}{x^2}\right) y = 0, \quad k = 0, 1, 2, \ldots,$$

$$y(a) = y(b) = 0, \quad b > a > 0,$$

show that $\phi_n(x) = U_k(\alpha_n x)$ given by (6.10.1), where $U_k(\alpha_n a) = 0$. Define the Sturm-Liouville transform corresponding to this system and find the inversion formula.

Ans. $$F^{-1}(n) = f(x) = \frac{\pi^2}{2} \sum_{n=1}^{\infty} \frac{\alpha_n^2 J_k^2(\alpha_n a) F(n)}{J_k^2(\alpha_n a) - J_k^2(\alpha_n b)}\, U_k(\alpha_n x).$$

15.3.8. Use the results of Exercise 15.3.7 to solve Exercise 8.5.12.

15.4 LEGENDRE TRANSFORM

In the discussion of the Sturm-Liouville system in Sec. 1.9 it was pointed out that when the function $r(x)$ vanishes at both end points it is not necessary to specify any boundary conditions to determine an orthogonal set of functions. Such is the case for the Legendre polynomials (see Chapter 4) since $r(x) = 1 - x^2$ and the interval is $(-1, 1)$. Also, we recall that $w(x) = 1$, $p(x) = 0$, and $\lambda_n = n(n + 1)$.

The *Legendre transform* of a function $f(x)$ is then defined by (15.1.4) as

$$P_n\{f(x)\} = F_P(n) = \int_{-1}^{1} f(x) P_n(x)\, dx, \quad n = 0, 1, 2, \ldots, \tag{15.4.1}$$

and its inverse is given by (15.1.5) as

$$P_n^{-1}\{F_P(n)\} = f(x) = \tfrac{1}{2} \sum_{n=0}^{\infty} (2n + 1) F_P(n) P_n(x), \quad -1 < x < 1. \tag{15.4.2}$$

Thus we see that the kernel is the Legendre polynomial $P_n(x)$, and again the transform is a set of functions.

The Legendre transform is useful when applied to a partial differential equation containing terms of the type

$$M\psi = \frac{\partial}{\partial x}\left[(1 - x^2)\frac{\partial \psi}{\partial x}\right],$$

as can be seen by comparing Legendre's equation with (15.1.6). Transforming this term by means of (15.1.9), we have

$$P_n\left\{\frac{\partial}{\partial x}\left[(1 - x^2)\frac{\partial \psi}{\partial x}\right]\right\} = -n(n + 1)P_n\{\psi\}.$$

For a convolution property of the Legendre transform, the reader is referred to [CD].

EXERCISES

15.4.1. We define *even* and *odd* Legendre transforms, respectively, by

$$P_{ne}\{f(x)\} = F_P(2n) = \int_0^1 f(x)P_{2n}(x)\,dx,$$

$$P_{no}\{f(x)\} = F_P(2n + 1) = \int_0^1 f(x)P_{2n+1}(x)\,dx, \quad n = 0, 1, 2, \ldots.$$

Obtain the inversion formulas (see Exercise 4.7.7).

Ans. $f(x) = \sum_{n=0}^{\infty} (4n + 1)F_P(2n)P_{2n}(x),$

$f(x) = \sum_{n=0}^{\infty} (4n + 3)F_P(2n + 1)P_{2n+1}(x).$

15.4.2. Show that, for $M\psi = (\partial/\partial x)[(1 - x^2)\psi']$,

$$P_{ne}\{M\psi\} = -P_{2n}(0)\psi_x(0) - 2n(2n + 1)P_{ne}\{\psi\}$$

and

$$P_{no}\{M\psi\} = (2n + 1)P_{2n}(0)\psi(0) - (2n + 1)(2n + 2)P_{no}\{\psi\}.$$

15.4.3. Solve Exercise 8.6.14 if $V = V(r, \theta)$ and $V(a, \theta) = f(\theta)$.

15.4.4. Find the steady-state temperature in a solid hemisphere $0 \leq r \leq 1$, $0 \leq \theta \leq \pi/2$, if the base is insulated and the hemispherical surface is kept at a temperature $f(\theta)$.

15.4.5. Solve for $\psi(x, t)$ if

$$\psi_{tt}(x, t) = \frac{\partial}{\partial x}\,[(1 - x^2)\psi_x(x, t)], \quad -1 < x < 1, t > 0,$$

$\psi(x, t)$ is continuous on $(-1, 1)$,

$\psi(x, 0) = f(x)$,

$\psi_t(x, 0) = 0.$

(See Exercise 8.3.11.)

15.5 LAGUERRE TRANSFORM

To vary our approach we define the *associated Laguerre transform* by means of the generalized Fourier series and coefficients for the Laguerre polynomials, which are given in Eqs. (10.5.12) and (10.5.13). From these we have the associated Laguerre transform

$$L_{n\alpha}\{f(x)\} = F_L(n, \alpha) = \int_0^\infty x^\alpha e^{-x} f(x) L_n^\alpha(x)\, dx \tag{15.5.1}$$

and the inverse transform

$$L_{n\alpha}^{-1}\{F_L(n, \alpha)\} = f(x) = \sum_{n=0}^\infty \frac{n!}{\Gamma(n+\alpha+1)} F_L(n, \alpha) L_n^\alpha(x). \tag{15.5.2}$$

Considering the associated Laguerre equation (10.2.6) in self-adjoint form,

$$[(x^{\alpha+1}e^{-x})y']' + nx^\alpha e^{-x} y = 0, \tag{15.5.3}$$

we note by a comparison with (15.1.6) and (15.1.9) that

$$L_{n\alpha}\{xf'' + (\alpha + 1 - x)f'\} = -nL_{n\alpha}\{f(x)\}, \tag{15.5.4}$$

and thus the Laguerre transform is suited for application to partial differential equations containing terms of the type

$$M\psi = x\frac{\partial^2 \psi}{\partial x^2} + (\alpha + 1 - x)\frac{\partial \psi}{\partial x}.$$

The case $\alpha = 0$ of the simple Laguerre polynomials may be obtained from the foregoing results. We will call these transforms the *Laguerre transforms* and use the notation,

$$L_n\{f(x)\} = F_L(n) = \int_0^\infty e^{-x} f(x) L_n(x)\, dx, \tag{15.5.5}$$

with the inverse transform,

$$L_n^{-1}\{F_L(n)\} = \sum_{n=0}^\infty F_L(n) L_n(x). \tag{15.5.6}$$

The interested reader is referred to [McC] where other properties of the Laguerre transform are considered.

EXERCISES

15.5.1. Find the Laguerre transform of

(a) 1, (c) $L_m(x)$,

(b) x, (d) $u(t - k)$.

15.5.2. Solve Exercise 10.5.14 by the transform method.

15.5.3. Solve Exercise 10.5.15.

15.5.4. Solve Exercise 10.5.16.

15.5.5. Solve for $u(x, t)$ if

$$u_t(x, t) = xu_{xx}(x, t) + (1 - x)u_x(x, t) + f(t), \quad x,t > 0,$$
$$u(x, 0) = g(x), \qquad |u(x, t)| < Me^{kx} \text{ as } x \to \infty, \quad k < 1.$$

15.6 HERMITE TRANSFORM

For the case of Hermite's equation

$$(e^{-x^2}y')' + 2ne^{-x^2}y = 0$$

we see that $w(x) = e^{-x^2}$, $\lambda_n = 2n$, $p(x) = 0$, and $r(x) = e^{-x^2}$. Using (15.1.4) and (15.1.5), we define the *Hermite transform*

$$H_n\{f(x)\} = F_H(n) = \int_{-\infty}^{\infty} e^{-x^2}f(x)H_n(x)\,dx \tag{15.6.1}$$

with inverse

$$H_n^{-1}\{F_H(n)\} = f(x) = \sum_{n=0}^{\infty} \frac{F_H(n)H_n(x)}{\sqrt{\pi}\,2^n n!} \tag{15.6.2}$$

For the Hermite transform, (15.1.9) takes the form

$$H_n\{f''(x) - 2xf'(x)\} = -2nH_n\{f(x)\}, \tag{15.6.3}$$

and thus we see that the Hermite transform is useful in solving equations having terms of the type

$$M\psi = \frac{\partial^2 \psi}{\partial x^2} - 2x\frac{\partial \psi}{\partial x}.$$

15.7 OTHER TRANSFORMS

We may use the results of Sec. 15.1 to obtain transforms corresponding to each of the special functions we have considered throughout the book, as well as to a variety of other functions. Since the procedures for each function are identical, we will give in this last section, and in the exercises to follow, the results for some of the other special functions.

Considering Chebyshev's equation in self-adjoint form,

$$(\sqrt{1 - x^2}\,y')' + \frac{n^2}{\sqrt{1 - x^2}}\,y = 0,$$

we may use Sec. 15.1 to define a *Chebyshev transform*,

$$T_n\{f(x)\} = F_T(n) = \int_{-1}^{1} \frac{f(x)T_n(x)}{\sqrt{1 - x}}\,dx, \tag{15.7.1}$$

with inverse

$$T_n^{-1}\{F_T(n)\} = f(x) = \sum_{n=0}^{\infty} \frac{F_T(n)T_n(x)}{\|T_n\|^2} = \sum_{n=0}^{\infty} \frac{\epsilon_n}{\pi}\,F_T(n)T_n(x) \tag{15.7.2}$$

(see Sec. 11.7). The operational property (15.1.9) for this case is

$$T_n\{(1 - x^2)f''(x) - xf'(x)\} = -n^2T_n\{f(x)\}. \tag{15.7.3}$$

As we shall see in the exercises, the Chebyshev transform is not new to us since it reduces to the finite Fourier cosine transform under the transformation $x = \cos\theta$.

Another useful transform is the *Jacobi transform*, given by

$$J_n^{(\alpha,\beta)}\{f(x)\} = F_J(n, \alpha, \beta) = \int_{-1}^{1} f(x)(1 - x)^\alpha(1 + x)^\beta P_n^{(\alpha,\beta)}(x)\,dx, \tag{15.7.4}$$

with inverse

$$f(x) = \sum_{n=0}^{\infty} \frac{F_J(n, \alpha, \beta)P_n^{(\alpha,\beta)}(x)}{\|P_n^{(\alpha,\beta)}\|^2} \tag{15.7.5}$$

and operational property

$$J_n^{(\alpha,\beta)}\{(1 - x^2)f''(x) + [\beta - \alpha - (\alpha + \beta + 2)x]f'\} = - n(n + \alpha + \beta + 1)J_n^{(\alpha,\beta)}\{f(x)\}. \tag{15.7.6}$$

We consider two other transforms, the *Gegenbauer transform* in Exercise 15.7.2, and a second Chebyshev transform in Exercise 15.7.12.

EXERCISES

15.7.1. Establish Eqs. (15.7.3) through (15.7.6) by the method of Sec. 15.1.

15.7.2. Derive the inverse transform and the operational property (15.1.8) for the *Gegenbauer transform*, defined by

$$C_n^\nu\{f(x)\} = F_G(n, \nu) = \int_{-1}^{1} f(x)(1 - x^2)^{\nu-1/2}C_n^\nu(x)\,dx.$$

See Exercise 13.4.8.

15.7.3. Find the Hermite transform of

(a) 1, Ans. $\sqrt{\pi}\,\delta_{n0}$.

(b) $H_m(x)$,

(c) $e^{-\alpha x}$. Ans. $(-\alpha)^n\sqrt{\pi}\,e^{\alpha^2/4}$.

15.7.4. Solve Exercise 9.7.3 by the methods of this chapter.

15.7.5. Solve Exercise 9.7.4.

15.7.6. Show that if $f(x)$ is continuous and $f'(x)$ is sectionally continuous on $(-\infty, \infty)$, and if $|f(x)| < Me^{\alpha x^2}$ as $x \to \infty$, $\alpha < 1$, then $H_n\{f'(x)\} = H_{n+1}\{f(x)\}$.

15.7.7. Extend the results of the previous exercise to obtain $H_n\{f^{(m)}(x)\} = F_H(n + m)$. What conditions are required on f and its derivatives?

15.7.8. Use the result of Exercise 15.7.7 to solve Exercise 15.7.3(c) and to find $H_n\{x^m\}$.

15.7.9. Solve Exercise 9.7.2 by transform methods. (See Sec. 9.7.)

15.7.10. Show that $T_n\{1\} = \pi\delta_{n0}$ and $T_n\{x\} = (\pi/2)\delta_{n1}$.

15.7.11. Using transform methods, find a solution of

$$(1 - x^2)y'' - xy' + ky = x, \quad k \neq 1.$$

Ans. $y = x/(k - 1)$.

15.7.12. Show that if $x = \cos\theta$, (15.7.1) and (15.7.2) become the finite cosine transform of $f(\cos\theta)$ and its inverse, respectively. Show also that if we employ the Chebyshev polynomials of the second kind $U_n(x)$, using Exercise 11.7.7, to define the transform

$$F_U(n) = U_n\{f(x)\} = \frac{2}{\pi}\int_{-1}^{1} f(x)U_n(x)\,dx,$$

then its inverse is

$$f(x) = \sum_{n=0}^{\infty} \sqrt{1 - x^2}\, F_U(n)U_n(x),$$

and these are the finite sine transform and its inverse if $x = \cos\theta$.

15.7.13. Solve Exercise 11.9.4 by transform methods.

15.7.14. To solve the system, $Lf + L_t f = 0$, $M_i(f) = A_i f'(a_i) + B_i f(a_i)$, $i = 1, 2$ (L_t independent of x), by transform methods, we must select $\{\phi_n(x)\}$ so that we can evaluate the terms in (15.1.8), which may be written

$$T\{Lf\} = r(x)\,\Delta(x)\Big|_{a_1}^{a_2} - \lambda_n T\{f\}.$$

(a) Show that if $B_i \neq 0$, then

$$B_i\,\Delta(a_i) = \begin{vmatrix} f'(a_i) & M_i(f) \\ \phi'_n(a_i) & M_i(\phi_n) \end{vmatrix}, \quad i = 1, 2,$$

and hence the ϕ_n may be determined from $L\phi_n + \lambda_n\phi_n = 0$ and conditions obtained from $\Delta(a_i)$, $r(a_i)$, or $B_i\,\Delta(a_i)$, $i = 1, 2$.

(b) Show that if $B_i \neq 0$, $r(a_i) \neq 0$, $i = 1, 2$, then $\{\phi_n\}$ is determined by $L\phi_n + \lambda_n\phi_n = 0$, $M_i(\phi_n) = 0$, $i = 1, 2$.

15.7.15. Use Exercise 15.7.14 to obtain the transform in Exercise 15.3.7.

15.7.16. Show that the transform suitable for solving Exercise 8.5.2(b) is

$$F(n) = \int_0^L f(x)\sin\frac{(2n - 1)\pi x}{2L}\,dx, \quad n = 1, 2, 3, \ldots.$$

Solve Exercise 8.5.2(b) by this method.

16

A General Class of Orthogonal Polynomials

16.1 A UNIFYING CONCEPT

We have seen that all our special functions arise as solutions of linear second-order differential equations. In particular, the orthogonal *polynomial* sets have been found to be solutions of the equation

$$f(x)y'' + g(x)y' + \lambda y = 0, \tag{16.1.1}$$

where, for $n = 0, 1, 2, \ldots,$

$$\begin{aligned} f(x) &= ax^2 + bx + c, \\ g(x) &= hx + k, \\ \lambda = \frac{n}{2}(1 - n)f''(x) &- ng'(x) = n[(1 - n)a - h]. \end{aligned} \tag{16.1.2}$$

Also, at least one of a,b,h is not zero and $f,g,\lambda \not\equiv 0$.

This equation may be put in self-adjoint form by multiplying it by the integrating factor

$$\mu(x) = \frac{1}{f(x)} \exp\left[\int \frac{g(x)\,dx}{f(x)}\right], \tag{16.1.3}$$

which results in

$$[r(x)y']' + \lambda\mu(x)y = 0, \tag{16.1.4}$$

with

$$r(x) = \exp\left[\int \frac{g(x)\,dx}{f(x)}\right]. \tag{16.1.5}$$

These results have been obtained in Sec. 1.11 where it was shown that (16.1.1) has a polynomial solution of degree n given by the Rodrigues' formula

$$y = G_n(a, b, c, h, k, C_n, x) = \frac{C_n}{\mu(x)} D^n\{[f(x)]^n \mu(x)\}$$

$$= C_n f \exp\left[-\int \frac{g\,dx}{f}\right] D^n\left[f^{n-1} \exp\left(\int \frac{g\,dx}{f}\right)\right]. \tag{16.1.6}$$

All the polynomial sets that we have considered are special cases of the polynomial G_n. For example, if $a = -1$, $b = 0$, $c = 1$, $h = -2$, $k = 0$, $C_n = 1/2^n n!$, then (16.1.1) becomes

$$(1 - x^2)y'' - 2xy' + n(n + 1)y = 0,$$

which is Legendre's equation, and (16.1.6) becomes

$$G_n\left(-1, 0, 1, -2, 0, \frac{(-1)^n}{2^n n!}, x\right) = \frac{1}{2^n n!} D^n(x^2 - 1)^n = P_n(x).$$

Other examples are considered in the exercises.

EXERCISES

16.1.1. Establish the following results by obtaining (16.1.6) in each case. Obtain also (16.1.1).

(a) $G_n(0, 0, 1, -2, 0, (-1)^n, x) = H_n(x)$ (Hermite polynomials).
(b) $G_n(0, 1, 0, -1, \alpha + 1, 1/n!, x) = L_n{}^\alpha(x)$ (associated Laguerre).
(c) $G_n(-1, 0, 1, -1, 0, (-2)^n n!/(2n)!, x) = T_n(x)$ (Chebyshev).
(d) $G_n(-1, 0, 1, -3, 0, (-2)^n(n + 1)!/(2n + 1)!, x) = U_n(x)$ (Chebyshev).
(e) $G_n(-1, 0, 1, -\alpha - \beta - 2, \beta - \alpha, (-1)^n/2^n n!, x) = P_n^{(\alpha,\beta)}(x)$ (Jacobi).
(f) $G_n(-1, 1, 0, n - \beta - 1, \gamma, C_n, x) = F(-n, \beta, \gamma, x)$ (hypergeometric).
(g) $G_n(1, 0, 0, 2, 2, 2^n, x) = y_n(x)$ (Bessel).
(h) $G_n(0, 1, 0, -2, -2n, 1/(-2)^n, x) = \theta_n(x)$ (Burchnall).

16.1.2. Obtain the set of polynomial solutions $\{h_n(x)\}$ of the equation

$$y'' - xy' + ny = 0.$$

Find an interval over which they are orthogonal, and the corresponding weight function. Ans. $h_n(x) = C_n e^{x^2/2} D^n e^{-x^2/2}$, $\int_{-\infty}^{\infty} e^{-x^2/2} h_n h_m\,dx = 0$, $n \neq m$.

16.1.3. Take $C_n = (-1)^n$ in the previous exercise and write out a few of the $h_n(x)$. Verify the orthogonality relation for some of these.

Ans. $h_0 = 1$, $h_1 = x$, $h_2 = x^2 - 1$, $h_3 = x^3 - 3x$,
$h_4 = x^4 - 6x^2 + 3$, $h_5 = x^5 - 10x^3 + 15x$.

16.1.4. Solve the equation in Exercise 16.1.2 by the method of Frobenius, and show that the polynomial solutions for $n = 1, 2, \ldots, 5$ check Exercise 16.1.3.

16.1.5. Show that $\phi_n = G_n(1, -2, 0, 1, -1, 1, x)$ is a polynomial solution of

$$x(x - 2)y'' + (x - 1)y' - n^2 y = 0,$$

and that

$$\phi_n = (-1)^n \sqrt{x(2 - x)}\, D^n[x^{n-1/2}(2 - x)^{n-1/2}].$$

16.1.6. Show that the set $\{\phi_n(x)\}$ in Exercise 16.1.5 is given by

$$\phi_n(x) = \frac{(2n)!}{2^n n!} \sum_{k=0}^{n} \binom{2n}{2k} (-1)^k x^{n-k} (2-x)^k.$$

16.1.7. Show that for $\{\phi_n(x)\}$ given in Exercise 16.1.5 we have

$$\int_0^2 [x(2-x)]^{-1/2} \phi_n(x) \phi_m(x)\, dx = 0, \quad n \neq m.$$

16.1.8. Show that the coefficient of x^n in Exercise 16.1.6 is

$$a_n = \frac{(2n)!}{2(n!)} (1 + \delta_{n0}).$$

Hence ϕ_n/a_n is a monic polynomial. Suggestion: Use the binomial expansion to show that

$$\sum_{k=0}^{n} \binom{2n}{2k} = 2^{n-1}(1 + \delta_{n0}).$$

16.1.9. Make up some differential equations which have polynomial solution sets other than those given in this section.

16.2 ORTHOGONALITY OF G_n

Since the function $G_n(a, b, c, h, k, C_n, x)$ is a solution of a Sturm-Liouville equation, the set $\{G_n\}$ will be orthogonal over an interval (α, β) if appropriate boundary conditions are specified, or if $r(\alpha) = r(\beta) = 0$ in (16.1.5), as was shown in Sec. 1.9. Following the procedure of that section we may show in general that the set $\{G_n\}$ is an orthogonal set, subject to the above conditions.

Comparing (1.9.5) with (16.1.4), we see that $w(x) = \mu(x)$, so that by (1.9.8) we have

$$(\lambda_m - \lambda_n) \int_\alpha^\beta \mu(x) G_n G_m \, dx = r(x)[G'_n G_m - G'_m G_n] \Big|_\alpha^\beta . \qquad (16.2.1)$$

If G_n satisfies the homogeneous boundary conditions (1.9.6) the right member is zero. As previously mentioned, this is true also if $r(x)$ vanishes at the end points. The latter case is the one we shall consider.

As we shall see, this latter case is $G_n(-1, \alpha + \beta, -\alpha\beta, h, k, C_n, x)$, $\beta > \alpha$, which satisfies the equation

$$(x - \alpha)(\beta - x)y'' + (hx + k)y' - n(h - n + 1)y = 0, \qquad (16.2.2)$$

and for which, by (16.1.5), we have

$$r(x) = (\beta - x)^{-(\beta h + k)/(\beta - \alpha)} (x - \alpha)^{(\alpha h + k)/(\beta - \alpha)}. \qquad (16.2.3)$$

Thus we see that $r(x)$ vanishes at α and β if

$$\beta h + k < 0, \qquad \alpha h + k > 0. \qquad (16.2.4)$$

We observe that these equations, for $\beta > \alpha$, require that $h < 0$, $\beta h \neq -k$, and $\alpha h \neq -k$.

For any $h < 0$ we may find an appropriate value of k, since (16.2.4) is equivalent to

$$\alpha h > -k > \beta h. \tag{16.2.5}$$

Therefore, given any finite values of α and β, we can construct a set of polynomials which are orthogonal over (α, β). The weight function will be

$$w(x) = \mu(x) = \frac{r(x)}{f(x)} = (\beta - x)^{-[\beta(h+1)+k-\alpha]/(\beta-\alpha)}(x - \alpha)^{[\alpha(h+1)+k-\beta]/(\beta-\alpha)}, \tag{16.2.6}$$

which we note can be written as

$$w(x) = (\beta - x)^{M-1}(x - \alpha)^{N-1}, \tag{16.2.7}$$

where

$$M = -\frac{\beta h + k}{\beta - \alpha} > 0, \qquad N = \frac{\alpha h + k}{\beta - \alpha} > 0. \tag{16.2.8}$$

For these values we will have

$$(G_n, G_m) = \int_\alpha^\beta w(x) G_n G_m \, dx = 0, \quad n \neq m.$$

We will consider the norm of G_n in Sec. 16.3.

As an example, we note that $\phi_n(x)$ in Exercise 16.1.5 has $\alpha = 0$, $\beta = 2$, $h = -1$, $k = 1$. For this case we have

$$w(x) = [x(2 - x)]^{-1/2}, \qquad r(x) = [x(2 - x)]^{1/2}.$$

16.3 NORM OF $G_n(-1, \alpha + \beta, -\alpha\beta, h, k, C_n, x)$.

The Rodrigues' formula for $G_n(-1, \alpha + \beta, -\alpha\beta, h, k, C_n, x)$ is given by (16.1.6) as

$$G_n = C_n(\beta - x)^{-M+1}(x - \alpha)^{-N+1} D^n[(\beta - x)^{M+n-1}(x - \alpha)^{N+n-1}], \tag{16.3.1}$$

where M and N are given by (16.2.8). We may use Leibnitz's formula to carry out the differentiation and arrive at a summation form of G_n. The result is

$$G_n = C_n \sum_{m=0}^{n} (-1)^m \binom{n}{m} (n - m + M)_m (m + N)_{n-m} (\beta - x)^{n-m} (x - \alpha)^m. \tag{16.3.2}$$

We observe from this expression that the coefficient of x^n is

$$a_n = (-1)^n C_n \sum_{m=0}^{n} \binom{n}{m} (n - m + M)_m (m + N)_{n-m}. \tag{16.3.3}$$

The norm of G_n is given by

$$\|G_n\|^2 = \int_\alpha^\beta \mu G_n^2 \, dx = C_n \int_\alpha^\beta G_n D^n(f^n \mu) \, dx. \tag{16.3.4}$$

If we integrate by parts, we have

$$\|G_n\|^2 = C_n\left[G_n D^{n-1}(f^n\mu)\Big|_\alpha^\beta - \int_\alpha^\beta G_n' D^{n-1}(f^n\mu)\,dx\right]$$
$$= -C_n \int_\alpha^\beta G'_n D^{n-1}(f^n\mu)\,dx,$$

since the integrated part is zero, as the reader may verify. We leave as an exercise the task of showing that

$$D^k(f^n\mu)\Big|_\alpha^\beta = 0, \quad k < n. \tag{16.3.5}$$

If we continue to integrate (16.3.4) by parts, and use (16.3.5), the nth integration will result in

$$\|G_n\|^2 = (-1)^n C_n \int_\alpha^\beta G_n^{(n)} f^n\mu\,dx$$

Since G_n is a polynomial of degree n with leading coefficient a_n given by (16.3.3), this expression may be written

$$\|G_n\|^2 = (-1)^n n!\,a_n C_n \int_\alpha^\beta (\beta - x)^{M+n-1}(x-\alpha)^{N+n-1}\,dx.$$

Letting $t = x - \alpha$ and factoring out $(\beta - \alpha)^{M+n-1}$, we may write this

$$\|G_n\|^2 = (-1)^n n!\,a_n C_n(\beta-\alpha)^{M+n-1}\int_0^{\beta-\alpha} t^{N+n-1}\left(1 - \frac{t}{\beta-\alpha}\right)^{M+n-1} dt.$$

Finally, letting $z = t/(\beta - \alpha)$, we have

$$\|G_n\|^2 = (-1)^n n!\,a_n C_n(\beta-\alpha)^{2n-h-1}\int_0^1 z^{N+n-1}(1-z)^{M+n-1}\,dz.$$

The integral on the right is the beta function (see Exercise 5.1.4), which is known. Substituting its value for this case, we have

$$\|G_n\|^2 = (-1)^n n!\,a_n C_n(\beta-\alpha)^{2n-h-1}\frac{\Gamma(n+M)\Gamma(n+N)}{\Gamma(2n-h)}, \tag{16.3.6}$$

where a_n is given by (16.3.3). We note that Eqs. (16.2.8) together with $h < 0$ ensure that (16.3.6) is well defined. The expression (16.3.3) for a_n contains a factor $(-1)^n C_n$ so that the norm is positive, as it should be.

Equation (16.3.6) may be put in a more useful form if we substitute for a_n from (16.3.3). This latter equation may be written

$$a_n = (-1)^n C_n \sum_{k=0}^{n}\binom{n}{k}\frac{\Gamma(M+n)\Gamma(N+n)}{\Gamma(M+n-k)\Gamma(N+k)}, \tag{16.3.7}$$

which when substituted into (16.3.6) yields

$$\|G_n\|^2 = n!\,C_n^{\,2}(\beta-\alpha)^{M+N+2n-1}\frac{\Gamma^2(N+n)\Gamma^2(M+n)}{\Gamma(M+N+2n)}$$
$$\times \sum_{k=0}^{n}\frac{\binom{n}{k}}{\Gamma(M+n-k)\Gamma(N+k)}.$$

Using the known formula (see Exercise 16.3.1)

$$\sum_{k=0}^{n} \frac{\Gamma(M+n)\Gamma(N+n)}{\Gamma(M+n-k)\Gamma(N+k)} \binom{n}{k} = \frac{\Gamma(M+N+2n-1)}{\Gamma(M+N+n-1)}, \qquad (16.3.8)$$

we finally obtain

$$\|G_n\|^2 = n!\, C_n^{\,2}(\beta-\alpha)^{M+N+2n-1} \frac{\Gamma(N+n)\Gamma(M+n)\Gamma(M+N+2n-1)}{\Gamma(M+N+2n)\Gamma(M+N+n-1)}.$$

This expression may be further simplified to yield the two cases,

$$\|G_0\|^2 = C_0^{\,2}(\beta-\alpha)^{M+N-1} \frac{\Gamma(M)\Gamma(N)}{\Gamma(M+N)}, \qquad (16.3.9)$$

and for $n \neq 0$,

$$\|G_n\|^2 = C_n^{\,2}(\beta-\alpha)^{M+N+2n-1} \frac{n!\,\Gamma(M+n)\Gamma(N+n)}{(M+N+2n-1)\Gamma(M+N+n-1)}. \qquad (16.3.10)$$

EXERCISES

16.3.1. Show that

$$\binom{n}{k} + \binom{n}{k-1} = \binom{n+1}{k}.$$

Use this result and show by induction on n that (16.3.8) holds.

16.3.2. Verify (16.3.5).

16.3.3. Using (16.3.9) and (16.3.10) find the norms of

(a) $P_n(x)$.
(b) $T_n(x)$.
(c) $U_n(x)$.
(d) $P_n^{(\alpha,\beta)}(x)$.
(e) $C_n^{\,\nu}(x)$.

16.3.4. Find the norm of $\phi_n(x)$ given in Exercise 16.1.5.

16.3.5. Make up a set of polynomials which are orthogonal over (1, 3) and find the differential equation, the Rodrigues' formula, and the orthogonality relation.

16.3.6. Show that $G_n(-1, 0, 1, -2m, 0, C_n, x)$, $m > 0$, is the ultraspherical polynomial $P_n^{(\alpha,\alpha)}(x)$ defined in Exercise 13.4.12. Work Exercise 13.4.13 by the method of this chapter.

16.3.7. Show that in Exercise 16.3.6 for $m = 1, \frac{1}{2}, \frac{3}{2}, \nu + \frac{1}{2}$, G_n is, respectively, $P_n(x)$, $T_n(x)$, $U_n(x)$, and $C_n^{\,\nu}(x)$.

16.3.8. Obtain the differential equation, the Rodrigues' formula, and the orthogonality relation for G_n in Exercise 16.3.6 for $m = 2, 3$.

16.3.9. Show that for $0 < c < 1$,

$$y_n^{\,c}(x) = G_n[-1, 0, 1, -1, 1-2c, (-2)^n n!/(2n)!, x]$$

satisfies the conditions

$$y_n^{1/2}(x) = T_n(x),$$

$$(1 - x^2)y'' + (1 - 2c - x)y' + n^2y = 0, \text{ where } y = y_n{}^c,$$

$$y_n{}^c = \frac{(-2)^n n!}{(2n)!}(1 - x)^{-c+1}(1 + x)^c D^n[(1 - x)^{c+n-1}(1 + x)^{-c+n}],$$

$$\int_{-1}^{1}(1 - x)^{c-1}(1 + x)^{-c} y_n{}^c y_m{}^c\, dx = \frac{2^{4n-1}(n!)^2\Gamma(n + c)\Gamma(n - c + 1)\delta_{mn}}{[(2n)!]^2}.$$

16.4 INFINITE INTERVALS

In the previous section we considered the possibility of constructing polynomial sets which are orthogonal on a given finite interval (α, β). In this section we will consider the cases where α or β or both may be infinite. We will again consider the cases where $r(x)$ in (16.2.1) vanishes at α and β.

First, for the case of an infinite upper limit, we consider $G_n(0, 1, c, h, k, C_n, x)$, which satisfies the differential equation

$$(x + c)y'' + (hx + k)y' - nhy = 0,$$

with

$$r(x) = \exp\left[\int \frac{g\, dx}{f}\right] = \exp\left[h\int \frac{x + k/h}{x + c}\, dx\right],$$

or

$$r(x) = e^{hx}(x + c)^{k-ch}. \tag{16.4.1}$$

Thus we see that if

$$h < 0, \quad k - ch > 0, \tag{16.4.2}$$

then $\{G_n\}$ is orthogonal over $(-c, \infty)$. If we are given $\alpha = -c$ as a prescribed lower limit, then we may select h, k to satisfy (16.4.2), which may then be written

$$h < 0, \qquad k > -\alpha h, \qquad c = -\alpha.$$

These conditions may always be satisfied for a given α.

The weight function for $G_n(0, 1, c, h, k, C_n, x)$ is given by

$$w(x) = \frac{r(x)}{f(x)} = e^{hx}(x + c)^M, \tag{16.4.3}$$

where

$$M = k - ch - 1 > -1. \tag{16.4.4}$$

The orthogonality relation is

$$\int_{-c}^{\infty} w(x)G_nG_m\, dx = \|G_n\|^2\, \delta_{mn}, \tag{16.4.5}$$

and the Rodrigues' formula is

$$G_n(0, 1, c, h, k, C_n, x) = C_n e^{-hx}(x + c)^{-M}D^n[e^{hx}(x + c)^{M+n}]. \tag{16.4.6}$$

To obtain the norm, we begin with the expression

$$\|G_n\|^2 = \int_{-c}^{\infty} wG_n^2\,dx = C_n\int_{-c}^{\infty} G_nD^n(f^n\mu)\,dx,$$

which is similar to (16.3.4). If we integrate by parts n times, we will find, as we did for (16.3.4), that the integrated part is zero, and that

$$\|G_n\|^2 = (-1)^n n!\, a_nC_n\int_{-c}^{\infty} e^{hx}(x+c)^{M+n}\,dx, \tag{16.4.7}$$

where again a_n is the leading coefficient of G_n. As we shall see in Exercise 16.6.1, it is given by

$$a_n = h^nC_n.$$

If we make the transformation $x = -t/h - c$ and substitute for a_n, (16.4.7) becomes

$$\|G_n\|^2 = n!\left(\frac{-1}{h}\right)^{M+1} C_n^2e^{-ch}\Gamma(M+n+1). \tag{16.4.8}$$

We notice that since $h < 0$, the norm is always positive.

To have G_n orthogonal over $(-\infty, -c)$ we have only to require in (16.4.1) that

$$h > 0, \qquad k - ch > 0. \tag{16.4.9}$$

We will consider this case in the exercises.

Also in the exercises we will consider the function $G_n(0, 0, 1, h, k, C_n, x)$, which satisfies the differential equation

$$y'' + (hx+k)y' - nhy = 0$$

and has

$$r(x) = \exp\left(\frac{hx^2}{2} + kx\right) = \exp\left[\frac{h}{2}\left(x+\frac{k}{h}\right)^2 - \frac{k^2}{2h}\right]. \tag{16.4.10}$$

The Rodrigues' formula is

$$G_n = C_n\exp\left[-\frac{h}{2}\left(x+\frac{k}{h}\right)^2\right]D^n\exp\left[\frac{h}{2}\left(x+\frac{k}{h}\right)^2\right], \tag{16.4.11}$$

and the orthogonality relation, for $h < 0$, is

$$\int_{-\infty}^{\infty}\exp\left(\frac{hx^2}{2}+kx\right)G_nG_m\,dx = n!\,C_n^2(-h)^{n-1/2}\sqrt{2\pi}e^{-k^2/2h}\,\delta_{mn}. \tag{16.4.12}$$

We note that $-h > 0$ so that the norm is positive.

16.5 GENERATING FUNCTIONS

If we do not restrict ourselves to real-variable methods, we may very easily obtain some other general results. For example, we may find a generating function

$$g(x, t) = \sum_{n=0}^{\infty} b_n G_n(a, b, c, h, k, C_n, x) t^n \tag{16.5.1}$$

by utilizing the known formula from the theory of functions of a complex variable (see, for example, [Kr], page 666),

$$D^n F(x) = \frac{n!}{2\pi i} \int_C \frac{F(z)\, dz}{(z - x)^{n+1}}, \quad n = 1, 2, \ldots. \tag{16.5.2}$$

Here C is a closed path in the x-y plane that encloses x but does not enclose singularities of $F(z)$; also, the integration is counterclockwise. The conditions on $F(x)$ are quite general and hold for all of our polynomial sets.

Substituting into (16.5.1) for G_n, using the Rodrigues' formula, we have

$$g(x, t) = \sum_{n=0}^{\infty} \frac{b_n C_n t^n D^n(f^n \mu)}{\mu},$$

which may be written, using (16.5.2),

$$g(x, t) = \sum_{n=0}^{\infty} \frac{b_n n!\, C_n t^n}{2\pi i \mu(x)} \int_C \frac{f^n(z)\mu(z)\, dz}{(z - x)^{n+1}}. \tag{16.5.3}$$

We may choose b_n such that B, defined by

$$B^n = b_n n!\, C_n, \tag{16.5.4}$$

is independent of n. Making this substitution and interchanging the orders of the operations in (16.5.3), we have

$$g(x, t) = \frac{1}{2\pi i \mu(x)} \int_C \frac{\mu(z)\, dz}{(z - x)} \sum_{n=0}^{\infty} \left[\frac{Bf(z)t}{z - x}\right]^n.$$

The series on the right is a geometric progression and may be summed if

$$\left|\frac{Bf(z)t}{z - x}\right| < 1, \tag{16.5.5}$$

resulting in the generating function

$$g(x, t) = \frac{1}{2\pi i \mu(x)} \int_C \frac{\mu(z)\, dz}{z - x - Bf(z)t} = \sum_{n=0}^{\infty} b_n G_n t^n. \tag{16.5.6}$$

To sum up the conditions under which this expression for $g(x, t)$ is valid, we note that the contour C must be chosen so that it does not enclose singularities of $\mu(z)$, it does enclose the point x, and Eq. (16.5.5) is satisfied. To satisfy these conditions will generally place restrictions on t also. The point x will be restricted to lie on the interval (α, β).

The integration in (16.5.6) may sometimes be carried out very easily by complex-variable methods. For example, if the integrand can be expanded in the form

$$\psi(z) = \frac{\mu(z)}{z - x - Bf(z)t} = \sum_{j=0}^{\infty} c_j(z - z_k)^j + \sum_{j=1}^{\infty} c_{-j}(z - z_k)^{-j}, \quad (16.5.7)$$

where not all of the c_{-j} vanish, then $\psi(z)$ has a singularity at $z = z_k$. If $c_{-N} \neq 0$ and if $c_{-j} = 0$ for every $j > N$, then $\psi(z)$ has a *pole* of order N at $z = z_k$. The coefficient $c_{-1} = R_k$ is called the *residue* of $\psi(z)$ at $z = z_k$, and by *Cauchy's residue theorem* (see [Kr], page 700), if the number of singularities z_k of $\psi(z)$ inside C is finite, say $k = 1, 2, \ldots, m$, then (16.5.6) may be written

$$g(x, t) = \frac{1}{\mu(x)} \sum_{k=1}^{m} R_k = \sum_{n=0}^{\infty} b_n G_n(x) t^n. \quad (16.5.8)$$

As an example, for the Laguerre polynomial $L_n(x)$, we have

$$\mu(z) = e^{-z}, \qquad f(z) = z, \qquad C_n = \frac{1}{n!}.$$

Observing (16.5.4), we see that we may take $b_n = B = 1$. Then we have

$$\psi(z) = \frac{e^{-z}}{z - x - zt},$$

which has a *simple* pole ($N = 1$) at $z_1 = x/(1 - t)$. The contour C must be chosen to include x and to satisfy (16.5.5), which for this case is

$$\left| \frac{zt}{z - x} \right| < 1.$$

This inequality can be satisfied for a path about x by making t sufficiently small, which in turn makes the pole z_1 arbitrarily close to x. Hence a path which encloses x and z_1 will suffice if t is sufficiently small.

We note from (16.5.7) that c_{-1} for the pole z_1 may be found from

$$R_1 = c_{-1} = \lim_{z \to z_1} (z - z_1)\psi(z) = (1 - t)^{-1} e^{-x/(1-t)}.$$

Hence (16.5.8) gives the generating function

$$(1 - t)^{-1} e^{-xt/(1-t)} = \sum_{n=0}^{\infty} L_n(x) t^n.$$

Other results, as we know, may be obtained from the generating function, as well as from the differential equation, the Rodrigues' formula, and the orthogonality condition.

16.6 SUMMARY

We have seen that all the orthogonal polynomials sets that we have considered in this book may be expressed as special cases of our function $G_n(a, b, c, h, k, C_n, x)$, and that their differential equations and Rodrigues' formulas may be obtained in general. In addition, all of these except the Bessel polynomials and the related Burchnall polynomials satisfy the conditions of Sec. 16.2, 16.3, or 16.4, and hence we may find their orthogonality relations by substituting into a formula.

The reader may observe that this last chapter is not applicable to the Bessel functions, Mathieu functions, associated Legendre functions, etc., because their differential equations are not of the type (16.1.1), the coefficient of y not being a constant in these cases. Also, it may be of some interest to see why the Bessel and Burchnall polynomials do not fit any of our formulas for the norm of G_n. The Bessel polynomials, for example, using (16.1.5), require that

$$r(x) = x^2e^{-2/x},$$

which does not vanish at two different values of x. The reader will recall, however, that the orthogonality relation for these polynomials is not of the orthodox type, but is one involving a contour integral.

Evidently it is quite conceivable that a book on special functions could begin with a chapter such as this, in which many of the important characteristics of the functions are obtained in general. Particular cases could then be given as exercises or worked out by substituting into the known formulas, any of which may be used as a beginning to obtain all the other characteristics.

Such an approach would have the merit of making the work much more compact, but would perhaps give the reader less insight and certainly much less experience in dealing with the special functions. For this reason, and to avoid giving the book too much of a "cook book" flavor, we have given the more conventional approach.

EXERCISES

16.6.1. Show that (16.4.6) may be written

$$G_n = C_n \sum_{m=0}^{n} \binom{n}{m} h^{n-m}(n + M - m + 1)_m(x + c)^{n-m},$$

and hence that $a_n = h^nC_n$.

16.6.2. Show that (16.4.5), (16.4.6), and (16.4.8) hold for the Laguerre polynomials $L_n(x)$.

16.6.3. Establish (16.4.10) and (16.4.11).

16.6.4. Find a_n for $G_n(0, 0, 1, h, k, C_n, x)$. Suggestion: Write $G_n(x)$ in (16.4.11) with $x = \sqrt{-2/h}\, y - k/h$, and use the relation obtained for $H_n(x)$ in Chapter 9:

$$e^{y^2} D^n e^{-y^2} = \sum_{m=0}^{[n/2]} \frac{(-1)^{n+m} n!\, (2y)^{n-2m}}{m!\, (n-2m)!}.$$

Ans. $a_n = h^n C_n$.

16.6.5. Show for $G_n(0, 0, 1, h, k, C_n, x)$ that

$$\|G_n\|^2 = n!\,(-1)^n a_n C_n \sqrt{\frac{-2\pi}{h}}\, e^{-k^2/2h},$$

and use Exercise 16.6.4 to obtain (16.4.12).

16.6.6. Show that (16.4.10), (16.4.11), and (16.4.12) hold for the Hermite polynomials $H_n(x)$.

16.6.7. Work Exercise 16.6.6 for the set $\{h_n(x)\}$ given in Exercise 16.1.3.

16.6.8. Find the differential equation, Rodrigues' formula, and orthogonality relation for $y_n = G_n(0, 1, -1, -1, 2, 1/n!, x)$. Write out the first four y_n's.

Ans. $(x - 1)y'' - (x - 2)y' + ny = 0$;
$y_n = (1/n!)e^x D^n[e^{-x}(x - 1)^n]$;
$\int_1^\infty e^{-x} y_n y_m\, dx = e^{-1}\delta_{nm}$;
$y_0 = 1$, $y_1 = -x + 2$, $y_2 = \frac{1}{2}(x^2 - 6x + 7)$, $y_3 = \frac{1}{6}(-x^3 + 12x - 39x + 34)$.

16.6.9. Obtain a set of polynomials that is orthogonal over $(2, \infty)$. Find the differential equation, Rodrigues' formula, and orthogonality relation.

16.6.10. Work Exercise 16.6.9 for the interval $(-\infty, \infty)$. Find a set other than $\{H_n(x)\}$ and $\{h_n(x)\}$.

16.6.11. Find a generating function for $\{P_n(x)\}$ using the method of Sec. 16.5. Suggestion: Show that the circle C with center at the origin and radius $1/|t|$ for $0 < |t| < 1$ encloses x for $-1 \le x \le 1$ and satisfies the condition of (16.5.5) for t sufficiently small. This circle contains only one of the two poles of $\psi(x)$.

16.6.12. Using (16.4.9), find the orthogonality relation for a set orthogonal on $(-\infty, \beta)$. Give an example.

Appendix A

Properties of Infinite Series

A.1 CONVERGENT SERIES

The infinite series

$$S(x) = \sum_{k=1}^{\infty} u_k(x), \tag{A.1.1}$$

converges to the sum $S(x)$ if given any $\epsilon > 0$, there exists an integer N_ϵ such that

$$|S(x) - S_n(x)| < \epsilon, \tag{A.1.2}$$

whenever $n \geq N_\epsilon$. $S_n(x)$ is the sum of the first n terms, given by

$$S_n(x) = \sum_{k=1}^{n} u_k(x).$$

In this section we give without proof (for a detailed treatment, see [Bu], Chapter 4) some of the better-known properties of convergent series. We abbreviate our series (A.1.1) as Σu_k:

1. If Σu_k converges, then $\lim_{k\to\infty} u_k = 0$.
2. (Comparison Test) If $0 \leq u_k \leq v_k$ for all k sufficiently large, the convergence of Σv_k implies the convergence of Σu_k.
3. (Ratio Test) If $\lim_{k\to\infty} |u_{k+1}/u_k| = r < 1$, then Σu_k converges.
4. If $\Sigma\, |u_k|$ converges, then Σu_k converges.

Definition: The series Σu_k converges *absolutely* if $\Sigma\, |u_k|$ converges.

5. If Σu_k converges absolutely, then the order of the terms in the summation is immaterial.
6. The *alternating* series $\Sigma(-1)^k u_k$, where $u_k > 0$, $k = 1, 2, 3, \ldots$, converges if $u_{k+1} < u_k$, $k = 1, 2, 3, \ldots$, and $\lim_{k\to\infty} u_k = 0$. Also,

$$\left|\sum_{k=N}^{\infty} (-1)^k u_k\right| < u_N.$$

In the case of double series, triple series, etc., the foregoing properties still hold with appropriate modifications. For example, the double series

$$S(x) = \sum_i \sum_j u_{ij} \tag{A.1.3}$$

converges if for any $\epsilon > 0$, there is an integer N_ϵ such that

$$\left| S(x) - \sum_{i=1}^{n} \sum_{j=1}^{m} u_{ij} \right| < \epsilon$$

whenever $n \geq N_\epsilon$ and $m \geq N_\epsilon$.

A.2 UNIFORMLY CONVERGENT SERIES

The validity of certain operations on infinite series such as termwise differentiation or interchange of "limit of a sum" with "sum of a limit" is closely associated with the concept of *uniform convergence*. The series

$$S(x) = \sum_{k=1}^{\infty} u_k(x), \quad a \leq x \leq b \tag{A.2.1}$$

is said to converge *uniformly* with respect to x if given any $\epsilon > 0$, there exists an integer N_ϵ, *independent of* x, such that

$$|S(x) - S_n(x)| < \epsilon, \tag{A.2.2}$$

whenever $n \geq N_\epsilon$, where again $S_n(x)$ is the sum of the first n terms of Σu_k. The reader will note that this definition of uniform convergence becomes the definition of ordinary convergence if the phrase "independent of x" is deleted. Thus, if a series converges uniformly, then it converges in the ordinary sense, though the converse is not true.

Uniformly convergent series have many useful properties, some of which we list as follows:

7. *Termwise integration*. If $S(x)$ and each $u_k(x)$ are integrable over $a \leq x \leq b$, then

$$\int_a^b S(x)\, dx = \Sigma \int_a^b u_k(x)\, dx. \tag{A.2.3}$$

Proof. Statement A: Given $\epsilon > 0$, there exists N_ϵ such that $|S(x) - S_n(x)| < \epsilon$ when $n \geq N_\epsilon$ for all x on $a \leq x \leq b$. Given Statement A, for $n \geq N_\epsilon$, then

$$\begin{aligned} \left|\int_a^b S(x)\, dx - \sum_{k=1}^{n} \int_a^b u_k\, dx\right| &= \left|\int_a^b S(x)\, dx - \int_a^b \sum_{k=1}^{n} u_k\, dx\right| \\ &= \left|\int_a^b S(x)\, dx - \int_a^b S_n(x)\, dx\right| \\ &\leq \int_a^b |S(x) - S_n(x)|\, dx < \epsilon(b - a). \end{aligned}$$

Therefore $|\int_a^b S(x)\,dx - \sum_{k=1}^n \int_a^b u_k(x)\,dx|$ can be made arbitrarily small for n sufficiently large, which implies (A.2.3).

8. If each $u_k(x)$ is continuous on $a \leq x \leq b$, then $S(x)$ is continuous on $a < x < b$.

   Proof. By Statement A (Property 7), for $n > N_\epsilon$ and for x_0 in $a \leq x \leq b$, $|S(x) - S_n(x)| < \epsilon/3$ and $|S(x_0) - S_n(x_0)| < \epsilon/3$. Since $S_n(x)$ is a finite sum and hence is continuous at x_0, there exists $\delta > 0$ such that

$$|S_n(x) - S_n(x_0)| < \epsilon/3$$

   for all x such that $|x - x_0| < \delta$. We may write

$$\begin{aligned}|S(x) - S(x_0)| &= |S(x) - S_n(x) + S_n(x) - S_n(x_0) + S_n(x_0) - S(x_0)| \\ &\leq |S(x) - S_n(x)| + |S_n(x) - S_n(x_0)| + |S_n(x_0) - S(x_0)| \\ &< \frac{\epsilon}{3} + \frac{\epsilon}{3} + \frac{\epsilon}{3} = \epsilon.\end{aligned}$$

   Hence for $|x - x_0| < \delta$, we may make $|S(x) - S(x_0)|$ arbitrarily small for n sufficiently large, and thus $S(x)$ is continuous at $x = x_0$.

   By a slight modification of the above argument it can be shown that $\lim_{x\to a+0} S(x) = S(a)$ and $\lim_{x\to b-0} S(x) = S(b)$. For example, instead of the condition $|x - x_0| < \delta$, we would use $0 < x - a < \delta$.

9. If for $a \leq x \leq b$, $\Sigma u_k = S(x)$ converges, each u'_k is continuous, and $\Sigma u'_k$ converges uniformly, then

$$S'(x) = \Sigma u'_k(x) \text{ for } a < x < b.$$

   Proof. Let $F(x) = \Sigma u'_k(x)$. Since this series converges uniformly, Property 1 enables us to write, for x on (a, b),

$$\int_a^x F(x)\,dx = \Sigma \int_a^x u'_k(x)\,dx = \Sigma[u_k(x) - u_k(a)].$$

   Since Σu_k converges, we may write this result as

$$\int_a^x F(x)\,dx = S(x) - S(a).$$

   By Property 8, $F(x)$ is continuous and hence

$$\frac{d}{dx}\int_a^x F(x)\,dx = F(x) = S'(x),$$

   or

$$S'(x) = \Sigma u'_k(x).$$

10. (Weierstrass M-Test) If the series of positive constants, ΣM_k, converges, and if for $k = 1, 2, 3, \ldots$, $|u_k(x)| \leq M_k$, for $a \leq x \leq b$, then the series (A.2.1) converges uniformly. The proof follows from the comparison test.

11. (Abel's Test) If $\Sigma u_k(x)$ converges uniformly on $a \leq x \leq b$ and the sequence $\{v_k(y)\}$, $k = 1, 2, 3, \ldots$, has for each k the properties, $|v_k(y)|$ is bounded and either $v_{k+1} \leq v_k$ or $v_{k+1} \geq v_k$, for $c \leq y \leq d$, then the series

$\Sigma u_k(x)v_k(y)$ is uniformly convergent with respect to x and y on $a \leq x \leq b$ and $c \leq y \leq d$. (For proof, see [C-1], pages 227–230.)

A.3 POWER SERIES

The special case of (A.1.1), known as a power series,

$$S(x) = \sum_{k=0}^{\infty} a_k x^k, \tag{A.3.1}$$

can easily be checked for convergence by the ratio test (Property 3). The application of the test to (A.3.1) results in an inequality $|r(x)| < 1$, which yields a solution $|x| < R$. The range $-R < x < R$ is the *interval of convergence* and if x is restricted to $|x| \leq R_0 < R$, where R_0 is fixed, the series (A.3.1) will converge absolutely and uniformly. In solving a differential equation by the method of Frobenius, the ratio test is particularly easy to apply since the essential part of the ratio has already been determined in the recurrence relation for the coefficients a_k.

Appendix B

Convergence of Fourier Series

B.1 SUFFICIENCY FOR CONVERGENCE

We shall restrict our attention to the case given by Eqs. (2.1.6) and (2.1.7), the first $2n + 1$ terms in the series being given by

$$S_n(x) = \frac{1}{2\pi} \int_{-\pi}^{\pi} \left\{ f(\xi) + 2 \sum_{k=1}^{n} f(\xi)[\cos k\xi \cos kx + \sin k\xi \sin kx] \right\} d\xi.$$

This equation may be written, by using a trigonometric identity, in the form

$$S_n(x) = \frac{1}{\pi} \int_{-\pi}^{\pi} f(\xi) F(\xi - x)\, d\xi, \tag{B.1.1}$$

where

$$F(u) = \tfrac{1}{2} + \sum_{k=1}^{n} \cos ku. \tag{B.1.2}$$

The sectional continuity of $f(x)$ (see Sec. 2.1) assures the existence of the integrals.

The latter series may be summed by multiplying both sides of (B.1.2) by $2 \sin \frac{1}{2}u$ and using the identity

$$2 \cos ku \sin \tfrac{1}{2}u = \sin (k + \tfrac{1}{2})u - \sin (k - \tfrac{1}{2})u.$$

All the terms in the right-hand side of (B.1.2) will cancel except one, and we may write

$$F(u) = \frac{\sin (n + \frac{1}{2})u}{2 \sin \frac{1}{2}u}, \tag{B.1.3}$$

from which we note that

$$F(u) = F(-u). \tag{B.1.4}$$

Equation (B.1.1) may be put in the two alternative forms

$$S_n(x) = \frac{1}{\pi} \int_{-\pi}^{\pi} f(x + \xi) F(\xi)\, d\xi$$

and

$$S_n(x) = \frac{1}{\pi}\int_{-\pi}^{\pi} f(x - \xi)F(\xi)\, d\xi,$$

the first by virtue of the periodicity of $f(x)$ and $F(x)$, and the second, which is obtained from the first, by virtue of (B.1.4). Adding these two expressions for $S_n(x)$, and noting that the resulting integrand is an even function of ξ, we obtain

$$S_n(x) = \frac{1}{\pi}\int_{0}^{\pi} [f(x + \xi) + f(x - \xi)]F(\xi)\, d\xi. \tag{B.1.5}$$

If we multiply (B.1.2) through by $(2/\pi)S_f(x)$, given by (2.1.5), and integrate with respect to ξ from 0 to π, we have

$$S_f(x) = \frac{1}{\pi}\int_{0}^{\pi} [f(x + 0) + f(x - 0)]F(\xi)\, d\xi. \tag{B.1.6}$$

We have used in the right member the expression for $S_f(x)$ given by

$$S_f(x) = \tfrac{1}{2}[f(x + 0) + f(x - 0)].$$

Also, we have made use of the orthogonality of the set $\{\cos ku\}$.

We now obtain, using (B.1.5), (B.1.6), and (B.1.3), the expression

$$S_n(x) - S_f(x) = \frac{1}{\pi}\int_{0}^{\pi} \left[\frac{f(x + \xi) - f(x + 0)}{2 \sin \frac{1}{2}\xi}\right] \sin [(n + \tfrac{1}{2})\xi]\, d\xi$$

$$+ \frac{1}{\pi}\int_{0}^{\pi} \left[\frac{f(x - \xi) - f(x - 0)}{2 \sin \frac{1}{2}\xi}\right] \sin [(n + \tfrac{1}{2})\xi]\, d\xi. \tag{B.1.7}$$

We note that

$$\lim_{\xi\to+0} \left[\frac{f(x + \xi) - f(x + 0)}{2 \sin \frac{1}{2}\xi}\right] = \lim_{\xi\to+0} \left[\frac{f(x + \xi) - f(x + 0)}{\xi}\right] \frac{\frac{1}{2}\xi}{\sin \frac{1}{2}\xi}$$

$$= \lim_{\xi\to+0} \frac{f(x + \xi) - f(x + 0)}{\xi}.$$

This latter limit, which is the right-hand derivative $f'_R(x)$, is postulated to exist (see Sec. 2.1). Hence the first integrand in the right member of (B.1.7) is sectionally continuous and the integral exists. A similar procedure can be applied to the second integral, in which the left-hand derivative will appear. Thus (B.1.7) may be written

$$S_n(x) - S_f(x) = \frac{1}{\pi}\int_{0}^{\pi} G(\xi) \sin [(n + \tfrac{1}{2})\xi]\, d\xi, \tag{B.1.8}$$

where $G(\xi)$ is sectionally continuous on $(0, \pi)$.

We may expand the right member of (B.1.8) into two integrals and make use of Exercise 2.4.18 to show that

$$\lim_{n \to \infty} |S_n(x) - S_f(x)| = 0.$$

Hence under the conditions assumed in Sec. 2.1 our series (2.1.6) converges to

$$S_f(x) = \tfrac{1}{2}[f(x + 0) + f(x - 0)].$$

Appendix C

Tables

Table I. Laplace Transforms

	$f(t)$	$F(s)$
1	$u(t)$	$\frac{1}{s}$
2	$t^n, \quad n > -1$	$\frac{\Gamma(n+1)}{s^{n+1}}$
3	e^{at}	$\frac{1}{s-a}$
4	$\sin kt$	$\frac{k}{s^2+k^2}$
5	$\cos kt$	$\frac{s}{s^2+k^2}$
6	$f(at)$	$\frac{1}{a} F\left(\frac{s}{a}\right)$
7	$e^{at}f(t)$	$F(s-a)$
8	$f(t-a)u(t-a)$	$e^{-as}F(s)$
9	$\int_0^t f(x)\,dx$	$\frac{F(s)}{s}$
10	$f^{(n)}(t), \quad n = 0, 1, 2, \ldots$	$s^n F(s) - \sum_{k=1}^{n} s^{n-k} f^{(k-1)}(0)$
11	$t^n f(t), \quad n = 0, 1, 2, \ldots$	$(-1)^n F^{(n)}(s)$
12	$f(t), \quad f(t+T) = f(t)$	$\frac{\int_0^T e^{-st} f(t)\,dt}{1 - e^{-sT}}$
13	$f(t) * g(t) = \int_0^t f(\tau) g(t-\tau)\,d\tau$	$F(s)G(s)$
14	$\operatorname{erf}(\sqrt{t}) = \frac{2}{\sqrt{\pi}} \int_0^{\sqrt{t}} e^{-x^2}\,dx$	$\frac{1}{s\sqrt{s+1}}$

Table I. (Continued)

	$f(t)$	$F(s)$
15	$J_n(t), \quad n > -1$	$\dfrac{(\sqrt{s^2+1}-s)^n}{\sqrt{s^2+1}}$
16	$\dfrac{k}{2\sqrt{\pi t^3}}\, e^{-k^2/4t}$	$e^{-k\sqrt{s}}, \quad k > 0$
17	$\dfrac{1}{\sqrt{\pi t}}\, e^{-k^2/4t}$	$\dfrac{1}{\sqrt{s}}\, e^{-k\sqrt{s}}, \quad k \geq 0$
18	$\operatorname{erfc}\left(\dfrac{k}{2\sqrt{t}}\right)$	$\dfrac{1}{s}\, e^{-k\sqrt{s}}, \quad k \geq 0$
19	$-e^{ak}e^{a^2t}\operatorname{erfc}\left(a\sqrt{t}+\dfrac{k}{2\sqrt{t}}\right) + \operatorname{erfc}\left(\dfrac{k}{2\sqrt{t}}\right)$	$\dfrac{ae^{-k\sqrt{s}},}{s(a+\sqrt{s})}, \quad k \geq 0$

Table 2. Finite Sine Transforms

	$f(x)$	$F_s(n)$
1	1	$\dfrac{L[1-(-1)^n]}{n\pi}$
2	x	$\dfrac{(-1)^{n+1}L^2}{n\pi}$
3	x^2	$(-1)^{n-1}\dfrac{L^3}{n\pi} - \dfrac{2L^3}{n^3\pi^3}[1-(-1)^n]$
4	$L-x$	$\dfrac{L^2}{n\pi}$
5	e^{ax}	$\dfrac{n\pi L}{n^2\pi^2 + a^2L^2}[1-(-1)^n e^{aL}]$
6	$\dfrac{2}{L}\arctan\dfrac{k\sin(\pi x/L)}{1+k\cos(\pi x/L)}$	$\dfrac{(-1)^{n+1}k^n}{n}, \quad \|k\| \le 1$
7	$\dfrac{2}{L}\arctan\dfrac{2k\sin(\pi x/L)}{1-k^2}$	$\left[\dfrac{1-(-1)^n}{n}\right]k^n, \quad \|k\| \le 1$

Table 3. Finite Cosine Transforms

	$f(x)$	$F_c(n)$
1	1	$0, \quad n = 1, 2, 3, \ldots; \quad F_c(0) = L$
2	x	$\dfrac{-L^2}{n^2\pi^2}[1-(-1)^n]; \quad F_c(0) = \dfrac{L^2}{2}$
3	x^2	$\dfrac{2L^3(-1)^n}{n^2\pi^2}; \quad F_c(0) = \dfrac{L^3}{3}$
4	$L-x$	$\dfrac{L^2}{n^2\pi^2}[1-(-1)^n]; \quad F_c(0) = \dfrac{L^2}{2}$
5	e^{ax}	$aL^2\left[\dfrac{(-1)^n e^{aL} - 1}{n^2\pi^2 + a^2L^2}\right]$
6	$\dfrac{2k}{L}\left[\dfrac{\cos(\pi x/L) - k}{1 - 2k\cos(\pi x/L) + k^2}\right]$	$k^n; \quad F_c(0) = 0, \quad \|k\| < 1$

Table 4. Summary of Properties of Polynomial Sets $\{\phi_n(x)\}$

Differential Equation: $f(x)\phi_n'' + g(x)\phi'_n + \lambda_n\phi_n = 0$; Self-Adjoint Form: $(\mu f\phi'_n)' + \lambda_n\mu\phi_n = 0$

Integrating Factor: $\mu(x) = \frac{1}{f(x)}\exp\left[\int \frac{g(x)\,dx}{f(x)}\right]$; Rodrigues' Formula: $\phi_n(x) = \frac{C_n}{\mu} D^n(f^n\mu)$

Orthogonality: $\int_a^b \mu\phi_n\phi_m\,dx = \|\phi_n\|^2\delta_{mn}$

$\phi_n(x)$	$f(x)$	$g(x)$	λ_n	μ	C_n	(a, b)	$\|\phi_n\|^2$	Section
$P_n(x)$	$1 - x^2$	$-2x$	$n(n+1)$	1	$\frac{(-1)^n}{2^n n!}$	$(-1, 1)$	$\frac{2}{2n+1}$	4.1
$H_n(x)$	1	$-2x$	$2n$	e^{-x^2}	$(-1)^n$	$(-\infty, \infty)$	$2^n n!\sqrt{\pi}$	9.1
$L_n^{\alpha}(x)$	x	$\alpha + 1 - x$	n	$x^{\alpha}e^{-x}$	$\frac{1}{n!}$	$(0, \infty)$	$\frac{\Gamma(1+\alpha+n)}{n!}$	10.1
$T_n(x)$	$1 - x^2$	$-x$	n^2	$(1 - x^2)^{-1/2}$	$\frac{(-2)^n n!}{(2n)!}$	$(-1, 1)$	π for $n = 0$; $\frac{\pi}{2}$ for $n \neq 0$	11.1
$U_n(x)$	$1 - x^2$	$-3x$	$n(n+2)$	$(1 - x^2)^{1/2}$	$\frac{(-2)^n(n+1)!}{(2n+1)!}$	$(-1, 1)$	$\frac{\pi}{2}$	11.1
$P_n^{(\alpha,\beta)}(x)$	$1 - x^2$	$\beta - \alpha - (\alpha+\beta+2)x$	$n(n+\alpha+\beta+1)$	$(1-x)^{\alpha}(1+x)^{\beta}$	$\frac{(-1)^n}{2^n n!}$	$(-1, 1)$	$\frac{2^{\alpha+\beta+1}\Gamma(\alpha+n+1)\Gamma(\beta+n+1)}{n!(\alpha+\beta+2n+1)\Gamma(\alpha+\beta+n+1)}$	13.2
$C_n^{\nu}(x)$	$1 - x^2$	$-(2\nu+1)x$	$n(n+2\nu)$	$(1 - x^2)^{\nu-1/2}$	$\frac{(-1)^n(2\nu)_n}{2^n n!(\nu+\frac{1}{2})_n}$	$(-1, 1)$	$\frac{2^{2\nu-1}(2\nu)_n\Gamma(\nu+\frac{1}{2})}{n!(\nu+n)\Gamma(2\nu)}$	13.4
$y_n(x)$	x^2	$2x + 2$	$-n(n+1)$	$e^{-2/x}$	2^n			13.5
$\theta_n(x)$	x	$-2(n+x)$	$2n$	$x^{-2n-1}e^{-2x}$	$(-2)^{-n}$			13.6

Table 5. Generating Functions

$$g(x, t) = \sum_{n=0}^{\infty} b_n \phi_n(x) t^n$$

$\phi_n(x)$	b_n	$g(x, t)$	Section
$P_n(x)$	1	$(1 - 2xt + t^2)^{-1/2}$	4.3
$P_n{}^m(x)$	1	$\dfrac{(2m)!\,(1 - x^2)^{m/2} t^m}{2^m m!\,(1 - 2xt + t^2)^{(2m+1)/2}}$	4.10
$J_n(x)$*	1	$\exp\left[\dfrac{x}{2}\left(t - \dfrac{1}{t}\right)\right]$	6.3
$H_n(x)$	$\dfrac{1}{n!}$	$\exp(2xt - t^2)$	9.2
$L_n{}^\alpha(x)$	1	$(1 - t)^{-\alpha-1} \exp\left(-\dfrac{xt}{1 - t}\right)$	10.1
$T_n(x)$	1	$\dfrac{1 - xt}{1 - 2xt + t^2}$	11.4
$U_n(x)$	1	$\dfrac{1}{1 - 2xt + t^2}$	11.4

Other Results

$J_n(x)$: $\quad x^2y'' + xy' + (x^2 - n^2)y = 0 \quad$ 6.1

$$\int_0^b xJ_k(\lambda_n x)J_k(\lambda_m x)\,dx = \frac{b^2}{2}[J_k'(\lambda_n b)]^2\,\delta_{mn}, \quad 6.7$$

where $J_k(\lambda_i b) = 0, \quad i = 1, 2, 3, \ldots$

$P_n{}^m(x)$: $\quad (1 - x^2)y'' - 2xy' + \left[n(n + 1) - \dfrac{m^2}{1 - x^2}\right]y = 0 \quad$ 4.8

$$\int_{-1}^1 P_n{}^m(x)P_k{}^m(x)\,dx = \frac{(n + m)!}{(n - m)!}\frac{2}{2n + 1}\delta_{nk} \quad 4.10$$

* The summation is from $-\infty$ to ∞.

Bibliography

[Ba] Bateman, H. *Partial Differential Equations of Mathematical Physics,* Cambridge University Press, London, 1959.

[Bo] Bowman, F. *Introduction to Bessel Functions,* Dover Publications, New York, 1958.

[Bu] Buck, R. C. *Advanced Calculus,* McGraw-Hill Book Co., Inc., New York, 1956.

[Bur] Burchnall, J. L. "The Bessel Polynomials," *Can. J. Math.,* 3 (1951): 62–68.

[CJ] Carslaw, H. S., and J. C. Jaeger. *Conduction of Heat in Solids,* 2d ed., Oxford University Press, London, 1959.

[C-1] Churchill, R. V. *Fourier Series and Boundary Value Problems,* 2d ed., McGraw-Hill Book Co., Inc., New York, 1963.

[C-2] ———. *Operational Mathematics,* 2d ed., McGraw-Hill Book Co., Inc., New York 1958.

[CD] ——— and C. L. Dolph. "Inverse Transforms of Products of Legendre Transforms," *Proc. Am. Math. Soc.,* 5 (1954): 93–100.

[Clt] Clement, P. R. "The Chebyshev Approximation Method," *Quart. Appl. Math.,* 11 (1953): 167.

[CL] Corson, D. R., and P. Lorrain. *Introduction to Electromagnetic Fields and Waves,* W. H. Freeman and Co., San Francisco, 1962.

[D] Dettman, J. W. *Mathematical Methods in Physics and Engineering,* McGraw-Hill Book Co., Inc., New York, 1962.

[E] Eringen, A. C. "The Finite Sturm-Liouville Transform," *Quart. J. Math., Oxford* (2), 5 (1954): 120–129.

[SCF] Fasenmyer, Sister M. Celine. "A Note on Pure Recurrence Relations," *Am. Math. Monthly,* 56 (1949): 14–17.

[F] Ford, L. R. *Differential Equations,* 2d ed., McGraw-Hill Book Co., Inc., New York, 1955.

[GM] Gray, A., G. B. Mathews, and T. M. MacRobert. *A Treatise on Bessel Functions and Their Applications to Physics,* 2d ed., Macmillan and Co., Ltd., London, 1952.

[Hn] Hobson, E. W. *The Theory of Spherical and Ellipsoidal Harmonics,* Cambridge University Press, London, 1931.

[Ht] Hochstadt, H. *Special Functions of Mathematical Physics,* Holt, Rinehart and Winston, New York, 1961.

[I] Ince, E. L. *Ordinary Differential Equations,* Dover Publications, Inc., New York, 1944.

[J] Jackson, D. *Fourier Series and Orthogonal Polynomials,* Carus Mathematical Monographs, No. 6, Mathematical Association of America, 1941.

[Ka] Karakash, J. J. *Transmission Lines and Filter Networks,* The Macmillan Co., New York, 1950.

[KF] Krall, H. L., and O. Frink. "A New Class of Orthogonal Polynomials: The Bessel Polynomials," *Trans. Am. Math. Soc.,* 65 (1949): 100–115.

[Kr] Kreyszig, E. *Advanced Engineering Mathematics,* John Wiley and Sons, Inc., New York, 1962.

[McR] MacRobert, T. M. *Spherical Harmonics*, Dover Publications, Inc., New York, 1948.

[McC] McCully, J. "The Laguerre Transform," *SIAM Review*, 2 (1960): 185–191.

[Mc-1] McLachlan, N. W. *Bessel Functions for Engineers*, 2d ed., Oxford University Press, London, 1955.

[Mc-2] ———. *Theory and Application of Mathieu Functions*, Oxford University Press, London, 1947.

[MM] Margenau, H., and G. M. Murphy. *The Mathematics of Physics and Chemistry*, 2d ed., D. Van Nostrand Co., Inc., Princeton, N.J., 1961.

[M] Miller, K. S. *Partial Differential Equations in Engineering Problems*, Prentice-Hall, Inc., Englewood Cliffs, N.J., 1953.

[MS] Moon, P., and D. E. Spencer, "Recent Investigations of the Separability of Laplace's Equation," *Proc. Am. Math. Soc.*, 4 (1953): 302–307.

[MF] Morse, P. M., and H. Feshbach. *Methods of Theoretical Physics*, Parts I and II, McGraw-Hill Book Co., Inc., New York, 1953.

[R] Rainville, E. D. *Special Functions*, The Macmillan Co., New York, 1960.

[Sa] Sansone, G. *Orthogonal Functions*, Interscience Publishers, Inc., New York, 1959.

[Sc] Scott, E. J. *Transform Calculus with an Introduction to Complex Variables*, Harper and Row, New York, 1955.

[Sn] Sneddon, I. N. *Special Functions of Mathematical Physics and Chemistry*, 2d ed., Interscience Publishers, Inc., New York, 1961.

[SR] Sokolnikoff, I. S., and R. M. Redheffer. *Mathematics of Physics and Modern Engineering*, McGraw-Hill Book Co., Inc., New York, 1958.

[St] Storch, L., "Synthesis of Constant-Time-Delay Ladder Networks Using Bessel Polynomials," *Proc. Inst. Radio Engr.*, 42 (1954): 1666–1675.

[Sz] Szego, G. *Orthogonal Polynomials*, American Mathematical Society Colloquium Publications vol. 23, New York, 1959.

[T] Tranter, C. J. *Integral Transforms in Mathematical Physics*, 2d ed., Methuen and Co., Ltd., London, 1962.

[VV] Van Valkenburg, M. E. *Introduction to Modern Network Synthesis*, John Wiley and Sons, Inc., New York, 1962.

[W] Watson, G. N. *A Treatise on the Theory of Bessel Functions*, 2d ed., Cambridge University Press, London, 1944.

[WW] Whittaker, E. T., and G. N. Watson. *Modern Analysis*, 4th ed., Cambridge University Press, London, 1950.

MATHEMATICAL TABLES

[1] *Handbook of Mathematical Functions with Formulas, Graphs, and Mathematical Tables*, National Bureau of Standards Applied Mathematics Series, vol. 55, U.S. Government Printing Office, Washington, D.C., 1964.

[2] *Tables of Chebyshev Polynomials* $S_n(x)$ and $C_n(x)$, National Bureau of Standards Applied Mathematics Series, vol. 9, U.S. Government Printing Office, Washington, D.C., 1952.

Index